AF651043

Striving for Sustainability
Environmental Stress and Democratic Initiatives in Kerala

About the Authors

Srikumar Chattopadhyay, M.Sc. and Ph.D. in Geography from the University of Calcutta, is a scientist at the Centre for Earth Science Studies, Thiruvananthapuram, Kerala. He earned his Post Graduate diploma in Applied Geomorphology and Environmental Survey from the International Institute of Aerospace Survey and Earth Sciences in the Netherlands (1986-87). He received a Post Doctoral Fulbright Fellowship to conduct research on sustainable development at the Environment and Policy Institute of the East-West Centre, Hawaii, USA in 1989-90. He was a visiting faculty member in the Department of Earth and Environmental Studies at Montclair State University, USA in 2002. In 1994, he received the Dr. S. Vasudev Award and in 2001 he received the R.N. Dubey Memorial Award.

Dr. Chattopadhyay has produced more than 90 research papers published in international and national journals of repute and has authored/edited five books. He has contributed significantly to develop the Kerala Panchayat Resource Mapping Programme. He is also the Chairman of the National Association of Geographers, India (NAGI) Commission on Sustainable Development and joint editor of the Annals, NAGI.

Richard W. Franke is a Professor of Anthropology at Montclair State University in New Jersey. He received his Ph.D. from Harvard University in 1972 and has worked in Surinam, Indonesia, West Africa, the Netherlands, France, and Kerala in India. With Barbara H. Chasin, he co-authored *Seeds of Famine : Ecological Destruction and the Development Dilemma in the West African Sahel* (1980). He lived in Kerala for eight months during 1986-87, carried out research for *Life is a Little Better : Redistribution as a Development Strategy in Nadur Village, Kerala (1993)*. Profs. Franke and Chasin have written *Kerala : Radical Reform as Development in an Indian State* (second edition, 1994). In 1998, Franke collaborated with Thomas Isaac and Pyaralal Raghavan on *Democracy at Work in an Indian Industrial Cooperative : The Story of Kerala Dinesh Beedi;* and in 2000 *Local Democracy and Development : People's Campaign for Decentralized Planning in Kerala.*

Striving for Sustainability Environmental Stress and Democratic Initiatives in Kerala

Srikumar Chattopadhyay
Richard W. Franke

CONCEPT PUBLISHING COMPANY, NEW DELHI-110059

Cataloging in Publication Data—DK
Courtesy: D.K. Agencies (P) Ltd. <docinfo@dkagencies.com>

Chattopadhyay, Srikumar, 1953-
Striving for sustainability environmental stress and democratic initiatives in Kerala / Srikumar Chattopadhyay, Richard W. Franke.
p. cm.
Includes bibliographical references (p.)
Includes index.
ISBN 8180692949

1. Environmental policy—India—Kerala. 2. Sustainable development—India—Kerala—Citizen participation. I. Franke, Richard W. (Richard William), 1944-, joint author. II. Title.

DDC 363.709 548 3 22

All rights reserved. No part of this work may be reproduced, stored in a retrieval system or transmitted in any form or by any means, electronic, mechanical, photocopying, microfilming, recording or otherwise, without the prior written permission of the copyright owner and the publisher.

ISBN : 81-8069-294-9

First Published 2006

© Srikumar Chattopadhyay and Richard W. Franke

Published and Printed by
Ashok Kumar Mittal
Concept Publishing Company
A/15-16, Commercial Block, Mohan Garden
New Delhi-110059 (India)

Phones : 25351460, 25351794
Fax : 091-11-25357103
Email : publishing@conceptpub.com

Preface and Acknowledgments

Kerala state in India is well known for its achievements in human development. The state enjoys a very favourable environmental resource base, which underpins many of the social development successes including easy access to environmental resources. In recent years, however, symptoms of environmental degradation have become widespread. The state is undergoing massive landscape changes. Deforestation, soil erosion, water quality deterioration, stagnation of fish production, shrinkage of wetlands, excessive extraction of river sand and clay, loss of biodiversity, fall in groundwater tables have the potential to cause irreparable damage to the natural resource base. How to arrest further degradation and restore environmental quality is debated in various circles particularly in the context of the recently popularised concept of sustainable development. Kerala is well known for its proactive civil society and its pro-people government policies. Kerala's mobilisation, structural changes such as land reforms, responsive scientific and technology institutions, trained manpower and mass movements for realising the ideals of democratic decentralization are some of the features of the state that can reverse the negative environmental trends. Significantly, Kerala has recently become a centre for local experimentation in sustainable and environment reviving development activities. These experiments can provide important lessons which can be scaled up as part of a more general strategy for sustainable development. We have written this book in order to bring to the Indian and international environmental audience, the scope of Kerala's environmental decline along with the original and promising responses the state's people are making towards possible alternatives that meet the basic criteria for sustainability.

The principle author, Srikumar Chattopadhyay has been working in Kerala for the past 25 years. His studies encompass many of the environmental and natural resource management problems indicated above. He was closely involved in the Panchayat Resource Mapping (PRM) programme and interacted frequently with the State Planning Board in the matters of decentralized planning, and the people's plan campaign. Richard W Franke took an interest in studying development in Kerala in the 1980s. He visited Kerala several times and produced four books on Kerala. He is keenly interested in understanding the Kerala mode of development in order to contextualize it in the larger international framework of sustainable development with social justice and equality.

During the 1990s, when Franke visited Kerala we had several discussions about the environmental problems faced by the state, the impending danger due to environmental degradation, and the potential to undermine Kerala's development and its many social justice achievements. We also discussed the local level initiatives to arrest environmental degradation. In the course of our discussions the idea crystallized to collaborate on a book dealing with these issues.

In January-May 2002, Srikumar Chattopadhyay was a visiting International Scholar at the Department of Earth and Environmental Studies at Montclair State University (MSU), where he worked with Franke drafting the main chapters. The MSU Global Education Centre sponsored Srikumar Chattopadhyay's visit. We take this opportunity to thank Marina Cunningham, Director of the Global Education Centre at MSU for extending support for Dr. Chattopadhyay's teaching and research.

In October 2002 Franke visited Kerala and worked on the manuscript. Further revisions were done mainly through e-mail.

We have freely drawn from the discussions with M. K. Prasad of the Kerala Sasthra Sahitya Parishad (KSSP), Dr. T. M. Thomas Isaac, Kerala MLA, Dr. John Kurien, Fellow, Centre for Development Studies, Thiruvananthapuram and T. Madhav Menon, Senior Fellow at the International School of Dravidian Linguistics in Thiruvananthapuram. M. Baba, Director, Centre for Earth Science Studies has always been cooperative and understanding of the demands for such a book. K. Soman, Head of the Resources Analysis Division whole heartedly supported the project. Mahamaya Chattopadhyay helped on various occasions while finalizing the manuscript. C. S. Lakshmi and Nitha Bhuvanan, Project Fellows, Resources Analysis Division, CESS processed the maps and graphs. We take this opportunity to thank all of them.

Franke's research visits to Kerala and a few other essential project costs were financed by the Montclair State University Global Education Centre, the MSU Faculty Scholarship Incentive Programme, and by Grant No. 99-61670-GSS from the John D. and Catherine T.

MacArthur Foundation. Among the many persons in Montclair who helped in putting the manuscript together, we thank Barbara H. Chasin, Richard Meyers, Nasir H. Naqvi, Laura Paynton and Agata Szwed. The 21 geography and anthropology students in Professor Chattopadhyay's MSU class on sustainable development created an atmosphere in which we could discuss mutual third world and first world environmental concerns while recognizing the primary responsibility of the first world and the primary suffering caused in the third world. Dr. Neeraj Vedwan provided valuable suggestions for improving the manuscript.

We take this opportunity to thank our publisher, Concept Publishing Company, New Delhi for undertaking the task of publishing this book.

Contents

PART II

Kerala's Initiatives for Sustainable Development: Employing the Power of an Awakened People

List of Tables

List of Figures and Boxes

Boxes

Abbreviations and Glossary

Adivasi	"Original"; member of a tribal group or Scheduled Tribe (ST)
Adivasi Gothra Sabha	Assembly of Indigenous People's Clans
African payal	Salvinia (water spangle or floating moss)
Alluvium (soil type)	Soil type mainly composed of sand, silt and clay deposited by running water
AWC	Available Water Capacity
Ayacut	The command or distribution area of an irrigation project
Basal value	Area covered by trees at ground level in a unit of forest area
Baseflow	River water flow during dry period/ non monsoon season without any addition by rain water
BG	Beneficiary Group
Biomass	Total vegetative matter in an area
Black soil	Type of soil, black in colour with high amount of clay particles, mostly developed in areas of lava flow
BOD	Biological Oxygen Demand
CDS	Centre for Development Studies
Cent	1/100 of an acre
CESS	Centre for Earth Science Studies
CGWB	Central Ground Water Board
Chagara	Suspended colloidal mud in sea water/ mud bank; also, the great shrimp catch along the Alappuzha coast
Charnockite	Dark coloured metamorphic rock widely found in Peninsular India
Cm	centimeter (one hundredth of a metre)
Conductivity	A measure of the ability of water to conduct electric current; related to concentration of total dissolved solids and major ions

COSTFORD	Centre of Science and Technology for Rural Development
CP:FC	Crop Production to Fertilizer Consumption ratio
CSE	Centre for Science and the Environment
CWC	Central Water Commission
CWRDM	Centre for Water Resources Development and Management
Dalit	Former untouchable caste member; Harijan
Dharna	Silent vigil; usually as a protest technique
DST	Department of Science and Technology
Ecovillage	Small intentional community where people build modest houses and use resources lightly with as much recycling as possible
Edaphic drought	Edaphic=soil related, indigenous
EU	European Union
Eutrophication	Loss of oxygen in a water body
FAO	Food and Agricultural Organisation
GALASA	Group Approach for Locally Adapted and Sustainable Agriculture
Geomorphic unit	Unit identified based on geomorphology
GIS	Geographic Information System
Gneiss	Metamorphic (transformed, usually by heat) rock with distinct banding
GOI	Government of India
GOK	Government of Kerala
Grama Sabha	Village (ward or neighbourhood) Assembly
GRP/LLP	Grassroots Planner/Local Level Planner
Ha	Hectare (=2.47 acres)
Harita Sangham	Green Group (vegetable cultivators working together)
HCBPL	Hindusthan Coca Cola Beverages Private Limited
Brown Hydromorphic	Type of soil developed under wet conditions in an enclosed basin
HYV	High Yielding Variety
ICAR	Indian Council of Agricultural Research

ICZM	Integrated Coastal Zone Management
IMU	Integrated Mapping Unit
IPCC	Intergovernmental Panel on Climate Change
IRDP	Integrated Rural Development Programme
IRS	Indian Remote Sensing (Satellite)
IRTC	Integrated Rural Technology Centre (in Palakkad)
ITK	Indigenous Technical Knowledge; Traditional Knowledge
Jala jatha	Water procession; village parade with displays about water use and conservation
Jalanidhi	Jala=water; nidhi=property. Programme for local water self-reliance.
JFM	Joint Forest Management
JRY	Jawahar Rozgar Yojana (Programme to Generate Rural Income)
KARA	Kerala Agrarian Relations Act
Kavu	Small Hindu shrine often situated in a grove of ancient trees and other plants
KERI	Kerala Engineering Research Institute
KFBOA	Kerala Fishing Boat Operators' Association
KFRI	Kerala Forest Research Institute
Khondalite (rock type)	Light coloured metamorphic rock, a variant of Gneiss widely found in South Kerala
KILA	Kerala Institute of Local Administration
Km	Kilometre
Km^2	Square Kilometre (= 100 hectares)
Kole	Low lying depression fields in Thrissur district
KSEB	Kerala State Electricity Board
KSI	Kerala Statistical Institute
KSLUB	Kerala State Land Use Board
KSPBC	Kerala State Pollution Control Board
KSSP	Kerala Sastra Sahitya Parishad; Kerala People's Science Movement

KWA	Kerala Water Authority
Lakh	100,000 (the Indian notation is 1,00,000)
Laterite (soil type)	Chemically weathered rock material red in colour with high content of iron oxides and aluminum hydroxide
Laterite with B horizon (soil type)	Laterite soil with B horizon in its vertical profile
Laterite without B horizon (soil type)	Laterite soil without B horizon in its vertical profile
Latosals	Laterite soils
Lithology	The study of rocks
Littoral	Coastal; pertaining to the seashore
Loam (soil type)	Soil composed of equal proportions of clay, silt and sand with admixture of organic matter
LSGD	Local Self Government Department
M^3	cubic metre
Malayalee	Person from Kerala; derived from Malayalam, the language spoken by 97% of Kerala's people
Mesa	A flat topped lateritic/ rocky hill with steep sides
Mg	Milligram (one thousandth of a gram)
Mg/l	Milligrams per litre
Moh	Micro molecules per litre, a measurement of the molecular concentration of a chemical
Mundakan (rabi)	The January harvest from the lesser or retreating monsoon of the previous October-November; also called the "winter crop." Rabi is the Hindi term.
NARP	National Agricultural Research Planning
NGO	Non Governmental Organization
NHG	Neighbourhood Group
NO_3	Nitrate
Onam	Kerala's main harvest and new year celebration based on a folk story about Mahabali, an ancient self-sacrificing raja.

Onattukara (soil type)	Sandy alluvium confined to a particular locality originally developed under sea water conditions
PAR	Photosynthetic Active Reduction
Parity Index	A number reflecting the costs of and returns to farming. An index of 100 means the farmer breaks even.
Patta	Land title paper in Kerala
PBR	People's Biodiversity Register
Peat	Semicarbonized vegetable tissue formed by partial decomposition in water of various plants; mosses of the gneiss Sphagnum
Pediplanation	Parallel slope retreat
Pedology	Soil science
Pokkali	A type of cultivation in which wet rice fields are reflooded part of the year to support prawn (shrimp) farming, then alternating back to paddy
Punja (boro)	Third harvest season for rice in the water logged lowlands in Kerala, it follows the mundakan harvest of January and occurs before the June monsoon rains begin. Boro is the Hindi term for rice cultivation in the pre monsoon season.
PWD	Public Works Department
Red loam (soil type)	Red coloured loamy soil marked in patches
Regolith	General term for unconsolidated weathered material including rock fragments and other superficial deposits or material usually lying on top of a formation
Runnel	Low lying area between two ridges
Sahyadri	Sanskrit name for the Western Ghats mountains
Salvinia	water spangle or floating moss
Sandy loam (soil type)	Loamy soil with higher portion of sand in its constituent
Shola	Vegetation concentrations along the water courses in areas above 1,000 metres

SVSS	Silent Valley Samrakshana Samithi; Silent Valley Defense Committee
TDS	Total Dissolved Solids
Thalweg	(German: "valley road") the line along the lowest elevation in a valley
TCA	Total Cropped Area
TDS	Total Dissolved Solids (in water)
Topo sequence	Topographic sequence spreading from hill tops through side slopes to the valley
UNEP	United Nations Environmental Programme
UNICEF	United Nations Children's Education Fund
USDA	United States Department of Agriculture
USLE	Universal Soil Loss Equation
Virippu (kharif)	September main rice harvest following the sowing in the June-August monsoon. Kharif is the Hindi term.
WCED	World Commission on Environment and Development

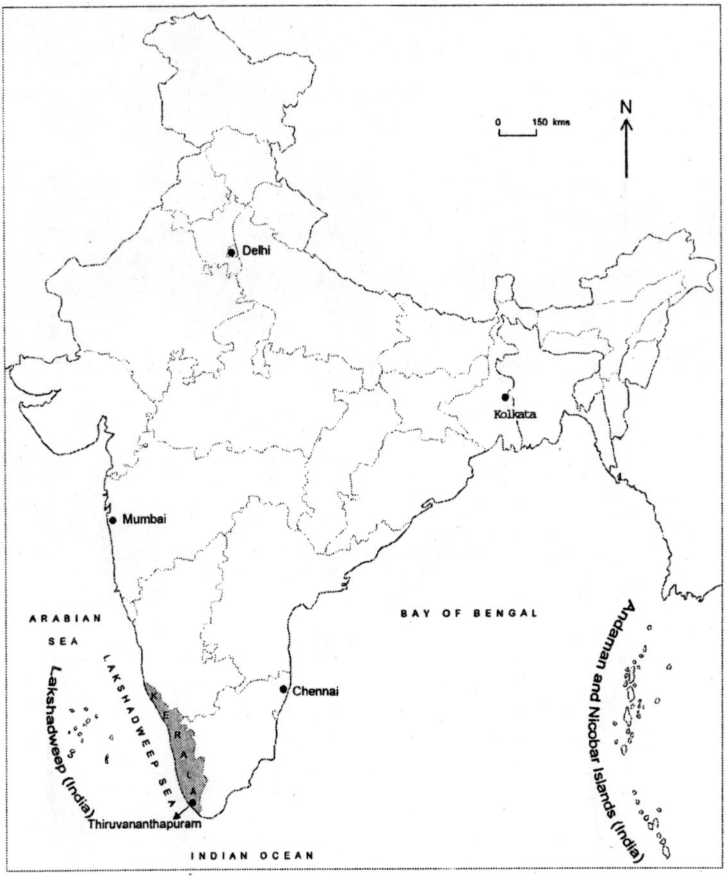

Fig. 1.1 : Location of Kerala

Part - I

Kerala and the World Environmental Crisis

Introduction

Our study is divided into two main parts. In part I, we consider the environmental changes in Kerala in the context of the world environmental crisis that everyone knows something about. In chapter 1, we give an overview of this worldwide crisis.

Following the introduction, we present five chapters on the ecological transformation of Kerala. In chapter 2, we describe Kerala's geographical setting and its environmental resource base. Chapter 3 examines the macro and micro level ecological zones. Chapter 4 is exclusively devoted to outlining the changes in land use while chapter 5 provides an overview of disturbing alterations occurring in the land, water and biomass base. Chapter 6 identifies some of the major human activities that account for the environmental changes analysed in chapters 4 and 5.

Part II of the book consists of four chapters dealing with the theme of "Kerala's initiatives for sustainable development: employing the power of an awakened people." We argue that the initiatives of Kerala's people in the context of the state's high level of activism and participation offer hope through many positive examples of local actions to foster sustainable development and environmental rehabilitation. Chapter 7 tries to bring out the state level struggles to protect Kerala's environment, their global implications and the possibility of creating a positive awareness among common people in the state. An overview of local panchayat activities, the experiment of the participatory panchayat resource mapping programme, and its impact will be treated in chapter 8. The 1996–2001 people's planning campaign will be discussed in chapter 9. Chapter 10 provides examples of recent micro level actions and project implementations and chapter 11 concludes our study with some thoughts on the sustainability challenges facing Kerala, India, and the world.

1 The World Environmental Crisis and the Kerala Experience

The Great Civilisational Change

"Humanity today is in the midst of a profound civilisational change." So begins Agenda 21, the document prepared after the Earth Summit held at Rio de Janeiro in 1992. The World Summit on Sustainable Development (WSSD) at Johannesburg in 2002 addressed many of the issues raised during the 1992 Earth Summit and at subsequent international conferences related to the environment.

Perhaps the most important contribution of the earth summit at Rio was to bring human actions to the center of the environment and development debate. Our environment is undergoing rapid change. While gradual change is a natural process, accelerated recent changes resulting from human activities are matters of serious concern. These changes carry the potential to jeopardise human welfare, particularly in the developing countries, whose people are more directly dependent on natural resources for their economic well-being than are those in the industrial nations. It is now widely acknowledged that the path of development so far followed is deepening the economic divisions within and between countries and leading to a deterioration of basic life support systems.

The Kerala Development Model

Among all the areas on the earth, Kerala has achieved great recognition for its human-centered approach to improve the lives of most of its people even at low levels of income and with low levels of economic growth. So unusually high are Kerala's levels of literacy and life expectancy; so unusually low its infant mortality, birth and maternal death rates; so effective its land reforms and its achievements in overcoming caste discrimination; so extensive its pension and worker protection systems; and so vibrant its political life and participation that experts for decades have discussed "The Kerala Development Model." It is a model of public action, of redistribution, of progressive change (Franke and Chasin 1994; Franke 1996; Jeffrey 1993; Ramachandran 1998).

From 1996 to 2001 Kerala became the focus of another internationally acclaimed action. The People's Campaign for Decentralized Planning attracted Indian, Third World, and Western intellectuals and development experts interested in the role radical decentralization and meaningful local democracy could play in seeking alternatives to corporate dominated globalization (Thomas Isaac and Franke 2000 and 2002). As it happens, Kerala's decentralization experiment is part of a larger Kerala tradition of environmental protection activities as well, and the follow up actions to the People's Campaign have generated experiments that offer the entire world examples and ideas worth consideration. At the same time, we acknowledge that Kerala suffers many ecological problems that threaten its own potential success. It is to bring out Kerala's potential contributions to the worldwide discussion on how to protect the earth's resources while maintaining an acceptable standard of living for all the world's people that we have written this book. We shall provide an overview of Kerala's environment, a summary of its main resources and the threats to them, and an account of how the Kerala tradition of political mobilization, an informed citizenry, and the use of democratic institutions and processes are being used to attempt to move towards sustainable development—a concept we shall define later in this introductory chapter. To more fully comprehend the achievements and challenges of Kerala's environmental movement, we must first return to the global crisis in which the Kerala Model of Development now finds itself compelled to operate.

Our Endangered Earth

Many experts use far stronger language than Agenda 21. U.S. scholar and former government advisor Lester Brown calls the earth a "planet under stress," with "a civilization in trouble." Until recently, Brown (2003:1; cf, Brown 2006) argues, "humans lived on the earth's sustainable yield—the interest from its natural endowment." Now, however,

> We are cutting trees faster than they can regenerate, overgrazing rangelands and converting them into deserts,

> overpumping aquifers, and draining rivers dry...We are taking fish from the ocean faster than they can reproduce.

Habitat destruction is combining with global warming to destroy

> ...plant and animal species far faster than new species can evolve, [thereby] launching the first mass extinction since the one that eradicated the dinosaurs 65 million years ago. (Brown 2003:1).

At present we confront at least five types of endangerment resulting from recent civilisational changes :

- global warming;
- impending water shortages;
- declining area and declining yields of productive land;
- pollution of land, fresh water, sea water and air; and
- biodiversity loss.

Global Warming. Most scientists today consider global warming a fact and its main cause to be human activities that release carbon dioxide (CO_2) into the atmosphere faster than it can be absorbed. From 316.7 ppm (parts per million) in 1960, CO_2 levels rose to 375.6 by 2003, an increase of 18.6 per cent in 43 years (Sawin 2005:40). Current concentrations of carbon dioxide are higher than they have been at any time in the last 400,000 years (Prugh, Flavin, and Sawin 2005:112). This "greenhouse effect" has resulted in a 3.6 per cent rise in average world temperatures from 14.01°C to 14.52°C in the same period (computed from Renner, Sheehan, et al. 2003:41). It is possible that negative feedback loops might reduce or slow global warming—such as increased carbon absorption by the oceans or the creation of carbon "sinks." It is possible, however, that positive feedback loops will intensify global warming—such as melting arctic tundra that could release huge amounts of methane, causing more melting, more methane release and so on (Meadows, Randers, and Meadows 2004:118).

Impending Water Shortages. Despite the enormous amounts of water visible in the oceans, only 1 per cent of the world's water supply is available as fresh water in the water cycle (Barlow 2002:5). In 1990, 131 million people were estimated to live in countries without adequate water. By 2025, the number could reach 817 million. In India in 1951, average water availability was 3,450 cubic meters per person per year. By the late 1990s it had fallen to 1,250 (Shiva 2002:1–2). As Vandana Shiva puts it: "In case after case, the story of water scarcity has been a story of greed, of careless technologies, and of taking more than nature can replenish and clean up" (2002:2). Her perspective seems especially appropriate for industrial agriculture. Agriculture absorbs 65–70 per cent of fresh water usage worldwide (Barlow 2002:8). It takes 1,000 tons of water to produce one ton of grain, 500 times as much as the 4 litres per day required for drinking purposes. When grain is consumed as animal products, the requirement is doubled (Brown 2003:25). Water tables are dropping across the globe: in parts of the major grain producing US states of Texas, Oklahoma, and Kansas it dropped more than 30 metres; in Pakistan's Punjab plains, it dropped from 1 to nearly 2 metres per year from 1982 to 2000; in North China it dropped 2.9 metres in the year 2000 alone; in wells in India's northern Gujarat, it has fallen from 15 metres to 400 metres below the surface in 30 years (Brown 2003:26–29). As water tables fall, getting the water up to the surface requires more and more mechanised pumping, which increases the use of fossil fuels and increases global warming.

Water diversion for agriculture and for urban use has so depleted rivers that many of the largest, including the Colorado, Yellow, Nile, Ganges, Indus, Chao Phraya, Syr Darya, and Amu Darya now run dry for some or all of the year (Meadows, Randers, and Meadows 2004:71). In chapter 5, we shall consider the problems facing Kerala's rivers. Despite the need for water in agriculture, as fossil water aquifers and rivers diminish, urban consumers win out in the water battle with farmers. And – as water supplies become more priced, multinational firms are drawn to water as an investment for profit, often with violent consequences through privatisation attempts as in Cochabamba, Bolivia or with harmful effects on local populations that are forced to fight back as in Plachimada, Kerala, which we will document in Chapter 7 of this study (Barlow and Clarke 2002; Shiva 2002).

Declining Productive Land. One of the foundations of modern civilization is the thin layer of topsoil on the earth's surface. Currently one-third or more of that layer is eroding faster than new soil is forming. The sequence is simple and devastating: forests turn to grasslands or fields; fields and grasslands turn to savannas; savannas turn to deserts. At the top of the sequence, tropical forest loss is now estimated at about 140,000 hectares per year, an area almost the size of Nepal (Bright 2003:8). More than half of all world forest loss has occurred since 1950. Between 1990 and 2000, 7 per cent of all tropical forest lands were converted to other uses – an area equal to the size of Mexico (Meadows, Randers, and Meadows 2004:75–76). On tropical agricultural lands, soil erosion may proceed at 34 times the rate in natural forests (Schücking and Anderson 1991:26). As soils erode, productivity falls. In the USA, studies suggest that for each inch of topsoil lost, wheat and corn yields decline by 6 per cent. India is thought to be losing 4.7 billion tons of topsoil annually through water-led erosion fed by the monsoons (Brown 2003:44). A large portion of the world's worst fed people live on marginal lands (Brown 2003:45). In marginal areas of Africa and Asia, deserts are spreading – the last stage in the land loss process. Meanwhile, in the still soil rich US, 16 million hectares of high quality cropland have been paved over to accommodate automobiles (Brown 2003:49). Overall, world cropland hit a peak of 732 million hectares in 1981; that had shrunk to 647 by 2002 (Brown 2003:53). Not surprisingly, per person grain production has also declined: it hit a high of 343 kg per capita in 1985, but was only 294 kg/capita in 2002, following a persistent drop (Halweil 2003:29).

Irrigated cropland such as that found in much of India, suffers from salinisation. Estimates are that about 20 per cent of the world's irrigated cropland has excessive toxic salts near the surface, reducing harvests to below their potential (Halweil 2002b :102). In chapter 4 we shall survey the problem of land degradation in Kerala.

Pollution. Various industrial processes introduce hundreds of new chemicals into the earth's environment each year. Acid rain, algal blooms, lead in gasoline, sulfur and nitrogen emissions, heavy metals, and pesticide residues are among the many elements assaulting the natural soils, water, and air of the earth. Recently, discarded computers and cell phones have become new sources of pollution, now referred to as "e-waste." This includes polyvinyl chloride (PVC), chlorine, and sulfur dioxide (Sarin 2004:44-45). In 1997 the average desktop computer required the production of 55 pounds of plastic, metal, glass, and silicon: drew 150 watts of electricity and generated 139 pounds of waste materials in its manufacture (Meadows, Randers, and Meadows 2004:41). Overall, it is estimated that 300–500 million tons of hazardous waste are spewed out each year. Between 50,000 and 100,000 synthetic chemicals are in production in the form of plastics, pesticides, lubricants, and solvents. Few have been studied for their toxicological consequences (Meadows, Randers, and Meadows 2004:113). Widespread evidence suggests that petrochemicals, heavy metals, and nitrates from fertilisers are filtering into underground aquifers that feed more than half the volume of the earth's lakes and rivers (Bright 2003:6–7).

Biodiversity Loss. The World Wide Fund for Nature (2002, cited in Meadows, Randers, and Meadows 2004:83) reports that its "Living Planet Index...an indicator of the state of the world's natural ecosystem...[a composite of] forest, freshwater, and marine species...shows an overall decline of about 37 per cent between 1970 and 2000." Dying out of species and birth of new species is part of evolution. Human-induced species loss, however, is now occurring faster than natural regeneration: some scientists have estimated the human-induced loss to be 40,000 times the rate at which natural selection alone would eliminate species from the earth's environment (Schückling and Anderson 1991:16). And this spectacularly rapid destruction occurs even as scientists have not even identified or fully described perhaps 95 per cent of the species in tropical rainforests (Schückling and Anderson 1991:15). British ecologist Norman Myers

Box 1.1 : Biodiversity

Biological diversity, or biodiversity, is the total number of genes, species, and ecosystems in a region and the variability between them. Biodiversity makes life itself possible. Not only do plants and animals provide actual and potential sources for human medicines and food, but also biodiversity has additional benefits that reach far beyond straightforward economic evaluations of utility. Scientists have shown that rich and diverse ecosystems improve water quality, reduce flooding, and absorb and clean wastes. They are also more resistant to environmental shocks and quicker to recover than regions depleted of genetic and species diversity. A group of scientists recently estimated the value of the services provided to humanity by the world's ecosystems—the pollination provided by insects, for example, and the water-cleaning capacity of healthy soils—as up to $54 trillion (Costanza et al 1997; Sutton and Costanza 2002), which is twice the size of the world economy.

McDonald and Nierenberg 2003:42; Definition of biodiversity from World Resources Institute 2002.

and the Washington-based environmental group Conservation International (CI) have defined 25 biodiversity "hot spots." These hot spots once covered 12 per cent of the earth's surface, but have now been reduced to 1.4 per cent. They contain rich varieties of plant and animal life but are threatened significantly by human activity. Gorillas, chimpanzees, forest buffaloes, and elephants are all threatened with extinction from the tropical forests of Central Africa (McDonald and Nierenberg 2003:44). Internationally, one fourth of the world's mammal and reptile species are threatened, along with 21 per cent of amphibians, and 30 per cent of fish. Twenty-eight percent of North American birds show significant decline in numbers. In Asia, 25 per cent of birds are thought to be in serious decline or are now limited to small, vulnerable populations. That is 664 species (Youth 2003:16). The Western Ghats mountain range that runs through most of Kerala has been designated as one of the twenty-five "hot spots" (McDonald and Nierenberg 2003:45). As we shall see in chapter 7, Kerala's people have fought hard to protect Silent Valley, one of Asia's remaining biodiversity islands. In chapter 8, we shall consider the Kerala Panchayat Resource Mapping Programme (PRM) that aims to create awareness and knowledge of biodiversity in the context of overall land use planning while in chapters 9 and 10 we shall examine several local programmes that attempt to integrate biodiversity directly into development planning.

Intensification, "Overshoot," Depletion, Collapse : A Theoretical Framework for Environmental Stress

Why have environmental concerns appeared on the historical stage at this moment? Humans have been using and sometimes destroying their resource base for millennia. But it is only in recent decades that two factors came together to generate a theoretical understanding of the interplay between humans and the earth's resources. With the publication of Rachel Carson's *Silent Spring* in 1962, the American and international reading public became aware of the dangers of particular chemicals and the possibility of environmental destruction. Then, as environmental movements grew around the globe, ecological theory developed. As data interacted with analyses and projections, the possibility of endangering the earth crept into the awareness of the scientific community and the public.

A synthesis of anthropology, geography, and ecological perspectives appeared in 1977 when anthropologist Marvin Harris published *Cannibals and Kings: The Origins of Cultures*. Drawing on the research from several disciplines, Harris suggested that much of human prehistory and history could be understood as a series of intensifications, depletions, innovations, and collapses. Population growth, increasing control over energy sources, inequality, and technological innovations were the driving forces of the sequence. Harris used this approach to explain the origins of agriculture, the rise of ancient states, the industrial revolution, and the collapse of several ancient cultures such as the Maya of Central America. The Maya collapse has particularly important implications for the present, because the archaeological and historical evidence now strongly suggest that they destroyed the fertility of their soils, leading to a civilisational collapse through widespread malnutrition that was itself intensified by the system of class inequality in which a ruling elite abused resources to satisfy its profligate lifestyle. More recently Jared Diamond's widely discussed *Collapse: How Societies Choose to Fail or Succeed* features a special chapter on the Maya rediscovering most of Harris's analysis. Diamond (2005:176) concludes that the Maya civilization collapsed primarily from internal mechanisms of intensification and depletion compounded by warfare that itself resulted in part from the declining resource base. Diamond finds similar sequences to account for ancient collapses on Easter Island, the Anasazi of the Southwest United States, the Viking settlements in Greenland, the recent genocide in Rwanda, extreme poverty coupled with violence in Haiti, and several other areas of interntional problems past and present. He identifies the same set of problems (2005:486–525; cf. Tainter 1988) as we have set forth at the beginning of this chapter.

The First Worldwide Depletion Crisis

The Maya, Anasazi, and other ancient collapses may carry valuable information for the present, but they also differ in one fundamental way – only now do humans face a potential collapse on a worldwide scale. Again, Marvin Harris suggests a framework for understanding the main components of the threat. He claims that the gains in living standards of the last two centuries are the result of a unique and perhaps non-replicable set of "extraordinary cultural events" (1977:186; cf. Wright 2004). The implementation of fossil fuel increased human energy control by thousands of times over that of the feudal or all previous periods. The discovery of safe contraception made possible conscious human control over reproduction. The rise of industrial jobs altered the cost/benefit ratio for children, dampening birth rates in the industrializing nations of Europe and North America. This combined with the first two factors to produce an "industrial bubble." The term industrial bubble refers to the idea that the high Western standards of living are based on a degree of resource drain that cannot be sustained if that standard of living is to be spread to the rest of the world. As Harris (1977:189) puts it: "the faster the underdeveloped world industrializes, the sooner the industrial world must develop a new mode of production." Short of that, current depletions will lead to some kind of collapse.

The Limits to Growth Debate

Harris was not alone in his concern. In 1972 a group in the System Dynamics Group of the Sloan School of Management at the Massachusetts Institute of Technology published their report of a project commissioned by the Club of Rome, a group of business persons, scientists, and politicians. Not having a theoretical framework or a comparative historical or anthropological perspective, they simply punched variables into a computer programme eventually called "World 3," which projected availability of a number of critical resources over various time spans as each variable increased or decreased by amounts selected for illustrative purposes. The results were the equivalent of a second "Silent Spring." All the runs indicated that humanity had about 50 years to a century to reduce its ecological footprint or one or more "Limits to Growth" would be bumped up against. The "ecological footprint" is an estimate of the "biologically productive space...necessary for a given population to produce the resources it consumes and to absorb the corresponding waste it generates, using prevailing technology" (Wackernagel and Silverstein 2000:392).

When the ecological footprint overshoots the available resource base through population growth, increased consumption per capita, and the spread of pollution, filling up the available sinks, eventually there must be catastrophic declines in human welfare or some other undesirable outcomes. World 3 estimates the declines to set in around 2050 at the latest. The models were never intended to predict events, just indicate general scenarios. The idea was to raise human consciousness to prevent such scenarios from ever occurring. In 1992 the same group issued an update entitled *Beyond the Limits*. Running a slightly modified World 3 programme with improved data, they discovered that their earlier predictions had been too optimistic: humanity's footprint had already "overshot" earth's resource base (see below in the next section).

In 2004 the team released a third edition called *Limits to Growth: The 30-Year Update*. Using further improved data, they project likely crisis conditions by the mid-twenty-first century, based on a logical model with four possibilities :

> *Unlimited growth* : All the evidence gathered so far is misleading or incorrect and the earth will sustain unlimited populations at unlimited levels of resource use. Needless to say, this option is not considered realistic.
> *Sigmoid growth* : Resource use somehow levels off smoothly to match population and consumption. This mathematical possibility is also considered unrealistic.
> *Overshoot and damped oscillation* : In this scenario threatened crises lead to corrective actions, which tend to go too far to the opposite extreme, leading to a period of unstable relations between resources and human welfare until some kind of steady state is achieved.
> *Overshoot and collapse* : Nature itself takes the corrective actions causing widespread harm to human welfare. A long period ensues with possibly: population decline, deterioration in living standards, increased warfare – this point emphasized by Diamond (2005) – spread of epidemic diseases, or other forms of collapse. The computer model does not predict any outcomes – it only predicts when certain variables will reach their limits. In chapter 11 we shall consider their "scenario 9," (2004:245) in which population is stabilised, poverty is eliminated, pollution is controlled, and human welfare remains at high levels while retracting from the current state of overshoot.

By the time Limits to Growth got its 30-year update, the scientific world on the whole already understood the general terms of the problem. The interconnected nature of the various forms of earth endangerment and the degree of overshoot make each factor potentially more destructive than they might appear separately. This point was emphasized in 1992 when 1,630 of the world's most eminent scientists, including 102 Nobel Laureates – a majority of the Nobel Laureates alive at the time – issued their "World Scientists Warning to Humanity." Their document (World Scientists 1992) includes the following observation :

> Our massive tampering with the world's interdependent web of life—coupled with the environmental damage inflicted by deforestation, species loss, and climate change—could trigger widespread adverse effects, including unpredictable collapses of critical biological systems whose interactions and dynamics we only imperfectly understand. Uncertainty over the extent of these effects cannot excuse complacency or delay in facing the threats.

First World Causes – Third World Consequences

The present day environmental crisis has much of its origins in the spread of capitalism, the industrial revolution, and the colonial empires and post-colonial exploitative relationships that are associated with both. While many first world observers continue to blame "population" for much of the environmental damages, there is growing awareness that inequality and especially the high-consumption life styles of those in the wealthy countries play equal if not more important roles.

Only 8 per cent of the world's people own cars, for example (Meadows, Randers and Meadows 2004:99), but most are in the industrial northern countries. Automobile emissions are considered a major contributor to global warming. Global warming threatens myriad consequences that have been widely discussed: melting glaciers and icecaps, rising sea levels, flooding of coastal areas, and the spread of infectious diseases. The short-term victims of global warming are mostly in the poor countries. Pacific islands and atolls are already going under the sea; tens of millions will be flooded out of their homes in Bangladesh; rising temperatures will cause thermal shock and dehydration to tropical crops. In effect, global warming will extend and exacerbate the tendencies inherited from the colonial era by virtue of the overlap between the political and economic inequality of the colonial system and the geographical distribution of colder rich countries and their former tropical colonies. The cause of this correlation can be seen most clearly in the statistics for the United States: with 5.2 per cent of the world's population, the U.S. consumes twice the commercial energy of Japan, its nearest competitor, and similar ratios of barrels of oil, and kilowatt hours of electricity per person (Sawin 2003:27). Americans generate twice the carbon emissions per person than do Germans, their nearest competitors (Renner and Sheehan 2003:21). Overall, the U.S. emits 24 per cent of the total world output of carbon from fossil fuel burning (Sheehan 2003:40). Eighty-four per cent of Americans belong to what the Worldwatch Institute calls the "consumer class," – individuals with more than $7,000 of purchasing power parity – compared with 12 per cent of people in India. As a result, U.S. consumers account for 31.5 per cent of total world private consumption expenditures (Gardner et al. 2004:6–7). And – apparently to maintain its privileged position despite the rhetoric of its leaders about spreading democracy – the U.S. in 2001 accounted for 48 per cent of world military spending, up from 24 per cent in 1980 (Chasin 2004:303). The expansion of US military power into the far reaches of the globe threatens to expand the violence that is regularly associated with inequality and with scarcity that we note starting on page 21. Military operations are among the most environmentally destructive of all human activities.

Because of its favourable climatic position and the low impact use of its soils by Native Americans over many centuries, up to the present the U.S. population has suffered few immediate consequences from its own misuse of environmental resources. But in low-income nations where many of the most marginal soils are found and where temperatures are already relatively high, the consequences of global warming can be immediate. In Kerala researchers have determined that for each temperature increase of one degree Celsius, rice yields decline by 6 per cent (Saseendran et al. 2000:504).

Rich country corporations and consumers account for most pollution, while the refuse is now more and more shipped off to third world locations where the land and water become poisoned. Perhaps only biodiversity is somewhat evenly under assault across the great power and income gaps that divide our world. Even here, however, as the internationally known Indian environmental advocate Vandana Shiva (Shiva et al. 1991:8) has forcefully argued, "the roots of the crisis of diversity in the South lie in the North." She goes on to point out that "The World Bank which has now launched the Biodiversity Action Plans, has for the past forty years been financing the destruction of genetic diversity in the Third World." She goes on to show how the very nature of the industrialized agriculture of the North reduces biodiversity in the South by patenting and privatising the latter's raw materials (Shiva 1991:54). The biodiversity issue involves the added dimension of tribal or other minority peoples, in whose territory the greatest diversity remains and who are the most vulnerable to the consequences of the loss of biodiversity (Gray 1991).

Saving the Endangered Earth : Sustainable Development and Ecological Democracy

Because people in countries like India are first to bear the brunt of world environmental decline, they have the most immediate interest in stopping it at least within their sphere of influence. Sustainable development is currently proposed as the conceptual basis for such actions. The concept of sustainable development emerged from the quest to create an alternative path of development. Attempts to define sustainable development started back in 1972 and continue today. Although sustainable development is a contested term, it has received wide application since 1987 with the publication of *Our Common Future* (WCED 1987). According to *Our Common Future*, "sustainable development is development that meets the needs of the present without compromising the ability of future generations to meet their own needs" (WCED 1987:43). This definition is now widely quoted. A set of selected definitions is provided in box 1.2. The definitional debate was quite intense until 1992, when the Earth Summit on the environment and development took place in Rio and Agenda 21 was brought out, spelling out a programme of action at different levels. The numerous definitions available in the literature (Pezzy 1989; Chattopadhyay and Carpenter 1990; Brown et al. 1987; Lele 1991) tried to capture a large number of issues in a single frame. As a result, the definition tends to be vague and subject to multiple interpretations. Some authors claim that this vagueness is a strength (Daly 1990a:32). Others (Lele1991) have expressed the view that the lack of precision in the definition prevents effective operationalisation. Although the term is loosely defined (Pretty 1995), it captures a widespread feeling that the state of the Earth is somewhat precarious (Wilbanks 1994); development is not synonymous with economic growth (Munasinghe and McNeely 1995) and finally there is a need for integration of ecology and economy (Carpenter and Harper 1989; Pearce et al. 1989).

Despite the differences in these definitions, we can note four features they share :

Box 1.2 : Sustainable Development : Selected Definitions

1. "The principal conditions of a stable society – one that...can be sustained indefinitely while giving optimum satisfaction to its members – are: (1) minimum disruption of ecological processes; (2) maximum conservation of materials and energy...(3) a population in which recruitment equals loss; and (4) a social system in which the individual can enjoy, rather than feel restricted by, the first three conditions" (Goldsmith et al. 1972:23).
2. "Sustainable society is one that lives within the self perpetuating limits of its environment. That society ... is not a no-growth society. ... It is, rather a society that recognizes the limits of growth... (and) looks for alternative ways of growing" (Coomer 1979:1).
3. "Sustainable development – development that is likely to achieve lasting satisfaction of human needs and improvement of the quality of human life..." (Allen 1980:23).
4. "Sustainability...a symbiotic relationship with nature. Development must operate within the constraints of local ecosystems. It must maintain and encourage biotic diversity as an insurance for the future. Essentially, it must be ecologically sustainable in terms of local ecosystems and the biosphere" (Dasmann 1984:20).
5. "The market does not distinguish an ecological sustainable scale of matter-energy throughput from an unsustainable scale, just as it does not distinguish between ethically just and unjust distributions of income. Sustainability, like justice, is a value not achievable by purely individualistic market processes" (Daly 1986:320).
6. "The primary objective [of sustainable economic development] is reducing the absolute poverty of the world's poor through providing lasting and secure livelihoods that minimize resource depletion, environmental degradation, cultural disruption, and social instability" (Barbier 1987:103).
7. "Sustainable economic development maximizes the goals across the biological and resource system, the economic system, and the social system" (Barbier 1987:104).
8. "A social definition of sustainability might include the continued satisfaction basic human needs¯food, water, shelter¯as well as higher level social and cultural necessities such as security, freedom, education, employment and recreation" (Brown et al 1987:716).
9. "Sustainability must be defined in terms of productivity and management..." (Carpenter and Harper 1989:44).

10. Sustainable development is "a pattern of social and structural transformation...which optimizes the economic and other societal benefits available in the present without jeopardizing the likely potential for similar benefits in the future" (Goodland and Ledec 1987:36).
11. "...development which allows all future generations to have a potential average quality of life at least as high as the average quality of life of the current generation....quality of life depends on traditional material consumption as well as on leisure, environment quality and so on" (Norwegian Research Council for Science and Humanities [NAVF] 1990:233 cited in Lafferty and Langhelle 1999:8).

- Humans are causing excessive stress to the biophysical environment, which is ultimately counterproductive to human activities as there is a limit to which productivity can be raised;
- Human survival can be ensured only by preserving the basic resource base;
- Even renewable resources can be exhausted, unless prudent management practices are followed; and
- Conventional economics does not appreciate or factor into its theories the actual status of resource utilisation. Vandana Shiva (2002:15) writes that "Market assumptions are blind to the ecological limits set by the water cycle and the economic limits set by poverty."

Physical Limits

Sustainable development is assumed to require that humans operate within physical, biological, ecological, social, and economic limits. These limits – taken together – are sometimes referred to as the "carrying capacity" of the earth or of a particular region or ecological zone (IUCN 1991:43–51). There are certain limits up to which an ecosystem can withstand perturbation. All ecosystems have self-ameliorating or assimilating capacities, also called the resilience capacity. For example, water can absorb up to a certain amount of pollutants and still purify itself. When the levels of pollutants cross the biophysical limit, the water becomes unable to self-purify and thus becomes unusable. Land, water and biomass are interdependent. Intervention in one results in changes in the others. According to steady-state economics theoretician Herman Daly (1991:187 and 225;

cf. 1987 and 1990a) the biophysical limit, which controls resource availability, resource use and economic viability, is a function of three components: (1) the finite nature of resources (2) entropy laws, and (3) complex ecological interdependence. We would add a fourth component: resilience or self ameliorating capacity. There is also a photosynthetic limit that ultimately restricts the output possible from plants and other biomass. Awareness of these limits led the World Commission on Environment and Development (WCED 1987:44) to add to its definition of sustainability "At a minimum, sustainable development must not endanger the natural systems that support life on Earth: the atmosphere, the waters, the soils, and the living beings" (quoted in Lafferty and Langhelle 1999:13).

Can we quantify the physical limits? A 2002 report presented as part of the Proceedings of the U.S. National Academy of Sciences suggest that humanity's "ecological footprint" now requires 20 per cent more than the available resources of the earth (Wackernagel et al. 2002). The study weights food crops, animal pasture, timber harvesting, fishing, infrastructure, and fossil fuel burning to generate a standardized per capita "global hectare area (gha):" the amount of the biosphere expressed as a land unit needed to support one human life. In 1999 the area available was estimated at 1.91 gha per person, but the usage was estimated at 2.33 (Wackernagel et al. 2002:9268). If 12 per cent of the biosphere was to be kept in reserve to protect biodiversity as recommended in the Brundtland Report, the "overshoot" becomes nearly 40 per cent. The conclusion to be drawn is that depletion is occurring and the capital reserves of the earth are being drawn down, but it cannot be stated with certainty how long the depletion can go on until an adaptation, oscillation, or collapse occurs.

Socio-cultural Limits

The socio-cultural limit is crucial but often overlooked in the development process. Values, beliefs and institutions develop within socio-cultural systems to meet human needs (Munasinghe and Menealy 1995). The social acceptability of any development process is important for its proper functioning. The management of resources is also a key factor. According to Freeman (1991:248) the economic value of natural resources depends in part on the institutions that manage the resources and ultimately, by the values embedded in the underlying socio-cultural matrix. Stephen Viederman (1995:37) has elevated this observation to create a specialized definition of sustainability :

> ...a community's control of capital, in all of its forms—natural, human, human-created, social, and cultural—to ensure to the degree possible that present and future generations can attain a high degree of economic security and achieve democracy while maintaining the integrity of the ecological systems upon which all life and production depend.

Apart from management, intra and inter generational equity cannot be ignored as economic disparities have grown to dangerous dimensions. A large section of people do not have access to basic human needs, while certain groups indulge in over consumption. Growing disparities can result in social discord with destabilising effects. The social limits involve distributive justice and cultural diversity.

Depletions, Inequality, and Violence

Our discussion of social limits and of inequality underscores the recent international acknowledgement of the importance of sustainable development as an essential component of world peace – itself a requirement for further sustainable development. Depletions lead quickly to scarcity and scarcity is associated with the onset of violence in many societies studied by social scientists. Substantial anthropological research suggests that otherwise peaceful hunting and gathering or early agricultural societies lapsed into endemic warfare and gender violence when they overshot their resource bases (Harris 1974:esp. 61–107). Unusually high rates of violence in the United States can be explained at least in part by the high degrees of inequality and by scarcity and competition (Chasin 2004). Jared Diamond (2005:311–28) offers a persuasive case that the Rwandan genocide of 1994 in which 800,000 people were massacred had depletions and scarcity compounded by a history of inequality among its main causal variables. The Nobel Prize Committee for 2004 honoured Kenyan environmentalist and women's rights activist Wangari Maathai with is annual Peace Prize, noting in its presentation statement (Mjøs 2004:1) that "Peace on earth depends on our ability to secure our living environment." In her Nobel Lecture Maathai (2004:4; cf. Maathai 2005) added that "there can be no peace without equitable development; and there can be no development without sustainable management of the environment in a democratic and peaceful space."

Another Nobel Peace Prize recipient, Nelson Mandela, while addressing the International Geographical Union (IGU) Regional Conference in Durban, South Africa focused on environmental sustainability at local level and observed :

> When I go to the place and area of my birth, so often as I do, the changed geography of the place strikes me with a force that I can not escape. And that the geography is not one of mere landscapes and topography, it is the geography of the people. Where once there were trees and even forests, we now see barrenness... one was saddened at the poverty of the people- poverty lived out in the geography of place... A devastated geography makes for a devastated people. It renders people vulnerable, and the traditionally vulnerable – women, children, the aged and disabled – will always be bearing the burnt of the suffering. Let us stand together to make of our world a sustainable source for our future as humanity on this planet (quoted in Wood 2004:2,4).

Finally, the Worldwatch *State of the World Report* for 2005 is subtitled *Redefining Global Security*. This report has become one of the international signposts of scientific thinking on sustainability for the past few decades. The 2005 report is entirely devoted to the connections between scarcity and violence versus sustainability and peace. Some important points they argue (Renner 2005:5) are that:

- "Weapons do not necessarily provide security;
- Real security in a globalizing world cannot be provided on a purely national basis; and
- Non-military dimensions have an important influence on security and stability."

These points have much relevance to our discussion of Kerala's experiments in egalitarian and sustainable development.

Economic Limits

Economic limits can be visualized through the concepts of efficiency, growth and stability. If a system does not function efficiently or is not cost-effective it can never be sustainable. However, we argue that the economic approach to sustainable development must be consistent with physical and biological systems. This requires the extension of economic theory to examine the biophysical resource limits and improve the valuation techniques to reflect degradation of environmental assets. The recent launching of the journal *Ecological Economics* is an indication of the expanding nature of economics in the present crisis period. International and intergenerational equity in resource allocation are also key questions. With respect to growth and distribution, Daly (1991) argues that growth must approach a steady state where qualitative improvement is given precedence over quantitative economic growth. A limit must be set on the use of scarce natural capital and on what is taken from the earth as a resource and what is dumped back into the global sink. Previously the argument against redistribution was that inequality is a necessary incentive for economic growth (Franke 1996:11–16). But if growth is stopped, the poor will no longer accept their poverty so there must be a mechanism for the international redistribution of income (Herman Daly as cited in Weaver et al. 1997:242).

Globalization has hardly benefited the poor or down trodden. The countries that are open to globalization are experiencing extreme forms of exploitation, the erosion of workers' rights, and increased vulnerability to global markets (Watkins 2002:25). While the developing countries are pressured to withdraw subsidies from agriculture the industrialised nations subsidise their farmers' overproduction and export dumping. Safeguarding of local interests in the face of this globalization is an important task, if development is to be truly sustainable.

The Strategy for Sustainable Development

Given this background, the challenge is to work out a development strategy that adheres to limits while also making increasingly efficient use of existing resources. To give just one (major) example: currently humans tap only one ten thousandth (0.0001) of the energy the sun delivers to the earth each day (Meadows, Randers and Meadows

2004:97). The question of production management in both agriculture and industry is of prime importance. In the sustainable development framework, the inputs and outputs of this production process should not reduce the biophysical components of the natural system including its resilience limits. Of crucial importance is to consider the environmental resource system as a whole. It is not just sustainable agriculture, sustainable forestry, or sustainable fisheries. It is sustainable development comprising all three, and more than that, there are synergies where each sector contributes to the other and gets enriched at the same time.

Local Sustainability

Creating sustainable development requires a strategic, cohesive, multidimensional and essentially interventionist approach. Translating this approach to regional sustainable development requires looking into the reality of the region in enough detail to appreciate local site differences and taking note of the policies that would suit the local environment (Chattopadhyay 1998; Li et al. 2001). Therefore, the local level processes are in aggregate important in both regional and global sustainability. The interactions and synergies of ecological, economic, and social processes are scale dependent and therefore change from the local to the regional, the national and global levels. Analysis of the processes and components in such a hierarchy cannot be accomplished in isolation. Both horizontal and vertical integration at different levels are called for to operationalise sustainable development.

The Global Environmental Change Programme (GECP 2001) pointed out that as global environmental change assumes a more and more central place in human affairs, scientists must accept the responsibility of developing and communicating the essential knowledge base that societies can use to debate, consider and ultimately decide how to respond to global change. Ultimately, preserving the global life support system depends upon science. The challenge of ensuring a sustainable future is daunting. It can be met only with a new and more vigorous approach to studying and managing the earth's integrated ecological systems.

Despite its global consequences, ecological transformation is primarily manifested at the local level. Aggregate data are not appropriate to comprehend many of the changes. Problems of environmental resource management can only be resolved at the level of their origins. This provides enormous scope and at the same time poses big challenges for local communities. Re-orientation of developmental programs depends upon various economic forces, such as resource availability and land tenure at the micro level and the main policy framework of rules and regulations at the macro level. To reverse current trends towards environmental degradation, we must not only understand the forces leading to change, but also be able to set the mechanisms, institutional and otherwise, that will enable resource users to strive for reorientation or alternatives. This means proactive intervention through empowerment of people at the lowest levels. Or, as Vandana Shiva (2002:14) puts it, "Ecological Democracy."

Kerala's Experiment in Local Planning for Sustainable Development

Among the places in the world where rapid environmental change and degradation are taking place, Kerala state in India has generated a number of experiments in local level planning that offer insights into how the local component of an environmental turnaround could be developed. The aim of this book is to describe Kerala's environmental resource base, illustrate the ecosystem changes that are taking place, and highlight the possibilities of local level planning in reversing the situation to move forward towards a sustainable world. The book will also try to assess the lessons from Kerala in the larger global context of growing environmental awareness of concern.

As we have noted near the beginning of this chapter, Kerala has drawn global attention for its unique achievements in improving the life of the people even at low levels of income. Perhaps the most environmentally connected component of the Kerala Model is the 1969 Land Reform Act. Kerala's land reforms are among the most noteworthy in India (Raj and Tharakan 1983). The tenants or sharecroppers got ownership rights and agricultural labourers received the site for their houses. Completely landless agricultural labourers, found in other states of India, are non-existent in Kerala.

This land reform gave marginal sections of the population access to land resources. Land productivity in Kerala is higher than the rest of India. The productivity of unirrigated rice is 1.5 times higher and overall agricultural productivity per ha of gross cropped area is 1.8

times higher. This high productivity helped even people with small holdings to raise their yields and live better than people with the same land holdings in other parts of the country.

The land reforms have interacted with Kerala's varied physical base to allow for a diversified crop culture even at the micro level. Zone specific production opportunities like fishing in the seas and backwaters, agriculture in the lowlands and midlands, and plantations and forestry in the highlands and mountains have provided the state a very favourable environment to raise productivity and to live life locally and also lightly, that is, without much consumption of energy. Power consumption for agricultural purposes is only 4.2 kwh in the state compared to the all India average of 24.3 kwh (KSLUB 1995). Kerala has recorded 10 times higher productivity from its forest areas than the rest of the country.

Because the higher productivity and high input-output ratio are determined principally by the ecological conditions of the state, maintenance of ecosystem integrity is essential for sustaining this high level of productivity and to raise it further. But changes in natural vegetation, land conditions, water quality and flow reported from various parts of the state are quite alarming. If these changes continue, Kerala's production base will be severely compromised which can jeopardise the state's earlier achievements in development.

Our study therefore highlights the necessity of local democratic interventions in ecosystem management, micro-level planning and sustainable development. These topics are taken up in part II of the book. We see in chapter 7 that activists and ordinary citizens have come together from the famous Silent Valley campaign of the 1970s to protect biodiversity to the Plachimada struggle to save a local water supply from the ravages of a multinational corporation. But beyond that : across Kerala numerous innovative projects are underway to enhance soil fertility, make more efficient use of water, protect sacred groves, and carry out other actions on behalf of sustainable development. Chapter 7 through 10 thus provides examples of the many ways in which local initiatives in an area of heightened awareness can bring enlightened change – change in the direction of sustainable development. In chapter 11, we attempt to situate the Kerala experience within the Indian national context and the international struggle for the sustainable world in which we all need to live.

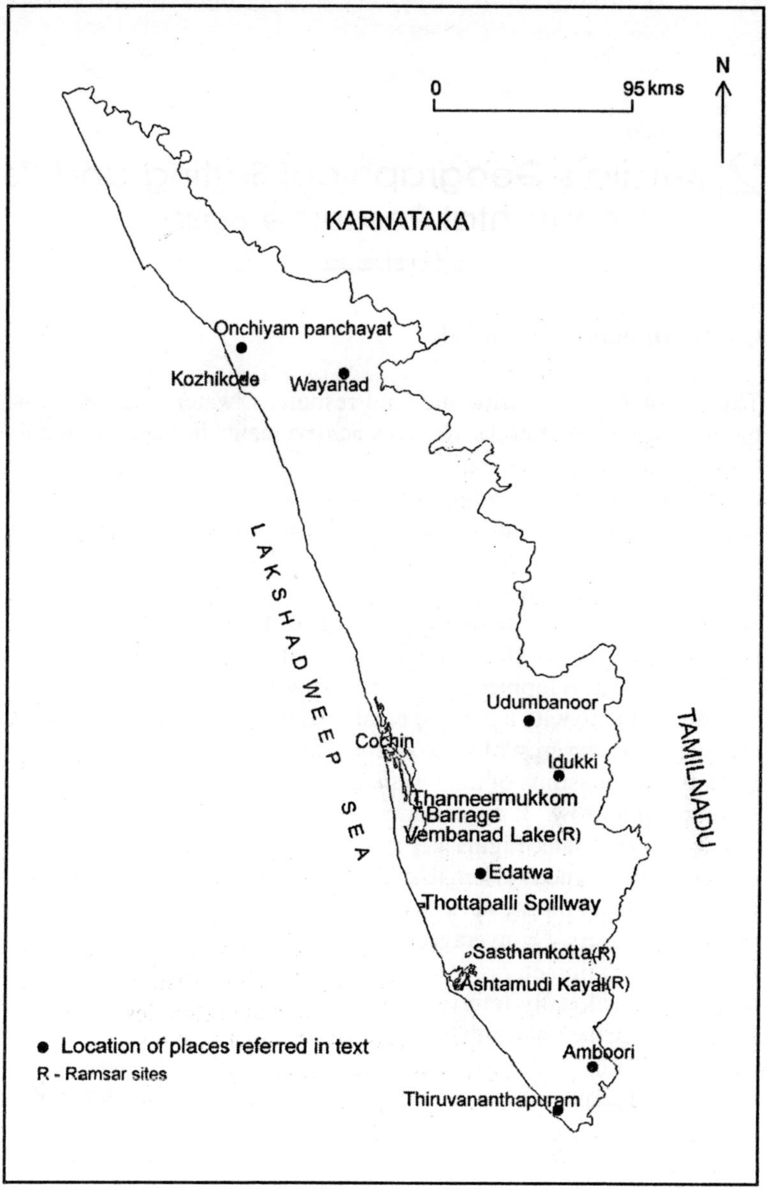

Fig. 1.2 Major Place Names Referred to in Part-1

2 Kerala's Geographical Setting and Its Environmental Resource Base

Coast, Midlands, Highlands

This chapter will describe the land resources, water resources, and biomass wealth of Kerala. Kerala's achievements in material quality of life that we summarized in chapter 1, are partly the result of its dispersed and rather evenly spread resources. Understanding the resource base is also important in order to properly appreciate the problems that the state will be confronted with if its biophysical foundation is weakened.

The geographical setting of Kerala provides a diversified resource base, which gives the state certain advantages conducive to decentralized development. The administrative boundaries of the state broadly coincide with a geographical region. The eastern border more or less follows the crest of the Western Ghat Mountains and the western border is the coastline of the Lakshadweep Sea (Arabian Sea). Kerala is long and narrow. It covers an area of 38,860 sq km. There are 14 districts in the state. Figure 2.1 indicates the administrative districts of Kerala. With a coastal length of 560 kilometers, the maximum width of the state from the western coastline to the eastern mountain border is 120 kilometers. On average, for every kilometer of coastline there are 70 square kilometers of landmass. A chain of water bodies parallel to the coast is locally referred to as *kayals*. Three distinct economic zones provide the state with diverse economic opportunities: (a) marine fisheries and coastal backwaters mostly located in the lowlands, (b) forestry and plantations along the eastern hills, plateaus and mountains, and (c) agriculture in the valleys lying mostly between these two areas.

Land Resources

We can discuss Kerala's land resources under three groupings : topography and relief, geology and geomorphology, and soil.

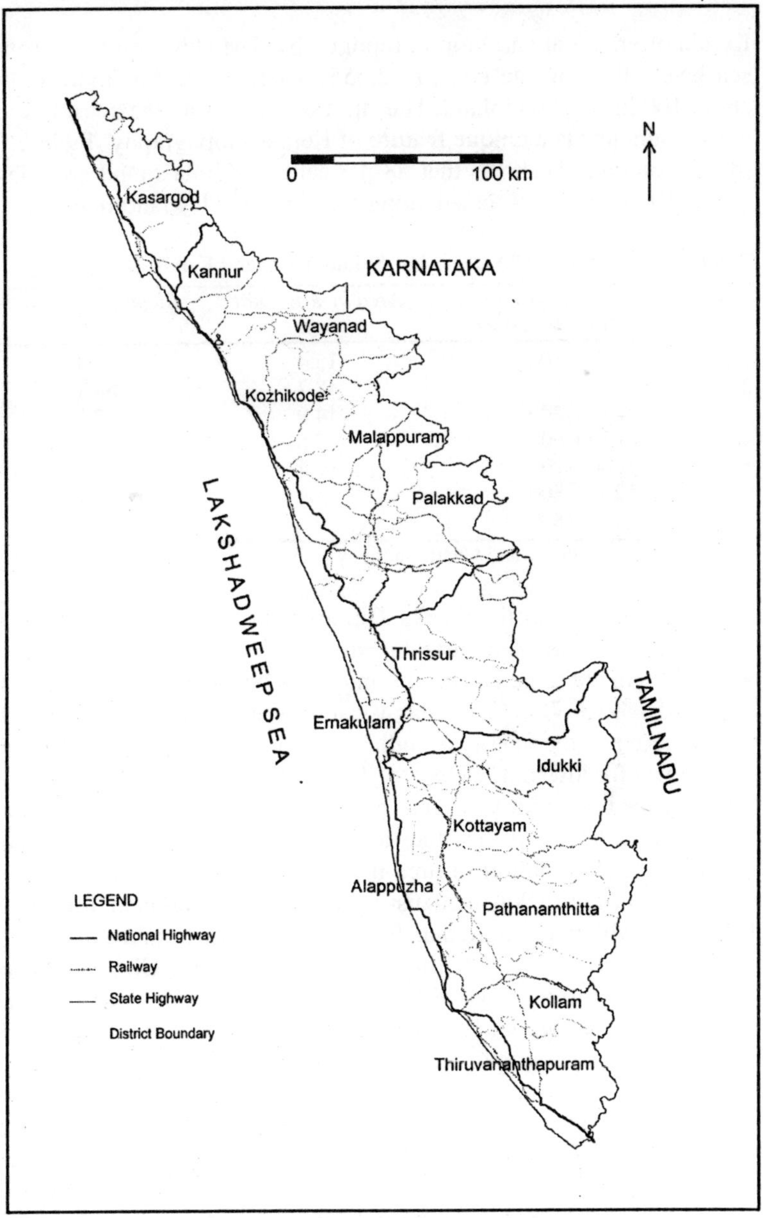

Fig. 2.1 Administrative Divisions of Kerala Showing Districts

Topography and Relief

Kerala offers great variation in topography. The elevation rises from sea level all along the coast to 2,655 meters at the Anaimudi only about 100 kilometers inland. The short distance from the coast to the high mountains is a unique feature of Kerala's topography. Table 2.1 (row 2, column 4) shows that 58 per cent of Kerala's land area lies below 100 meters, dispersed along the coast and lowland regions.

Table 2.1 Distribution of Area within Altitudinal Zones: Kerala, India

Row No.	*Altitude above sea-level in metres*	*Area in % to total area*	*Cumulative area %*
1.	0–100	15.4	15.4
2.	10–100	42.7	58.1
3.	100–300	14.0	72.1
4.	300–600	7.8	79.9
5.	600–1200	16.4	96.3
6.	1200–1800	3.4	99.7
7.	Above1800	0.3	100.0

Source : Chattopadhyay and Chattopadhyay (1995:19)

The Kerala Public Works Department (PWD) developed a classification of the physiographic areas of Kerala in which lowlands were recognized to be below 8m, midlands between 8 and 75m, and highlands above 75m. This classification scheme had been evolved earlier based on the Survey of India old series maps. The demarcation was 25 feet for lowland, 25 feet to 300 feet for midland and more than 300 feet for highland. Lowland, midland and highland are the most widely used terms in the state to denote geographical areas. The district-wise distribution of area under these three units is given in table 2.2. The entire Wayanad district comes under highland, followed by Idukki, which has more than 95 per cent of its area in the highland category. Both of these districts are in the Western Ghats and are characterised by plateaus with altitudes of 600–700m. Apart from these two, in Kasargod, Palakkad and Pathanamthitta more than 50 per cent of the geographical area falls under highland. Conversely, only one district, Alappuzha, has more than 75 per cent of its area as lowlands. Kollam is the district where distributions of midland and highland are almost equal. Each of the remaining seven districts has more than 50 per cent of its area under midland. This area distribution indicates that all districts have relief diversity. Being heavily influenced by physiography both in climate and agriculture, Kerala's diversity in relief at the district level signifies that most of the districts enjoy comparable topographic variations resulting in equal opportunities for diverse cropping.

Table 2.2 Highland, Midland and Lowland Area in Percents by District in Kerala

Districts	*Area in %*		
	Lowland	*Midland*	*Highland*
Kasargod	9.79	37.72	52.49
Kannur	5.30	53.71	40.99
Wayanad	0	0	100.00
Kozhikode	15.55	57.63	26.82
Malappuram	2.84	78.26	18.90
Palakkad	0	35.59	64.41
Thrissur	15.55	51.67	32.78
Ernakulam	24.29	66.78	8.93
Idukki	0	3.58	96.42
Kottayam	18.14	58.66	23.20
Alappuzha	76.66	23.34	0
Pathanamthitta	2.73	19.72	77.55
Kollam	12.84	44.72	42.44
Thiruvananthapuram	5.21	68.52	26.27
Kerala State Totals	10.09	41.76	48.15

Source : Chattopadhyay and Chattopadhyay (1995:18)
Note : Districts are listed from North to South.

A more intricate relief and topography classification scheme (Chattopadhyay and Chattopadhyay1995) suggests that the state could be divided into five geographic zones. Figure 2.2 is a map of the geographic zones of Kerala. These are: (1) the high ranges of the Western Ghats (above 600m), (2) a Western Ghat foothill region (300–600m), (3) an upland region (100–300m), (4) an undulating midland (10–100m), and (5) a coastal area and adjoining lowland (less than 10m). Each of these units can be further divided into two or more of eleven possible sub-units. Identification of these zones and sub-units is to capture the differential production potential at the local level, which is necessary to work out an area plan for sustainable natural resource management.

The eastern high ranges are clearly identifiable in two separate blocks: to the north and south of the Palghat Gap. This 30 km wide gap with an altitude ranging from 100 to 300m is a major break within the otherwise monolithic Western Ghat range. The gap connects Kerala

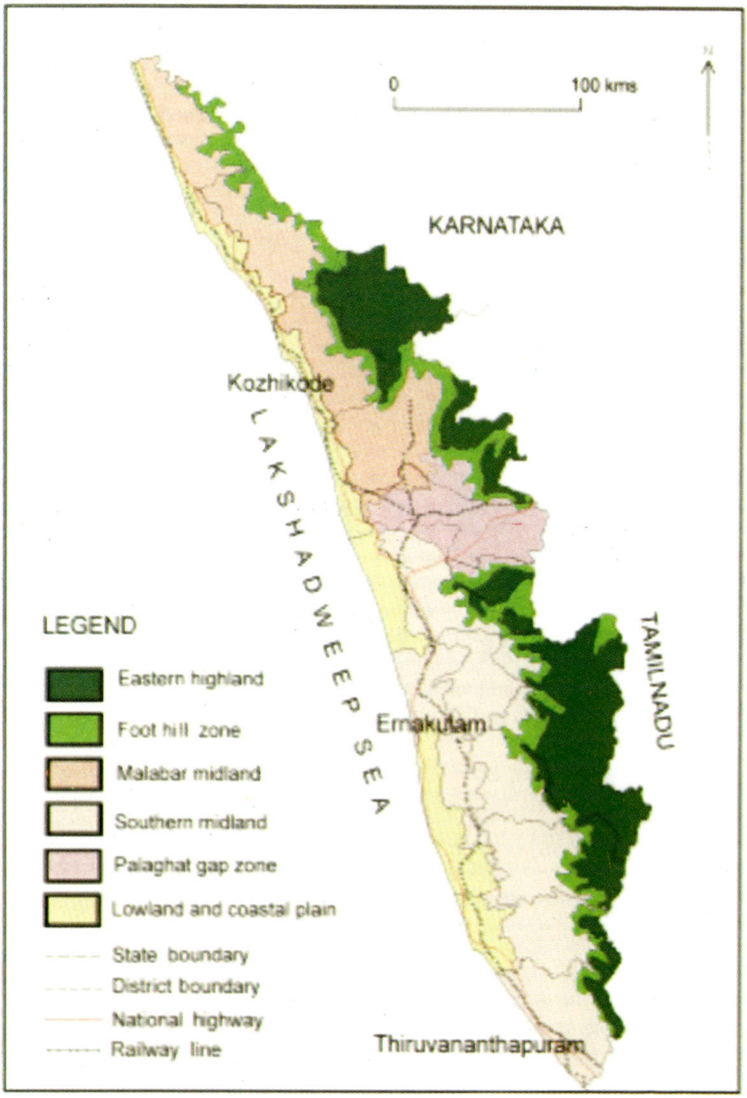

Fig. 2.2 Geographic Zones of Kerala

with the rest of India through the Tamil Nadu plain. The gap exerts considerable agro-climatic influence, which we will discuss below in the section on agro-climatic zones.

The foothill zone (300–600m) abuts the high ranges and is identified in two patches to the north and south of the Palghat Gap. The upland region (100–300m) below foothills can be divided into three sub-units: (1) the Malabar upland, dominated by low-level lateritic plateaus, (2) the Palghat Gap zone and (3) the Thrissur-Thiruvananthapuram rolling plain. These three units have different agricultural production characteristics. The undulating midland (10m–100m) is traceable throughout the state as a single unit placed longitudinally between the upland region and the coastal plain. The coastal plain also breaks into three segments (a) the Malabar coastal tract mostly comprised of laterites, (b) the Ponnani-Alappuzha coast with a dominance of recent sediments and (c) the Kollam-Thiruvananthapuram coast with occasional rock outcrops and intervening stretches of recent sediments.

One of the most conspicuous features of the Kerala coast is a backwater system parallel to the entire coastline. This backwater system gives the state resources not only for fishing but also to facilitate coir retting, an essential process for extracting coir fabric. Detailed discussion on the backwaters will be taken up in the section on water resources in this chapter.

Undulation starts almost at the coastline except in central Kerala, where a flat coastal plain extends about 30 km inland. The mountain ranges lie more or less parallel to the coast in a NNW to SSE direction and 41 of the 44 rivers flow towards the west. The descending topography generally follows the river courses in the central area. Alternating valley ridge topography has evolved in the process for most parts of the state. The low relief from the coastal plain extends towards the east within the hilly eastern part along the rivers and streams. The variation of altitudes within a short distance results in a high drainage texture – that is, a large number of drainage lines per unit of area.

Geology and Geomorphology

Geologically, Kerala is characterised by four major formations: crystalline rocks of the Archean Age, sedimentary rocks of the Tertiary Age, laterite capping on the crystalline and sedimentary rocks, and recent sediments forming the low areas and river valleys (GSI 1976; Soman 2002). The crystalline rocks are hard and cover the major part of the state. These crystalline rocks are chiefly composed of charnockitic, the khondalite suite of rocks, granite and granitic gneisses. There are basic dykes traversing these rock bodies.

These rock types, depending on their mineral composition, have differential hardness and are subject to variable weathering. The relatively hard rocks form the ridges while the weaker parts erode quickly to form the valleys. The monolithic hard rock body also experiences variable erosion rates due to structural deformation through geological processes. Sedimentary rocks are present in the coastal belt.

Laterites are a product of chemical weathering and are unique features in the tropical countries. They occur as cap rocks. The principle geological mechanism involved in laterite formation is horizontal and vertical movement of water in a near peneplained surface and chemical alteration of the parent rocks. It is in the Angadipuram region of Malabar where Francis Hamilton Buchanan, a medical officer in the services of the East India Company first reported laterite during his travels there in 1800–01 (Soman 2002:155).

Recent sediment occupies the valleys and coastal plain. Apart from the coastal plain the depth to bedrock is limited to 10 to 15 meters. Bedrock is exposed in many valleys, indicating that the sediments are not very thick there.

The lithology and structure have a pronounced impact on land and water distribution and landform patterns. Many of Kerala's rivers flow in straight lines along the geologically weaker planes known as lineaments in earth science literature. The ridge and valley alterations are coordinated by the presence of these weaker planes occupied by rivers or streams. These weaker planes are mostly lineaments. They are conducive to water percolation and thus facilitate ground water recharging.

The geological set up, high rainfall and location with respect to sea and mountains have given rise to a diversified geomorphology. Four main processes are operative: marine, fluvio-marine/estuarine, fluvial and denudational. There are units developed purely under erosion, or accretion/accumulation, or, a combination of both. A major part of the state is governed by denudational processes. As many as six units can be identified under this process. These six units together cover around 80 per cent of the total area of the state. All the districts except Alappuzha exhibit development of all six categories. Landforms in Alappuzha are dominated by fluvial, fluvio-marine and marine processes. Five districts in Kerala do not extend to the coastline. The remaining nine districts exhibit most of the units. In all, 22 geomorphic units have been identified in the entire state (Chattopadhyay and Chattopadhyay 1995). Kollam district in the south and Kasargod, Kannur and Kozhikode districts in the north each has 14 units. The diversity appears to be extremely high. The diversity brings important contributions in enhancing the biological productivity of the state.

Soil

Climate, geology, relief and other factors influence soil formation. It is the top-most part of the land, which is used for all activities. Traditionally, soils in Kerala had been classified into ten broad groups based on morphological features and physico-chemical properties (Gov. of Kerala–GOK 1978). These are red loam, lateritic soil, coastal alluvium, greyish *onattukara* soil, brown hydromorphic soil, hydromorphic saline soil, acid saline soil, black soil and forest loam. The lateritic soil covers around 60 per cent of the total surface area of the state. Forest loam has considerable areas in the eastern part. This classification has been further improved. Based on texture, depth, soil drainage, slope and erosion, 38 soil categories have been identified under five physiographic divisions (KSLUB 1995). There are seven soil types in the lowlands, which cover 16 per cent of the total geographical area of the state. There are limitations due to drainage and salinity. Soils in the midlands have been classified into six types. Soils are very deep, well drained, gravelly and occur on various types of slopes on laterites. The combined area of these six types accounts for 34 per cent of the total area of Kerala. Central Sahyadri (Sahyadri is the Sanskrit name for the Ghats) stretching from 100m to 2300m in the high hills is characterized by 14 different soil types, covering 20 per cent of the total area of the state. These soils vary widely from moderately shallow to very deep and from imperfectly drained to well drained. The soil developed in the Wayanad Plateau, Palghat Gap and under thick vegetation cover in this part of the Western Ghats is included in these types. There are six types of soil in the Nilgiris hill area. These types occupy the eastern part of Malappuram district. Area coverage is slightly over 12 per cent of the state total. The Southern Sahyadri extends from the southern flank of the Palghat Gap to

Thiruvananthapuram district. Elevations range from 300m to 2,695m. Soils are deep to very deep, well drained and with moderate to severe erosion. Five types of soil have been recognized. Total coverage of these types is around 18 per cent of the total area of the state.

The major limitations recorded for these soils are erosion and runoff, stoniness, and wetness/drainage. It has been estimated that only about 37 per cent of Kerala's area is suitable for irrigation – and this is with some limitations. The three major factors influencing irrigation potential are soil type, soil drainage and topography. Soil drainage primarily depends upon its textural composition. The irrigation potential of ill drained soils (soils with a high proportion of clay) and excessively drained soils (soils with a high proportion of coarse sand) is generally low.

Soils are generally deep in Kerala. As much as 64 per cent of the area has soil depths of more than 150cm. Loamy texture covers 59 per cent of the total area. Only 4 per cent of the area falls under the severely eroded category while 69 per cent of the area falls under the moderately eroded category. The capacity of soils to hold water in reserve expressed as AWC (available water capacity) is an important soil characteristic for agricultural purposes and it is measured in millimetres of water in 100cm of soil or in the entire soil column if the soil is shallower. Thirty five percent of the soils in the state have high AWC values (150-200mm) (KSLUB 1995:115).

Kerala's soils are productive and have a capacity to generate high biomass. We will elaborate this point in the section on ecological zones.

Water Resources

As we noted in chapter 1, water is one of the important natural resources that determines the productivity of land. The importance of fresh water in every major aspect of development is well realised, particularly in the developing countries, where the availability of potable water is often a major problem. Water security, like food security, is a major concern today (Biswas 1993). Kerala is well endowed with water resources (Nair and Chattopadhyay 2005). It also had traditional water harvesting systems. Our discussion of water resources will cover rainfall, surface water and ground water.

Rainfall

Rain is the source of all water. Once precipitated, part of the rain evaporates back into the atmosphere, while another part flows as surface runoff and the remaining water infiltrates. A certain portion of infiltrated water remains as soil moisture, and the rest recharges the sub-surface water or ground water.

Kerala enjoys a southwesterly, as well as a northeasterly monsoon. On average, Kerala receives over 3,000 mm of rain each year. In general, precipitation increases from the coast inland and the maximum is recorded along the foothills around the Neriamangalam-Palai stretch (5,883 mm) in the south and the Kuttiadi area in the north. The Palghat Gap, lying almost in the central part of the Western Ghats of Kerala disrupts the rainfall trend due to its sudden reduction in relief, resulting in considerably less rainfall. The eastern slopes of the Anaimudi around Chinnar and North Marayur, being on the leeward side receive less than 1000 mm of rainfall annually.

As mentioned above, Kerala experiences two monsoons – southwest and northeast. More than 60 per cent of all rainfall occurs during the southwest monsoon period. The contribution of the northeast monsoon is high in certain parts. The effect of the northeast monsoon is found to be more towards the south. As a result, the rainfall station in the southern part of the state records two peaks in the rainfall curve, one in June-July and the other in October, whereas the part of Kerala lying to the north of the Palghat Gap is characterised by a single peak during June-July.

Normal rainfall patterns indicate that the state receives some rainfall throughout the year, although amounts are considerably less during the dry months. Table 2.3 shows the average monthly rainfall in all the districts while figure 2.3 contains graphs of the monthly distribution at selected rainfall stations.

Rainfall variability worked out for the period 1961–79 showed that there are considerable variations between normal predicted and actual rainfall in several places (Shravankumar et al.1988). The general variability for the state is between 20 and 40 per cent above or below the expected values. Trends of variability change according to season. The northern part, especially the Kasargod-Kannur area records variability around 50 per cent from year to year.

The annual moisture index, a measure of water availability calculated from rainfall and temperature data for the state as a whole

Table 2.3 1963–1993 Average Kerala Rainfall in Millimeters by District

District	*J*	*F*	*M*	*A*	*M*	*J*	*JLY*	*AU*	*S*	*O*	*N*	*D*	*Annual*
Thiruvananthpuram	22	19	45	120	225	377	279	157	170	280	182	67	1943
Kollam	24	31	75	166	242	487	428	273	212	350	216	60	2554
Pathananthitta	28	41	99	202	306	585	550	372	272	339	242	62	3098
Allapuzha	25	27	56	129	323	637	547	327	271	333	228	61	2965
Kottayam	24	24	52	125	310	697	640	400	285	331	188	55	3137
Idukki	19	23	47	129	228	638	831	518	305	310	167	50	3265
Ernakulam	15	17	41	134	318	740	721	451	288	312	172	45	3254
Thrissur	7	10	27	86	296	769	759	443	257	301	144	31	3130
Palakkad	8	9	27	87	161	477	633	349	165	249	136	28	2329
Malappuram	5	7	21	86	221	688	797	399	205	291	157	30	2907
Kozhikode	9	7	21	93	297	925	1034	547	257	284	157	36	3667
Wayanad	7	9	23	99	183	762	1261	644	252	226	99	23	3588
Kannur	4	4	8	64	244	891	1085	570	249	226	99	23	3467
Kasargod	5	3	9	51	226	1004	1081	610	269	209	91	23	3581
All Kerala	14	17	39	112	256	691	760	433	247	288	163	42	3063

Source : Farm Guide, 1994 (KSLUB 1995:57).

is 80 per cent. From December until February the value ranges between 60 and 90 per cent, and during the March-May period, the index value comes down to 60 to 80 per cent. While this lower value restricts seasonal crop culture, it does not affect tree crops that have sufficient root penetration capacity.

Orographic influence on rainfall is pronounced. Moisture-laden air condenses in the hills, and thus the foothills receive maximum rainfall. As there is another level of high altitude over the plateau, rainfall is quite high – even on the plateau surface. Thus, both the Idukki and Wayanad plateaus receive high rainfall.

This year-round rainfall and the high precipitation in the foothills have contributed to the perennial water supply in the rivers. It is generally observed that if rainfall is low in the southwest monsoon, it is partly compensated for during the northeast period. Monsoon failure for both periods is very rare. The state therefore enjoys a certain amount of assured rainfall each year.

There is, however, a decreasing tendency in rainfall patterns. Soman et al. (1988:10–11) reported a 15–20 per cent decrease in rainfall in the foothill region during the 1941–80 period compared to 1901 to 1940. We shall elaborate this point in chapter 5 while discussing micro climate changes.

Surface Water

Kerala is drained by 44 rivers. Some of the rivers join together before their outfall; therefore, there are 34 river basins drained by the 44 rivers. Out of these 44 rivers three flow east, while the rest flow towards the west and join the Lakshadweep Sea through backwaters and estuaries. The catchments of some of the west-flowing rivers extend beyond the state boundary to the east. The total annual yield of the west-flowing rivers is estimated at 64,263 million cubic meters, out of which the annual utilisable yield is around 50 per cent (GOK 1974). On average, one square kilometer of catchment area yields 1.81 million cubic meters of water, of which 60 per cent is usable.

The east-flowing rivers located in the Wayanad Plateau, in Attappadi and in the northeastern part of Munnar plateau have a combined and fully utilisable annual yield of 6,060 million cubic meters. The total annual utilisable water in the state is 42,772 million cubic meters, about 61 per cent of the total yield.

River length varies from 16 km for the Manjeshwar River to 244

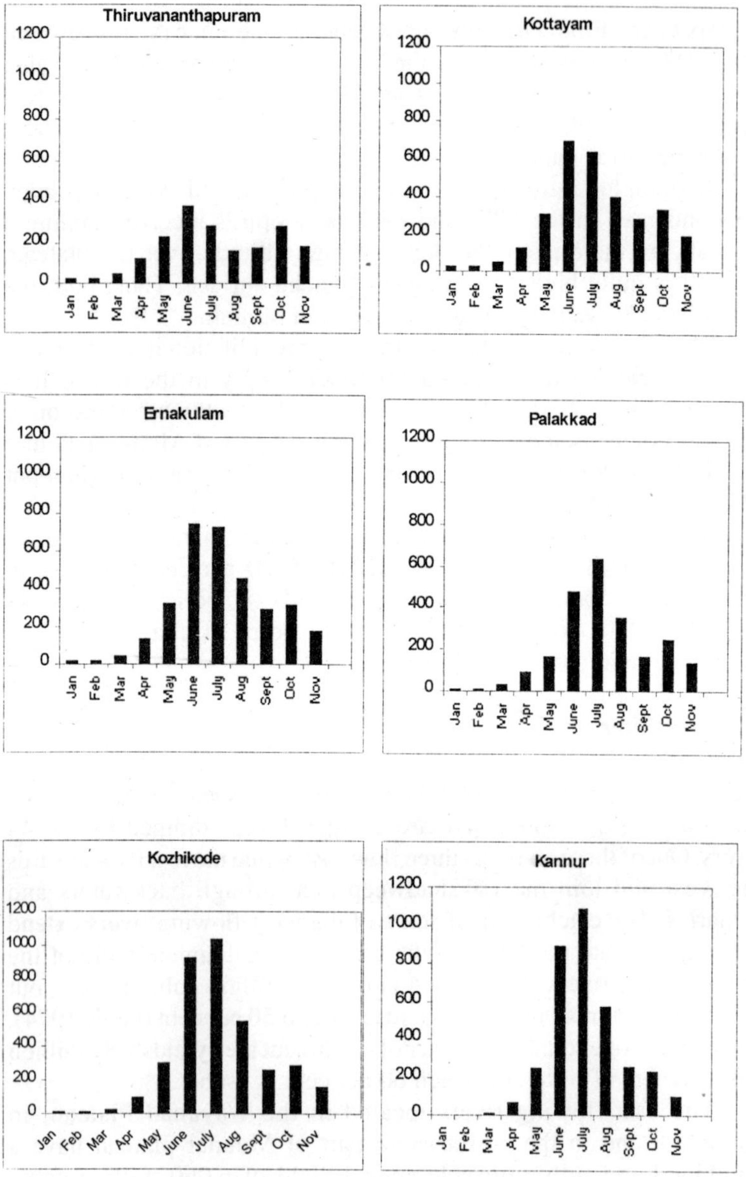

Fig. 2.3 Monthly Distribution of Rainfall (mm) at Selected Stations of Kerala

km for the Periyar. The Bharathapuzha River has the largest catchment area, at 6,186 sq km, with 1,786 square kilometers being located in Tamil Nadu. The Periyar River drains a catchment area of 5,398 sq km of which 5,284 are in Kerala. Rivers of more than 100 km length are: the Chandragiri (105 km), Valapattanam (110 km), Chaliyar (169 km), Kadaluandi (130 km), Bharathapuzha (209 km), Chalakudi (130 km), Periyar (244 km), Muvattupuzha (121 km), Pamba (176 km), Achankovil (129 km) and Kallada (121 km). The combined length of all the major rivers is around 3,200 km, which gives a figure of 12 sq km of catchment for every kilometer of the major rivers. In other words, 1 km of main river is fed by 12 sq km of catchment area. This indicates a higher capacity of the land to sustain a river and also high water yield compared to adjacent states. Figure 2.4 is a map showing the drainage patterns of the main rivers of Kerala.

Most of the rivers in Kerala are perennial. Many of the rivers and tributaries originate in the foothills, where the rainfall is very high. The distribution of rainfall throughout the year, combined with deep soil and high biomass density contribute to the perennial water availability of Kerala's rivers. Despite high fluctuations between monsoon and non-monsoon months in drainage discharge, there are base flows in the rivers throughout the year. The river level controls the water table in the adjoining dug wells, which are the main sources of domestic water in the rural areas. Along with fluctuation of water level in the rivers, the level of the ground water table also fluctuates. Maintenance of the base flow in the rivers during the non-monsoon months is therefore very significant for water security.

Many of the rivers have been dammed in their upstream sections for irrigation and hydroelectric projects. There are 30 reservoirs covering an area of 296 sq km. Water is diverted through canals for irrigation during the dry season. It also helps recharging wells in certain areas. But the reservoirs are mixed blessings. On the one hand they facilitate generation of electricity, provide irrigation water for agriculture and therefore contribute to development. On the other hand they restrict silt movement and the coastal area is deprived of its natural sediment supply from the upland through these rivers. The distribution of sediment is affected in the process, which in many cases leads to coastal erosion. Furthermore, impounded water may cause reservoir induced seismicity (RIS) bringing on minor earthquakes. The adverse impacts of the large reservoirs raise doubts about the viability of the canal irrigation system.

Fresh Water Lakes

There are nine fresh water lakes in Kerala. Three of them are the most significant: Sasthamkotta Lake in Kollam district, Vellayani Lake in Thiruvananthapuram district and Pukkot Lake in Wayanad district. While Pukkot Lake is used for pisciculture and irrigation, Vellayani Lake and Sasthamkotta Lake supply water for urban areas. Kollam municipality, with a population of 361,441 in the 2001 census depends upon the Sasthamkotta Lake for its drinking water. Natural depressions, locally known as *'chira'* distributed surrounding this lake, function as water harvesting devices during the monsoon months.

Detailed studies were conducted by the Centre for Earth Science Studies (CESS) in 1998 and 2000 concerning the water potential of Sasthamkotta and Vellayani lakes. Water budgeting for Sasthamkotta Lake indicates that the annual yield of this lake is to the tune of 4.7 million cubic meters, contributed by direct rainfall, runoff and groundwater recharge. There is a significant inflow of water from a vast sub-surface catchment, comprising the tertiary sedimentary formation. The palaeo channels that connect the lake to the adjacent river and the flood plain of the Kallada river used for paddy cultivation also contribute to the lake's storage capacity. With slightly over 4 sq km, the surface catchment area of the lake does not appear to indicate the actual area from which the lake is being recharged. In the case of Vellayani Lake, an inflow of 15 million cubic meters was estimated for an eight-month period (end of April to December 1999). Surface runoff contributed to two-thirds of the inflow.

Ponds and Tanks

Ponds and tanks are traditional rainwater harvesting structures, used widely in tropical countries. India has a long tradition of utilising ponds and tanks for household water and for irrigation. Excavating a pond or replacing an existing one used to be considered a virtuous act. In the precolonial times concern of the native rulers (Rajas or Maharajas) for their subjects was expressed partly through excavating ponds or building tanks, along with laying roads and constructing places of worship.

Ponds and tanks are an effective traditional rainwater harvesting technology, and are well established throughout South Asia. In many areas the ponds are quite large, but in Kerala these structures were

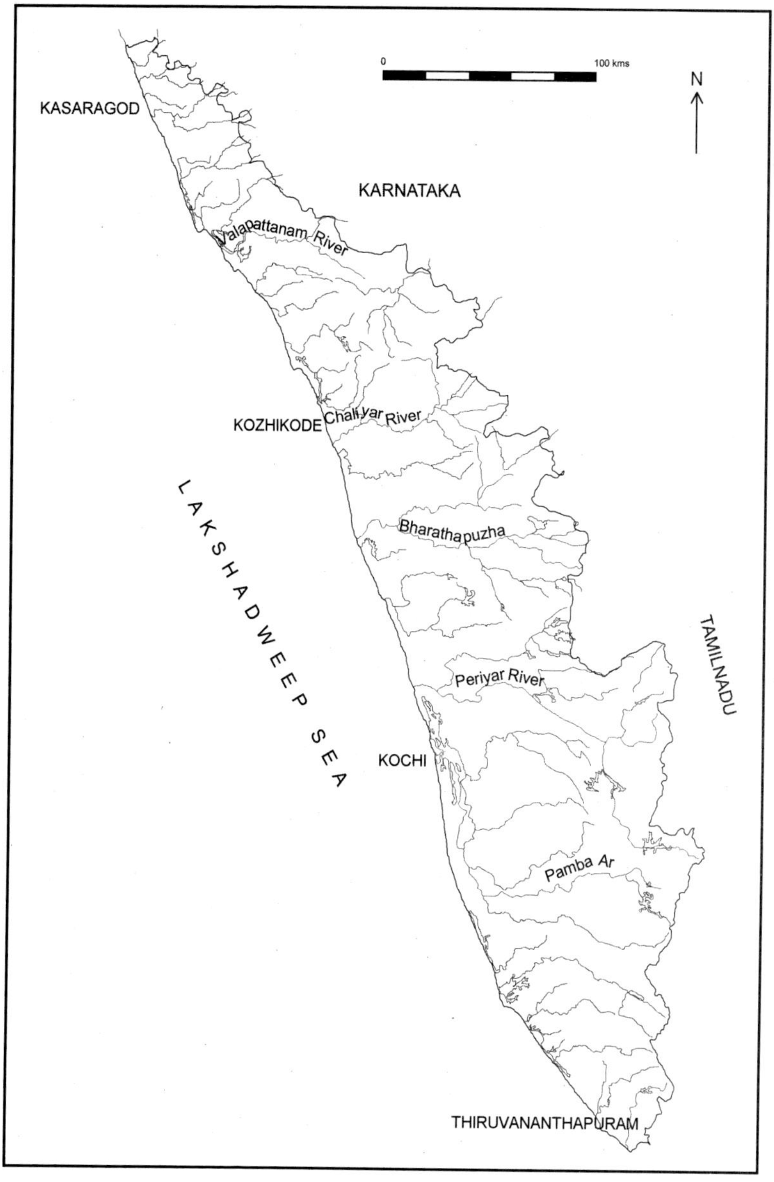

Fig. 2.4 Drainage Pattern in Kerala

developed on a small scale, conforming to the region's more variegated topography. Nevertheless, they have helped the state to augment its water resources from early days.

Tamil Nadu has a large network of ponds and tanks to hold spill water at different levels of flooding. The ponds and tanks observed in Kerala are not similar to those of Tamil Nadu with the possible exception of those in Palakkad district. The ponds in Kerala can be grouped into three types (Chattopadhyay 1999): natural depressions locally known as *chiras* in some parts of the state, excavated and man-made ponds or tanks, and valley-head ponds or 'Talakulam' (tala = head, kulam = pond).

While the first two types are located in the low lying areas the talakulams are located at the valley head. Water is impounded by constructing a barrier or dam across the valley. On both sides of the valley, small channels are excavated following the slope to carry water. Valleys of minor streams are converted to agricultural fields pushing the stream channel from the centre of the valley to the side along the break-in-slope that demarcates the boundary between the valley and the side slopes. This is a common sight throughout Kerala. Figure 2.5 is a photograph of a landscape showing a flat bottomed valley, a channel along the break in the slope and a side slope in Thiruvananthapuram district.

Fig. 2.5 Midlands Landsape in Kerala. Photo by Srikumar Chattopadhyay

Valley-head ponds are used mainly for irrigation, supplementing rainwater to grow rice or other crops after the monsoon. Similar structures are found in Sri Lanka. Optimisation of water use from these structures has also been attempted in Sri Lanka (Mahendrarajah et al. 1993). Many of the valley-head ponds in Kerala are associated with springs. Spring fed ponds are perennial, however all springs do not yield large quantities of water.

Some of the main items mapped by the volunteers under the Panchayat Resource Mapping Programme (CESS 1991; Chattopadhyay et al. 1999) were water structures such as ponds, tanks, wells, rivers and canals. Survey results brought out the existence of large numbers of ponds in the rural and also in the urban areas. Table 2.4 indicates the status of ponds in five selected panchayats. These panchayats are distributed in different parts of the state (north, central and southern Kerala) with various types of landscapes, ecological conditions and rainfall patterns.

Table 2.4 Status of Ponds in Five Selected Panchayats, 1992

Panchayat	*Area of the panchayat (km^2)*	*Location*	*No. of Ponds*		*Uses*	*Problems*
			Total	*Perennial*		
Mayyil	48	North Kerala midland	27	17	Irrigation	Siltation
Onchiyam	9	North Kerala coastal plain	84	39	Irrigation and domestic	Siltation and quality
Madakkathara	25	Central Kerala midland/ high land	26	7	Irrigation	Siltation
Kumarakom	52	South Kerala lowland	120	120	Domestic/ Irrigation	Quality/ siltation
Mezhuveli	14	South Kerala midland/ Lowland	47	27	Irrigation/ Domestic	Quality/ siltation

Source : Panchayat Resource Mapping Project, Centre for Earth Science Studies, 1993

The average density of ponds in these five panchayats is two per sq km or one pond for every 0.5 sq km of area. The distribution varies from Onchiyam with one pond in every 0.10 sq km area, to Mayyil, where one pond can be found in every 1.8 sq km area. Mayyil, being in the hard lateritic area has fewer ponds. Siltation is a common problem encountered everywhere. Problems of water quality were also reported in these panchayats. From these samples and from our observations in every district, it can be reasonably assumed that all the panchayats in the state are endowed with ponds and tanks.

A state level survey conducted by the Centre for Water Resource Development and Management (CWRDM 1989) identified 995 large ponds with a minimum water storage capacity of 1500 cubic meters each. Of these, 287 ponds are public and 137 belong to temples. They are found throughout the state, but a sizable number are concentrated in Palakkad district, which is drought-prone due to the effect of the Palghat Gap. Rainfall in this area is less compared to other parts of the state. Even the Thiruvananthapuram Urban Corporation area can boast of 80 ponds, some of which are still in use.

The Backwaters

Kerala has a chain of backwaters running from the north to south. The backwaters add enchanting beauty to the coastal region and have become a major tourist attraction. Backwaters can be categorised into lagoons, estuaries and lakes, connected by channels and also broken up by man-made lands. The brackish water (mixed salt and freshwater) forms part of the coastal wetlands system. When taken together, backwaters, estuaries and brackish water lakes, cover an area of 2,426 sq km (GOK 1992). There are 46 estuaries and backwaters. The Vembanad is the largest backwater system in the state. Vembanad Lake, along with the Kochi backwaters, covers an area of 233.50 sq km. The Ashtamudi estuary has an area of 64.24 sq km. The Government of India's Department of Environment has designated both of these as wetlands of international importance (See figure 1.2 for locations). Accordingly these wetlands come under the purview of certain conservation measures with required fund allocations from the Government of India. This is part of the international effort to protect wetlands initiated following an international convention, also known as the Ramsar Convention, held at Ramsar Iran in 1971. An inter-governmental international treaty was signed to protect wetlands.

Wetlands identified under this category are designated as Ramsar sites or Ramsar wetlands.

The backwaters perform a variety of functions that directly or indirectly benefit humans. They provide fish; hold excess monsoon runoff; facilitate water percolation; purify water; maintain the nutrient cycle; provide sites for biodiversity, easy transport, retting of coconut husks; and act as pollutant absorbing sinks. As most of Kerala's rivers pass through the backwater system before emptying into the Lakshadweep Sea, the backwaters tend to be enriched with nutrients brought down by the rivers. The sediment and nutrient loads flowing down with the river water first settle (or are naturally panned) in the backwaters before they move to the sea. As a result, the productivity of the coastal waters is linked to the nature of the settling process in the backwaters. The presence of the backwater system in the coastal plain has added to the diversity of the resource base of the state.

Ground Water

Kerala has good ground water potential. Crystalline rocks occupy about 90 per cent of the total geographical area of the state and the rest has sedimentary formations. There is widespread laterite cover throughout the state. Ground water in the sedimentary formation occurs in both water table and confined conditions (CESS 1984). In crystalline terrain it is available in the fractures and contact zones, through which ground water recharging takes place. The potential of ground water in this terrain depends upon rock type and fracture pattern.

Average annual ground water recharge in the state is 7,900 million cubic meters, of which 85 per cent can be utilised for irrigation purposes every year. The net draft for irrigation as of 1992 was 1,006 million cubic meters or 15 per cent of utilisable ground water. However, there are wide spatial variations in utilisation patterns. Block level utilisation patterns provided by the Ground Water Department (GOK 1995) indicate that in some Blocks, such as Chirayinkkil, net draft has reached 60 per cent. In the Western Ghats districts, the level of ground water use centers around two to three per cent. It is important to note that ground water is available throughout the state. Even the lateritic wasteland that dominated Kasargod had utilisable water equal to 85 per cent of the annual recharge. This provides more or less equal opportunity to all segments of the state in utilising ground water resources.

Dug wells are widely used in Kerala for drawing sub-surface water. Depth to the water table in dug wells varies from one to two meters in the coastal plain to more than 15 metres in the hill areas. In general, the water level changes from lowland to highland, and in particular, it varies from valley bottom to ridge top. In the north, with thick laterite cover, the depth can reach more than 20m and sometimes the wells go dry in the late dry season months of April and May. Such areas have more than 200 wells per sq km. In some places even a density of 400 has been reported. Dug wells are still in use for drinking water purposes in rural areas. Some households have more than one well in their compound.

Flora

Kerala is well endowed with natural vegetation and wildlife. The climatic variability and physiographic diversity are well manifested in the range of vegetation types. Moist tropical forest dominates the natural vegetation, covering the largest area. Dry deciduous sub-tropical montane vegetation can be marked in patches having low temperatures and meagre rainfall. The main vegetation types are wet evergreen, semi-evergreen, moist deciduous, dry deciduous, sub-tropical montane and temperate shola forests. Sholas are vegetation concentrations along the water courses in the high mountain areas, usually above 1,000 metres. Between the water courses dry savanna areas predominate (Nair et al. 2001). Besides this, there are forest plantations. Mangrove patches are found along the coast and the estuarine margins.

According to the Forest Department, Government of Kerala (1992), the total forest area in 1991–92 was 11,223 sq km of which 7,860 sq km were dense forest, 1,538 sq km in plantations, and the remaining 1,825 sq km were given to various agencies, such as the Kerala State Electricity Board (KSEB) and Public Works Department (PWD) for the river valley projects, the Forest Development Corporation, and the Kerala Livestock Development Corporation. There are disputes about the figures for the dense forest cover. Through field observation and satellite image analysis it has been found that there is considerable thinning of vegetation in many areas.

Natural virgin forests in Kerala today are confined to rugged slopes and high altitude areas. The state witnessed sizable deforestation in the present century. This point will be discussed in the section on changes in land use patterns in chapter 4. Figure 2.6 is a photo of forest vegetation along rugged slopes in Kerala.

Fig. 2.6 Forest Vegetation along Rugged Slope in Kerala.
Photo by Srikumar Chattopadhyay

Forests are classified by legal status as reserve forest (83%) and vested forests (privately owned forest subsequently acquired by the government (17%). Before reorganisation of the state in 1956, parts of the forests in the Travancore-Cochin area were privately owned. Following reorganisation, all forests were taken over by the government and at present there is no difference between reserve forest and vested forest. All forest lands are treated under the same laws. There are certain regulations to follow in the matter of using forestlands.

A brief description of the vegetation types, irrespective of their legal status, is provided here to clarify the nature of Kerala's forest wealth. Figure 2.7 shows the various forest types in Kerala. The tropical wet evergreen and semi-evergreen types cover 37 per cent of the forest area. These types are characterised by tall trees in the upper storey followed by dense second storey and undergrowth of numerous ferns and tall herbs. The tropical moist deciduous forests cover 44 per cent of the effective forest area. Here, the vegetation is less dense, but commercially valuable trees such as teak and rosewood abound. The

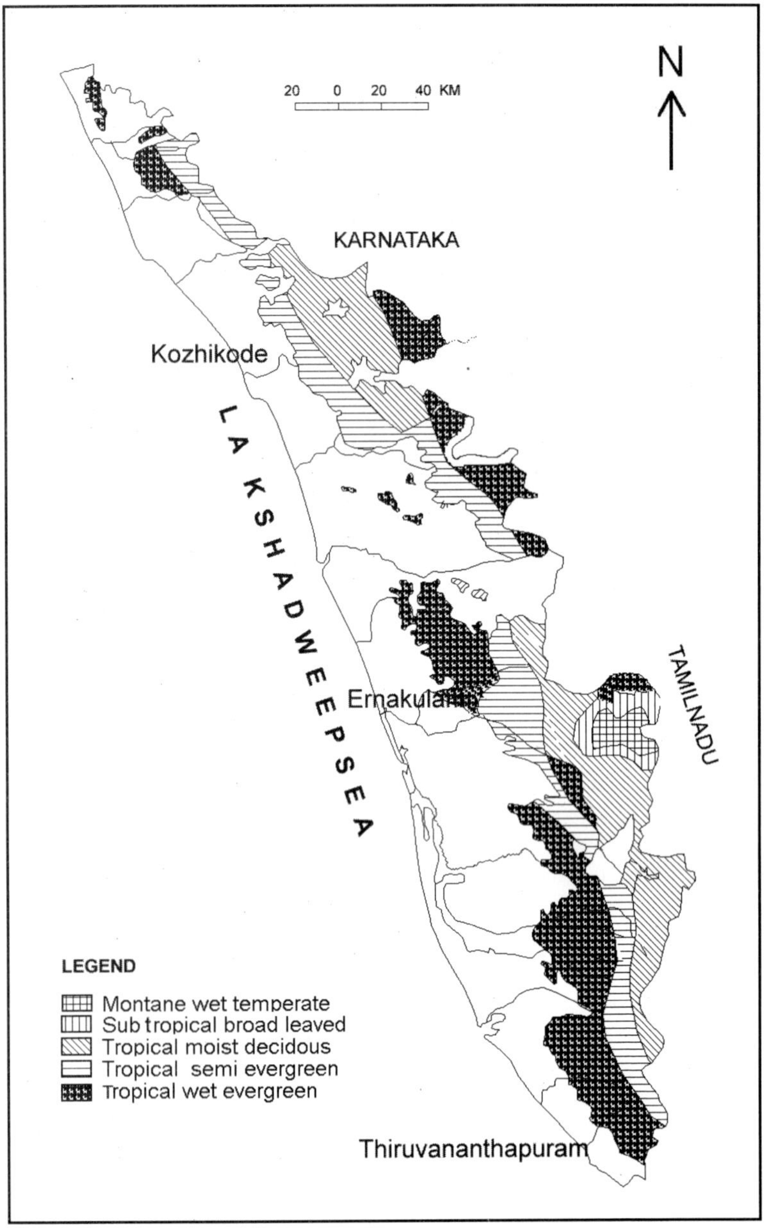

Fig. 2.7 Types of Forest in Kerala

evergreen, semi-evergreen and deciduous forest types are located in the rainfall zone of 250 cm to 350 cm and above and at a temperature of more than 20°C (Shravankumar et al. 1988). They are mostly located above 300m altitude. In the northwestern flank of the Wayanad Plateau in north Kerala, these forests are found at lower levels. The dry deciduous type, which accounts for 1 per cent of the effective forest area is located in the Pambar valley to the northeast of the Anaimalai mountain complex. This is a rain shadow region with annual rainfall of less than 100 cm, and winter temperatures below 20°C.

Montane sub-tropical and temperate sholas cover 2 per cent of the effective forest area and are found along streams in the valleys within high ranges. Grasslands are found in isolated patches throughout the state. They mostly occupy the ridges. Some of the grasslands are due to climatic climax and some others are due to human intervention. Grasslands along the Western Ghats crest developed in stretches of forest clearance where roads were laid for plantation development in the nineteenth century. The remaining 16 per cent comes under forest plantations.

There are 21 forest divisions and 6 wildlife divisions to administer the forest areas. Forests are found in all districts except Alappuzha. Kottayam District has only 101 sq km of forest. Idukki District has a forest area of 2,991 sq km, followed by Palakkad with 1,604 sq km. Wayanad District has 884 sq km of forest area. The Western Ghats part of the state along the eastern border is under forest cover of various intensity, except for the Palghat Gap where more grasslands are found.

Fauna

Kerala provides home to a variety of wild animals: elephants, various types of deer, wild boars, tigers, panthers, nilgiri taur, nilgiri langur and lion tailed macaques. To protect wildlife, and to promote forest conservation, 2,315 sq. km or 21 per cent of the forest area has been earmarked as either wildlife sanctuaries or national parks. There are seven wildlife divisions in the Forest Department to manage fourteen wildlife protected forest area. These are at Thiruvananthapuram (Thiruvananthapuram district), Agasthyavanam (Thiruvananthapuram district), Thekkady (Pathanamthitta and Idukki districts), Eravikulam (Idukki district), Parambikulam (Palakkad district), Silent Valley (Palakkad district), and Wayanad (Wayanad district). This distribution throughout the state is significant for decentralised management, and promotes habitat-specific development.

There are eleven wildlife sanctuaries, two national parks and one bird sanctuary in Kerala. The Silent Valley National Park is nationally and internationally known. Periyar Wildlife Sanctuary is a well-known site globally with a four-star hotel and a steady stream of visitors eager to view the wildlife. The environmental movement in Kerala centered on the Silent Valley during the 1970s. This 89 sq km virgin tropical rainforest is still pristine. There was a proposal to construct a hydel power project in this area. The movement mobilised against this proposal. The site is now protected as a national park. The wildlife sanctuaries and national parks attract tourists from various parts of the country and abroad. The people's movement to save Silent Valley is described in detail in chapter 7.

Resource Zones

Kerala has three distinct natural resource zones distributed longitudinally parallel to the coast. These are fisheries and aquaculture, including marine, backwater and fresh water bodies, agriculture, and forest.

Fisheries and Aquaculture

Kerala's long coastline and large backwater areas have provided for ample fishery development. The swift flowing small rivers bring down rich nutrients to the backwaters and coastal waters. As indicated earlier, on average, one km of main river is fed by 12 sq km of catchment area and one km length of coastline is supported by 70 sq km of land area. This coupled with the high rainfall experienced by the state has resulted in high nutrient delivery in the backwaters and the sea. The Kerala coastal waters are extremely rich. The marine fisheries are important for the state economy and provide livelihood support to a good section of the people.

Nine out of fourteen districts in the state have coastal access. Alappuzha and Kannur districts have the maximum coastline length. The continental shelf area extends to 39,139 sq km (GOK 1992) where fishing can be done. The backwaters and estuaries used for brackish water fish culture also make an important contribution to the state and local economy. The state has 652 sq km of area in this category.

The Fisheries Department is promoting backwater fish and prawn culture farms. These farms cover 11.42 sq km and are distributed over all nine coastal access districts. Ernakulam district has the maximum area of 6 sq km. There are traditional prawn filtration fields in the state. Waterlogged paddy fields are used for prawn culture. These are mostly single-crop paddy fields. Prawn paddy alternation has become an important practice in many parts of India, especially along the east coast. Kerala had 125 sq km under such practices in 1992. Ernakulam district had the maximum area of 106 sq km in this sector.

Prawns are important export items in the fisheries sector. Data for 2003–04 indicate that export of marine products was 76,627 metric tons. The export value of fish products was worth Indian Rupees 10,990 million (GOK 2005:103).

Kerala has well developed marine and backwater fishing infrastructures. There are nine fishery harbours and landing centres for mechanised fishing boats. Neendakara fishing harbor at the mouth of the Ashtamudi estuary is one of the most modern and important. There is also scope for development of inland fisheries. The 30 reservoirs, 9 lakes, and the rivers and ponds in the rural areas can be used for inland fish farming. There are 111 villages (out of 991, total villages in Kerala) where inland fishing is prevalent. Out of 222 coastal panchayats involved in marine fishing, many of the panchayats are in the sector of inland fishing by virtue of their location between sea and backwater.

The widely dispersed distribution of resources throughout the state has given a good section of the people easy access to resources. The problems emanate mostly from inappropriate management, rather than the resource base. It is here that the advantage of the decentralized resource distribution has to be maximised. We shall describe an innovative coastal area project in chapter 10.

Agriculture

The main agricultural systems comprising seasonal crops, annual crops, tree crops and plantation agriculture are prevalent throughout the state, and cover lowlands, midlands and parts of the highlands. The cultivators in lowland and midland regions in all the districts resort to more or less similar practices. The crop system in the state can be classified under four broad groups: seasonal crops, annual crops, tree crops and plantations. The plantations include tea, coffee, cardamom and rubber. Another important crop raised in the state is black pepper. Kerala was internationally well known in ancient times as a land of spices or 'land of black pepper.' Oil palm plantations are now being introduced in some parts of the state.

Coconut has surpassed all other crops in terms of individual crop coverage. In 1996, it had 1,005 thousand hectares (one million and 5,000) or 33 per cent of the total cropped area of 3,067 thousand hectares. This is followed by rubber (15%), rice (14%), pepper (6%), tapioca (5%) and cashew (3%).

Apart from tea, coffee and cardamom, which require a certain temperature and rainfall, the other crops can be raised throughout the state. This provides an advantage to the people who depend on agriculture in terms of cultivation as well as marketing. There is no undue demand in any single zone to raise a specific crop.

To bring out the picture, district-wise percentage distributions of particular crops in terms of total area under those crops in the state have been worked out. The data are shown on Table 2.5. This gives the spatial concentration of the crops at the district level. For each crop, the difference between the highest percentage and the lowest percentage is calculated. Dividing this value by the number of occurrences will provide the degree of concentration. The low value indicates a low degree of concentration.

It should be noted that none of the districts accounts for more

Table 2.5 : Concentration Index for Selected Crops in Kerala – 1991

Crops	*Per cent coverage at district level*		*Difference (D) (=Col 2 minus Col 3)*	*Index (D/14) (=Col 4 divided by 14)*
	*Highest**	*Lowest**		
Total food grain	28.57	2.01	26.56	1.9
Sugarcane	49.25	0.50	48.75	3.5
Condiments and spices	22.82	1.27	21.55	1.5
Total fruit trees	17.00	2.75	14.25	1.01
Vegetables	18.38	2.09	16.29	1.15
Total food crops	16.19	3.25	12.94	0.92
Total oil seeds (including coconut)	13.52	1.93	11.59	0.83
Total non-food crops	10.44	4.57	5.89	0.42

Basic Data Source: Agricultural Statistics of Kerala, 1992–1993.

* The highest and the lowest have been worked out by accounting percentage coverage of a particular crop in a district with respect to total area under that crop in the state.

than 20 per cent for any major crop except for Palakkad, which accounts for 28.57 per cent of the total area under food crops and Kottayam for rubber (20.19%). Almost all the crops are raised in all the districts with certain variations in proportion. The concentration index is found to be the highest for total food grains (1.9), while the lowest is for total non-food crops (0.42). Sugarcane is a minor crop. It covers less than 0.05 per cent of the total cropped area, and is grown mainly in Palakkad. As it is concentrated, the index value is high. Food grains, which are dominated by rice, also have relatively high concentrations.

It can also be inferred that apart from seasonal crops and tree crops, non-food crops or commercial crops are also distributed throughout the state. This reinforces the picture of agro-climatic diversity that exists even at the district level.

Forest Resources

The earlier discussion on vegetation highlighted the type of biomass with which Kerala is endowed. A brief discussion on forest resources is provided here. Around 18 per cent of the state's forest area was under plantations as of 1992, as compared to 12.6 per cent in 1973. Teak is the main plantation crop covering a little less than 50 per cent of the total plantation area. Mixed plantations and eucalyptus follow. In all, there are 11 types of plantations. The Forest Development Corporation, which maintains a plantation area of 106 sq km, raises pulpwood, teakwood, softwood, in addition to cardamom and tea. Most of the cardamom and tea plantations are privately owned. Forest products include logs, poles, charcoal, firewood, sandalwood, honey, reeds and bamboo.

Honey is one of the main minor forest products. There are wild ginger, herbs and medicinal plants. These products are regularly collected from the forests by the tribal people (*adivasis*). Reeds and bamboo are used for basket making, a cottage industry that thrives along the peripheries of the forest areas. As was pointed out in the discussion on flora and fauna, forest resources are found in almost all the districts.

Productivity and Access to Resources

The foregoing discussion on resource zones brought out the decentralised nature of the resource distribution pattern in Kerala.

This section attempts to highlight questions of productivity and access to resources that are enjoyed by the people of the state. Productivity will be discussed in terms of crop yields and income generation.

Rice is the principal crop in India and also in the state, although coverage-wise, rice is now the third ranking crop in Kerala as we noted above. The average yield of rice in Kerala is 2,018 kg/ha, against the all India figure of 1637 kg/ha. But for the high yielding varieties, the all-India average yield rate is 2310 kg/ha compared to Kerala's 2129 kg/ha. For unirrigated rice, Kerala's yield rate is 1.5 times the national average (1090 kg/ha). The proportion of gross irrigated area to gross cropped area is less in Kerala (12.3%) than that in India (32.6%). Per capita power consumption for agriculture in Kerala is six times lower than that in all-India. Use of power in agriculture is primarily for lifting irrigation water. Kerala consumes less power in this sector due to various factors such as sufficient rain water, canal irrigation, spring water and river water.

The high yields in Kerala can be attributed to the availability of rainwater for longer periods, inherent soil quality, and comparatively better management. Since the rice fields in Kerala are located in the valleys, they receive sediment and water from the slopes. Due to high biomass coverage, both water and silt are nutrient laden and thus enrich the fields. Although canal irrigation is limited, ponds, springs and river water are commonly used for irrigating rice fields. There are fields in Palakkad district where rice can be grown three times in a year with in-situ spring water. These types of wet paddy fields can be found in many places in the state. Fertiliser use in Kerala is higher at 81.5 kg/ha than the all-India average of 72.4 kg/ha.

The high productivity of the state is further reflected in the yield of total food grains: 1,784 kg/ha in Kerala against an all-India average of 1,384 kg/ha. Total food grains include rice, millets, and pulses. Kerala also has a much higher proportion of area under non-food crops. Income generated in agriculture from one hectare of gross cropped area in the state is 1.8 times the all India average.

Like the agricultural sector, forests in the state are also highly productive. It is globally well known that tropical rainforests are highly diverse in biota and also highly productive. Not only biomass but also value-added products are quite high in the tropical forest area. The value of forest products per hectare of forest area in Kerala is around 10 times higher than the all-India average.

Kerala accounts for 20 per cent of the total fish production of

India, although the state operates in little more than 10 per cent of the coastal waters.

Access to resources can be discussed at two levels: location and possession. Location is related to settlement. Kerala has a unique settlement pattern. Linear settlement in the coastal plain along the sandy ridges provides easy access to the sea. Inland settlements are along ridges, and agriculture spreads along the valleys. The dispersed nature of the settlement has given easy access to the people for harvesting local resources. This also provided space for household waste disposal and better sanitation. As there are few limitations in obtaining water, the basic structure of the settlement system is decentralised.

Historians and geographers have noted Kerala's unusual settlement pattern. Most Indian villages fall into one of three types. *Nucleated* villages have houses close together in a well-defined site. Such villages are common in western Uttar Pradesh, Delhi, Punjab, and some parts of southern India and are associated with wheat, barley, and millet production. *Hamleted* villages are found in the middle and lower Ganges plain and some other areas. They are composed of "a central settlement, several hamlets, and satellite settlements scattered over the fields" (Cohn 1971:142). Hamleted villages generally occur where rice and other grain agriculture are mixed (Cohn 1971:143). In Kerala, villages correspond to a third type: dispersed (Mencher 1966:144; Franke 1993:19–20). This pattern is also found in some of the hill regions and in the lower delta of the Ganges, and in other areas of intensive rice agriculture. Dispersed villages have homesteads spread across the landscape in such a way that one cannot always tell where one village begins and another ends. In many parts of Kerala, villages are more like extended suburbs of towns with no clear boundaries between urban and rural areas. The dispersed village patterns in Kerala have traditionally been linked to the dispersed access to water, which we documented in this chapter.

Kerala's dense and fairly even population distribution may be a factor in its high quality of life achievements. It is easier and more cost-effective to provide health and educational services to evenly dispersed populations than in nucleated or hamleted villages where the logistics of roads and other infrastructure are more complicated. The dispersed settlement pattern in Kerala may also help reduce infectious and parasitic diseases because of the distance between one house compound and the next (Franke and Chasin 1994:23). In this chapter we saw how the dispersed settlement pattern grows out of the resource base.

Land reform has given possession of the land to the people. Given Kerala's high land productivity even a small holding can meet food demands to a certain level. Due to the high biomass, fuel and fodder availability at the local level are higher. Now, open grazing is very limited due to conversion of pastures and grazing land for other purposes. Still high biomass availability helps the rural people in the fringe areas to raise livestock and collect fuel wood.

To conclude this discussion on the resource base, we stress the following points since they are extremely important in understanding why there should be overriding concern about the degradation of the environmental resource base in the state. The general productivity level of the state is high, and in the case of forests it is much higher than for all India. Long growing seasons add to its agricultural productivity. High biomass production, high assimilating capacity of the biosphere and very high species diversity combine to increase the production base of the state. This has given Kerala a higher carrying capacity than other Indian states. There is greater access to resources both in terms of location and possession. All these may be the underlying reasons for the co-existence of the highest population density and the highest quality of life, despite a low level of industrialisation when compared to the rest of India.

The basic pattern of the resource system is decentralised or dispersed, as is the case with settlements. All these natural advantages have pushed the state up front to utilise the opportunities at the individual and community level. These advantages have to be protected and enhanced by wise use with limited injury to the environment. The need for local level intervention and local self-government emanates from this basic requirement in Kerala's unique circumstances.

3 Macro and Micro Ecological Zones

Why Micro Matters

Chapter 2 gave an overview of the general features of Kerala's environment and their links to the nature of Kerala's population patterns – the demography. The layout of communities in terms of access to water is one important element in explaining Kerala's unusual population dispersal of non-nucleated villages. Kerala's ecological diversity in terms of soil types, altitude variations, rainfall regimes and biomass – plant and animal resources – is also important. Classifying the state's resources into zones improves our understanding of the production potential and further links the basic geographical features to the issue of decentralisation that we will investigate in detail later in this book. In this chapter we identify and discuss both macro and micro level ecological zones in Kerala. The ecological zones result from physical and biological factors that give rise to combinations of land, water and biomass patterns. These patterns in turn produce the production possibilities to sustain human life. Identifying these zones helps us to recognise the potential for various levels and types of activity. In earlier years ecology had focused on paradigms that assumed spatial homogeneity for long periods of time, relegating heterogeneity to unexplained variants. But today it is not possible to ignore such heterogeneity in spatial patterns (O'Neill et al. 1995). Landscapes can be classified to capture significant aspects of the pattern. These are mostly temporal and spatial change-sensitive, and can be interpreted in terms of ecological processes (O'Neill 1992). The prime objective of identifying the zones is to find the range of variability for a given territory. This raises the possibility that the landscape can be managed in order to produce items for human use without compromising its ecological integrity.

Macro Level Zones

Ecological zones can be recognised both at the macro and micro levels.

The variables having regional dimensions are considered for demarcation of macro level zones and local variabilities are captured through micro level zonation. It is also possible to design a hierarchic classification system linking micro level zones to the macro level zones. This section intends to provide an understanding of the macro level zones that exist in Kerala.

Climatic Zones

Climate is perhaps the most widely used parameter to classify the earth into production zones. The climatic classification schemes proposed by Köppen (1936) and Thornthwaite (1948) are widely used throughout the world. These schemes considered rainfall and temperature as the basic parameters for classifying the world into climatic zones. Regional varieties of temperature and rainfall, seasonal wetness, dry conditions and the prevalence of monsoons are also taken into account. Applying their classification scheme to India and applying regional data, Subramanyam and Murthy (1982) divided Kerala into two zones : tropical monsoon and tropical savanna. Parts of Kannur and Kozhikode districts in the north, more than half of Kottayam and Alappuzha districts, the entire Kollam, Pathanamthitta and Thiruvananthapuram districts are designated as tropical savanna. The rest of the state falls under the tropical monsoon type. The difference between tropical monsoon and tropical savanna is determined by the amount of rainfall; the savanna areas receive comparatively less rainfall. However the entire state enjoys a monsoon climate and it is designated as Amw (tropical monsoon variety) in Köppen's classification scheme. The A stands for tropical, m for monsoon and w for wet (Jarrett 1977). According to the scheme of classification proposed by Miller (1971) the entire state of Kerala falls under the tropical continental monsoon type (Shravankumar et al.1988). Miller's classification extended the limits of the humid tropics farther north considering the influence of trade winds. Monsoon areas have been subdivided into equatorial, maritime and continental types. All these classifications position Kerala in the global context with the national and international level data bases. It is imperative to consider climatic data available at the state level to delineate comparable macro level zones.

One such attempt was by Prabhakaran et al. (1991). They grouped 77 rainfall reporting stations into nine clusters based on principle component analysis of rainfall and potential evapotranspiration. These clusters are not comparable to climatic zones, but indicate climatic variations within the state. Climatically these clusters range from moist humid regimes to dry tracts. With data from seventy-seven reporting stations, it was found that forty-three stations distributed throughout the state belong to the climate category of 'normal monsoon' by receiving rainfall during the southwest monsoon months of June, July, August and September.

Bio-climatic Zones

Bio-climatic classification has gained in importance since it has been realised that vegetation types reflect climate. Recorded data on rainfall and temperature, the two most important components of climate study, are not available for most remote areas. It is felt, therefore, that the incorporation of vegetation data will improve our understanding of climate.

In the 1980s, the French Institute in Pondicherry prepared bio-climatic maps of India (Gaussen et al. 1986). Kerala was shown in the tropical monsoon and tropical savanna zones on this map. A detailed analysis of bio—climate characteristics was attempted, taking into consideration rainfall, precipitation effectiveness index (an index to assess amount of monthly effective rainfall), temperature, temperature effectiveness index (an index to assess degree of effective temperature in every month), evaporation loss and vegetation (CESS 1984; Shravankumar et al. 1988). This analysis established certain spatial relationships between altitude, rainfall, temperature and vegetation.

Considering the amounts and patterns of rainfall, temperature and vegetation, it is found that there are 12 bio-climatic zones in Kerala. Rainfall patterns are somewhat unique in Kerala, a situation that needs additional explanation before going further. Some parts of Kerala receive rainfall both during south-west monsoon and the north-east or retreating monsoon. A graph showing monthly rainfall for those areas shows two peaks—one during July/August and another during October/ November. The zones to the north and south of the Palghat Gap (a break within the Western Ghats) differ in rainfall patterns. Rainfall south of the Gap is characterised by two peaks (high rainfall in two months separated by low rainfall months), whereas in the north, there is a single peak (see Fig. 2.3). Despite high total rainfall, the north is therefore relatively dry compared to the south.

Agro-climatic Zones

Another type of macro zone classification is called agro-climatic zones. This concept combines elements of climate with the potential for particular types of agricultural production. Because agroclimatic studies are important for agricultural planning, the Government of India's National Commission on Agriculture initiated a programme to delineate agro-climatic regions at the national level and agroclimatic zones at the regional level. They identified 15 agro-climatic regions in the country (ICAR 1991). According to this classification, the entire state of Kerala falls in what they call the West Coast plain and Ghat region. Another government agency, the National Agriculture Research Planning (NARP) reported to have identified eight agro-climatic zones in Kerala. These are the northern zone, central zone, southern zone, high altitude zone, Onattukara zone, Kuttanad and Kole zone, Pokkali and a low rainfall zone (KSLUB 1995). The last four areas are sometimes clubbed together as the problem area zone since they face excessive drainage, high salinity or high alkalinity, water logging and water deficiency. Onattukara is a local name for grayish sandy areas that are supposed to have been deposited under marine conditions. The Kuttanad, Kole and Pokkali lands represent low lying, water logged areas with deep clay soil.

Prior to the NARP and ICAR classification and following directives from the Government of India, the Kerala government brought out a report on agro-climatic zones and cropping patterns (GOK 1974). According to this report, thirteen agro-climatic zones of Kerala could be identified, based on topographical patterns, rainfall patterns and soil types. Data gaps limited the precision of the zonal boundary demarcations to approximately that of the administrative boundaries, such as blocks – the sets of usually 4 to 5 panchayats used by the central government to administer certain antipoverty and employment programmes. The areas, soil types and principle crops for these 13 zones are provided in table 3.1. It may be observed that rice is one of the principle crops in all the zones except in the high ranges. Coconut is also widespread. Tea, coffee and cardamom are limited to the high ranges. River bank alluvium cuts across different zones and therefore is not computed separately for area measurement. However, it is important to note here that river bank alluviums are highly suitable for seasonal crops such as rice and vegetables.

Table 3.1 : Agro-climatic zones in Kerala

Name of Zone	*Area in %*	*Soil Type*	*Principle Crops*
Highland	27.4	Laterite without B-horizon	Rubber, coconut, cashew nut, arecanut
High ranges	15.9	Red loam	Tea, coffee, cardamom
North midland (Malappuram type)	13.1	Laterite	Coconut, rice, cashew, arecanut, rubber
North midland	11.6	Laterite	Coconut, rice, cashew-nut, rubber, arecanut
Southern midland	10.0	Laterite with B-horizon	Rubber, coconut, rice, tapioca, arecanut
Central midland	8.2	Laterite	Rubber, rice, coconut, tapioca, arecanut, banana.
Coastal Sandy	4.8	Sandy loam	Coconut, rice
Palakkad	3.9	Red loam	Rice, cotton, groundnut, millet
Chittoor Block	1.6	Black soil	Rice, sugarcane, cotton, groundnut, millet
Onattukara	1.6	Sandy loam	Rice, coconut, tapioca, arecanut
Red loam	1.0	Red loam	Coconut, tapioca, rubber, rice
Kuttanad	0.9	Peat	Rice, coconut
Riverbank alluvium	-	Alluvium	Rice, coconut, sugarcane, arecanut

Source : Report of the Committee on Agro-climatic Zones and Cropping Patterns, Department of Agriculture, GOK (1974).

Note : Total area covers 32,382 sq km. The rest is under forest and not considered for this zoning. For details of the terms used under soil type see the Glossary.

Table 3.1 shows the main cropping pattern prevalent in these zones. Rubber is cultivated in all the zones with the few exceptions of the coastal alluvium and Onattukara zone. Groundnut is confined to the black soil in Palakkad, as it can not survive in other pedological (soil) conditions. Pepper is an important crop raised in almost all zones.

Having thirteen agro-climatic zones in such a small region as Kerala implies a high degree of diversity and therefore great scope for raising diverse crops. The climate data thus reinforce the observations we made in chapter 2 about Kerala's diverse productive capacity. Crops like rice and coconut are adaptable to a wide range of climatic conditions, but groundnut, sugarcane, coffee, tea and cardamom are zone specific. Fortunately these sites are also spatially distributed, giving rise to diversified cropping patterns in all parts of the state.

Landscape Ecological Zones

Another way to classify ecological zones is according to features of the landscape. The concept of landscape ecological zones is of recent origin, but gained wide application due to a growing understanding about the importance of landscape ecological analysis for monitoring large geographical areas for environmental sustainability (O'Neill 1995). The Canadian Committee on Ecological (Biophysical) land classification emphasised it as a new methodology and proposed a hierarchical system with three levels of classification, namely, ecoregion, ecodistrict and ecosection (Davidson 1982). Ecoregion refers to a large area, which can be subdivided into ecodistricts and ecosections at the lower levels. The parameters considered to delineate these units are not the same and the levels of details are also different.

Chattopadhyay (1985) proposed a landscape ecological classification for Kerala employing a combined set of land attributes: topography, geology, climate, soil, vegetation, ground water, existing land use, human activities not connected to land use, and the degree of human interference in the natural system. Each of these attributes is classified according to its spatial characteristics. This classification is based on types of geology, soil, and land use, and for other attributes, a scale of gradation has been followed.

Three primary zones have been identified depending upon the nature of activities and the dominant processes operating therein. These are termed: (1) *coastarea,* (2) *cultivarea*, and (3) *naturarea*. The coastarea covers 5.6 per cent of the total area of Kerala and extends about 4 to 5 km inland on average. The maximum width of this zone is 30 km in the Alappuzha Kottayam segment. The coastarea as a whole is heavily populated and has industrialised pockets. Apart from fisheries and coastal agriculture, modern and traditional industries contribute to the economy and to land use change. The *cultivarea* covers 52.07 per cent of the total area. It lies parallel to the *coastarea* and is dominated by agriculture. It extends along the entire state. Intensive use of the land from lowland agriculture to tree crops results from centuries-long human activities to exploit the landscape. Thus, the *cultivarea* denotes a landscape of predominantly human-induced factors. The *naturarea* lying farther east is found in two patches, one to the north and one to the south of the Palghat Gap. This zone covers 42.33 per cent of the total area of the state. A large part of this area is still dominated by natural processes. Virgin forests with intact pristine ecosystems cover around 12 per cent to 14 per cent of the total geographical area of the state.

Based on landscape attributes, each of these three areas can be divided so that a set of 24 zones delineates Kerala's ecology as shown in Figure 3.1. There are 6 subunits of the coastarea, 13 subunits of the cultivarea, and 5 subunits of the naturarea.

The Coastarea Subzones

Coastarea 1 : This northernmost coastal tract lies in Kasargod district. It is a narrow strip composed of recent sediments. Topographically, it is gently rolling hills with coastal alluvium and brown hydromorphic soils. Remnants of palaeo sand dunes occur in patches. Littoral vegetation is also noted. This part of the coastal zone is lateritic in patches. The depth to water table is 1–2 metres. Annual rainfall varies between 3000 and 4000 mm. Fishing and agriculture are the main activities. Tree crops such as coconut dominate agriculture, although seasonal crops like tobacco are also grown. The level of industrial activity is low and the overall human intervention in the natural ecosystem is medium.

Coastarea 2 : This stretch extends from Kasargod to Kannur. It is partly lateritic and sandy with gently rolling topography. Soils are coastal alluvium, brown hydromorphic and lateritic. Rainfall is within 3000–4000 mm. Agricultural land use is dominated by coconut cultivation. Depth to water table is mostly less than 2 metres; however, in the lateritic area and in elevated areas the depth is greater. Marine fishing occurs in pockets. Cottage industries have been well developed. The Ezhimala Hill Complex lies within this zone. This is an isolated hill right on the coast surrounded by backwaters on the other three sides. Mangrove vegetation is well established in the southeastern part of this hill complex. There is also a well developed beach along the coast in the northern segment and a cliff section on the southern sea face. Human intervention in nature's domain is medium.

Coastarea 3 : This zone covers the Kozhikode and Malappuram coasts. It is lateritic, interspersed with sandy stretches. Rainfall is 2000–3000 mm. Depth to water table is within 2 metres. Coconut is the dominant crop; fishing is the main productive activity. The level of

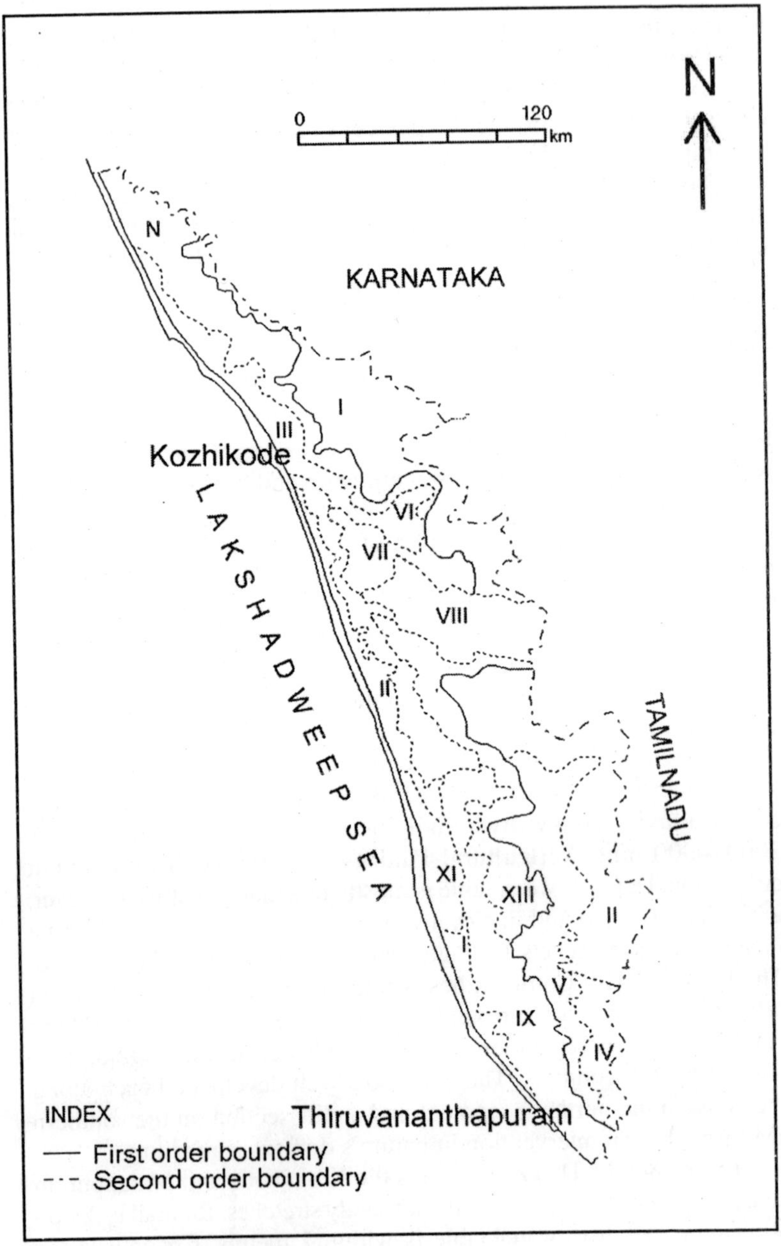

Fig. 3.1 : Landscape Based Ecozones of Kerala

industrialisation is low except around the urban centers. A couple of wood industries are found. Overall human intervention on the natural ecosystem is medium.

Coastarea 4 : This is a long stretch extending from the mouth of the Ponnani river in Malappuram district in the north to the Thottapally spillway in Alappuzha district in the south. Coastal alluvium and riverine alluvium dominate the soil types. The topography is a flat basin. The depth to soak water table is less than 1metre. Rainfall is 2000–3000 mm. Extensive backwaters are conspicuous in this area. Coconut gardens as well as seasonal crops such as paddy, banana and vegetables constitute the agricultural land use. This coastal tract has a very high population density and hosts well developed modern industrial activities around Kochi city. Traditional industries are also scattered throughout the zone. Fishing is an important activity. There are problems of water pollution. The level of human intervention is very high.

Coastarea 5 : This zone occupies the Kollam-Thiruvananthapuram coast. It is mostly flat with coastal alluvium, and patches of laterite exposure. Rainfall is below 2000 mm. The depth to water table varies between 1 and 2 metres. Agriculture, industry and fishing activities are prominent. Water pollution is reported from many parts. The level of human intervention is high. Kollam municipal town (population 341,441 in 2001) and Thiruvananthapuram urban agglomeration (population 889,191 in 2001) are located in this zone. Coconut cultivation dominates, but arecanut, banana and rice are also found.

Coastarea 6 : This zone is formed out of the southernmost part of the coast. Its topography is undulated. The depth to water table is more than 2 metres in the laterite landforms. The seacoast is rocky with pocket beaches. It stretches from the Kovalam beach and tourist complex in the south of Thiruvananthapuram city to the state boundary of Kerala and Tamil Nadu farther southeast in Thiruvananthapuram district. Population pressure is very high. Tourism is a major activity along with fishing. Human intervention is high, although the level of industrialisation is not very high. Coconut cultivation is extensive.

The Cultivarea Subzones

Cultivarea 1 : This zone lies to the east of Coastarea 5 and spreads across Kollam and Thiruvananthapuram districts. Geologically it is composed of recent sediments, tertiaries and crystalline rocks.

Undulating topography with laterite and sandy soils characterises this unit. Rainfall is 2000–3000 mm. Groundwater depth is 1–2 meters. Industry forms part of the activity of the region, along with agriculture, which comprises seasonal, annual and tree crops. Rice, tapioca and coconut are all found here. The cashew industry has been developed. The level of human intervention is high. The coastal backwaters are affected by extensive and intensive coconut husk retting. This is an age old practice to soak coconut husks under brackish water in the coastal plain for several months. Coconut fibre can be extracted only after proper retting. Water quality deteriorates in this process and many of the dug wells around the backwaters report water pollution and the smell of hydrogen sulphide (H_2S).

Cultivarea 2 : This zone is located to the east of Coastareas 3 and 4, and surrounds the backwater system of Lake Vembanad. Topographically this zone is a flat basin with soils of alluvium, brown hydromorphic and acid saline. The lateritic landscape borders this unit to the east. Industrial activity is high around Alwaye (Aluva) and Kalamassery in Ernakulam district. Human intervention is very high, water pollution being the main problem. Rice, coconut, arecanut, banana and other seasonal and tree crops dominate agriculture. Rainfall is 3000 to 4000 mm, and well water depth varies between 3 to 5m below ground level.

Cultivarea 3 : This zone lies farther north, stretching up to Kasargod. The topography is undulating with lateritic and crystalline rocks. Rainfall is 3000 to 4000 mm. The water table in the region's open wells is more than 5m below the surface. Soil types are dominated by laterite, brown hydromorphic and red loam. Agriculture, particularly coconut, fruit trees and annual crops such as tapioca dominate. The level of industrialisation is medium, and is mainly cottage industries of *beedis* (local cigarettes) and handloom. The level of human intervention is medium.

Cultivarea 4 : This zone lies in the midland of Kasargod and Kannur districts and extends up to the foothills of the Wayanad Plateau. The topography is undulating. Lateritic mesas (table-like landforms) are the major landscape features. Crystalline rocks form the lithology. Rainfall is high at 3000–4000 mm. Water table depth exceeds 5m on average, but there are places where it is more than 20m. Seasonal and annual crops such as rice and tapioca, and tree crops such as coconut, arecanut, and fruit trees, as well as cashew and rubber plantations dominate the agricultural landscape. The level of industrialisation is low. Human intervention is also low. Extensive grasslands are found on hard crusted lateritic mesas, which are otherwise wastelands.

Cultivarea 5 : This forms the foothill zone of the Wayanad Plateau. Rock types are dominated by Charnockite, and Gneisses. The landform is highly undulating with long slopes. Soils are lateritic and residual. Residual soils develop in situ on weathered rock. This typically occurs where the gradient is gentle but the location is not a valley. Rainfall is 3000 to 4000 mm or more in places. Agriculture is dominated by coconut and fruit trees along with rubber, cashew and pepper plantations. There are patches of natural vegetation towards the east. Human intervention is low with a very low level of industrialization. This is a transitional zone between *cultivarea* and *naturarea.*

Cultivarea 6 : This zone covers parts of Malappuram district. It is similar to *Cultivarea 5,* but is devoid of natural vegetation. The topography is undulating, the soil is primarily lateritic, and rainfall is between 2000 and 3000 mm. Well water depth varies between 3 to 5m. Tree crops such as coconut and fruit trees, and cashew, rubber and pepper plantations dominate. Industrial development is very low, but human intervention is high on the agricultural front.

Cultivarea 7 : This zone occupies a major part of Malappuram district. Unclassified Gneiss and Charnockite are the main rock types. As a cap rock, laterite is widely developed. Topographically undulated, this zone is characterised by lateritic and hydromorphic saline soil. Industrial activity is very low. Agriculture is dominated by rubber plantations. Pepper is also cultivated. Seasonal crops in the valleys and tree crops on the slopes are found between plantations. Rainfall is 2000–3000 mm and water table depth goes beyond 5m.

Cultivarea 8 : The Palghat Gap and adjoining areas constitute this zone. Climatically, it is distinct from the rest of the state. The gap is a zone of wind convergence. Dry winds from the Tamil Nadu plain have profound influences on the climate here. Rolling topography developed on Charnockite and Gneiss characterises the area. Rainfall is less than 2000 mm, and diurnal and seasonal temperature variations are relatively high. Diversified soils are found: alluvium, laterite, brown hydromorphic and black soils. The cropping pattern is dominated by seasonal agriculture including paddy. This is one of the main rice growing areas in addition to groundnut and sugarcane. Tree crops are mainly coconut with palmyra in places. Rubber plantations are found in patches. The water table in some places is 10m below ground level (Ramachandran et al. 1987:12). This zone is noteworthy for canal irrigation facilities.

Cultivarea 9 : This zone covering the midland between coast-lowland and highland extends from Kottayam to Thiruvananthapuram. The rock types are dominated by Khondalites with lateritic caps. Topographically undulated, this area receives 2000 to 3000 mm of rain annually. Well water depth varies between 3 and 5m. Rubber plantations dominate the land use pattern. Coconut trees grow on the slopes, and seasonal rice is cultivated in the valleys. The level of industrialization and the level of human intervention are medium. Widespread terracing of the landscape for rubber plantations is found in this zone, particularly in the Kottayam area. The productivity of rubber in this zone at 830 kg/ha surpasses all other areas and the state average of 749kg/ha (Chattopadhyay 1996b:5).

Cultivarea 10 : Lying east of Cultivarea 9, this zone in the Western Ghat foothills of central Kerala is transitional to the naturarea. It is confined mainly to Kottayam district. Charnockite dominates the rock type. The topography is highly undulated. Rainfall varies between 3000 and 4000 mm. The soil is lateritic. The level of industrialization is low and human intervention is medium in terms of land use and land cover change. Rubber plantations dominate cultivation. Coconut and pepper are also prominent. There are patches of degraded forest in this zone.

Cultivarea 11 : This zone covers the "rice bowl" of Kerala. With basin type topography and with deep clay and alluvium soils, this zone spreads over the districts of Ernakalum, Kottayan and Alappuzha. Rainfall is 2000 to 3000 mm. The depth to the water table for wells is less than 3m. This is an intensely cultivated rice cropping area. There are areas where water logging is a regular feature. The prawn and paddy combination is popular in waterlogged areas. Part of this zone is below mean sea level and is protected by embankments. A sizable area has been reclaimed from the Vembanad lake bed. Intense human intervention is manifested in fertiliser and pesticide pollution of water bodies. The Vembanad lake water is also polluted due to the construction of a barrage that converted the southern part of the lake into a fresh water body. This barrage has restricted the inflow of sea water during high tides. As a result the normal mixing process between brackish water and fresh water has been jeopardised. Waste water effluents from the surrounding areas dominated by agriculture, settlements and to some extent industries contribute to the high pollution levels of the southern segment of Vembanad lake. The level of industrialization is low, but the level of human intervention is high in the land and water systems.

Cultivarea 12 : This zone extends from Kozhikode to Ernakulam districts and borders the eastern side of Cultivarea 2. Topographically rolling, the soils are predominately laterite. Some of the enclosed valleys are characterised by water logging and wet soil for a considerable part of the year. Rainfall is 2000 to 3000 mm and water table depth goes beyond 5m. The industrial activity level is medium. Tree crops such as coconut and arecanut dominate agricultural land use. Pepper plantations are found. Rice cultivation takes place in the valleys. Rubber plantations are also making inroads. The level of human intervention is medium.

Cultivarea 13 : This is the foothills region, extending from the southern flank of the Palghat Gap to the foothills of the Munnar Plateau. This zone can be considered as the northern segment of Cultivarea 10. Charnockite is the main rock type. The topography is undulated with laterite in the slopes and ridges and alluvium in the valleys. Rainfall varies between 2000 and 3000 mm, but in the southern end it goes much above 5000 mm near Nariamangalam along the boundary of Kottayam and Idukki districts. The level of industrialization is low. Agriculture is dominated by rubber plantations. Coconut, rice and pepper plantations are common. Human intervention is medium.

Naturarea

The naturarea covers the entire Western Ghats region at altitudes above 300 metres. This is the eastern border area of Kerala, next to Tamil Nadu. It is dominated by forest ecosystems, although there are places of human activities within. All of Kerala's rivers originate in this zone. Biosphere reserves and wildlife sanctuaries are distributed throughout this ecoregion. Table 3.2 gives details about the wildlife sanctuaries and national parks while Fig 3.3 provides a map showing their locations.

Five zones can be identified within this ecoregion.

Naturearea 1 : This zone covers the Wayanad Plateau, the Kunda Hills and the northern flank of the Palghat Gap. It is coterminous with the northern segment of the Western Ghats in Kerala. Topographically hilly, this area receives more than 3000 mm of rainfall annually. Temperatures drop at the higher elevations. Forests dominate land use in most places, while tea, coffee, and cardamom plantations are widely found. Pepper and ginger are the other two commercial

Table 3.2 : Wildlife Sanctuaries and National Parks of Kerala

No.	*Name of Sanctuary*	*Area in km²*	*Year of Formation*
1.	Periyar Wildlife Sanctuary	777.54	1950
2.	Wayanad Wildlife Sanctuary	344.46	1973
3.	Parambikulam Wildlife Sanctuary	285.00	1973
4.	Neyyar Wildlife Sanctuary	128.00	1958
5.	Peechi - Vazhani Wildlife Sanctuary	125.00	1958
6.	Shendurney Wildlife Sanctuary	100.32	1984
7.	Eravikulam National Park	97.00	1978
8.	Chinnar Wildlife Sanctuary	90.44	1984
9.	Silent Valley National Park	89.52	1984
10.	Chimmony Wildlife Sanctuary	75.00	1984
11.	Idukki Wildlife Sanctuary	70.00	1976
12.	Aralam Wildlife Sanctuary	55.00	1984
13.	Peppara Wildlife Sanctuary	53.00	1983
14.	Tattekkad Bird Sanctuary	25.16	1983
	Total	2315.44	

Source : Kerala Forest and Forestry Handbook, 1995.

crops. Paddy is grown in the valleys. Tapioca, coconut and arecanuts are also found near human settlements.

The forest vegetation is comprised of wet evergreen, moist deciduous and semi-evergreen types. Bamboo is also found. Teak plantations cover around 4 per cent of the forested area. Landslides in the scarp slope have been recorded during monsoon months associated with several continuous rainy days.

Naturarea 2 : This zone forms the core of the southern part of the naturarea. It covers the Nelliyampathi, Munnar, and Periyar plateaus. Being located on a plateau surface this sub zone is topographically hilly. Rainfall is 2000–3000 mm. The water table depth ranges from 1 to 3 metres with greater variations between valleys and hilltops. Forest loam and laterite are the main soil types.

Kerala's highest peak, Anamudi (2,695m), is located in this zone. Tea, coffee and cardamom plantations are found here. Dry land paddy cultivation was introduced in the past, but has been almost completely discontinued. A sizable area remains forested with evergreen, moist deciduous, dry deciduous and temperate shola types of vegetation. The area is considerably deforested (40–50% of total forest area has been lost during the last 100 years) and hosts a number of hydroelectric projects.

Naturarea 3 : This zone is located at the western fringe of

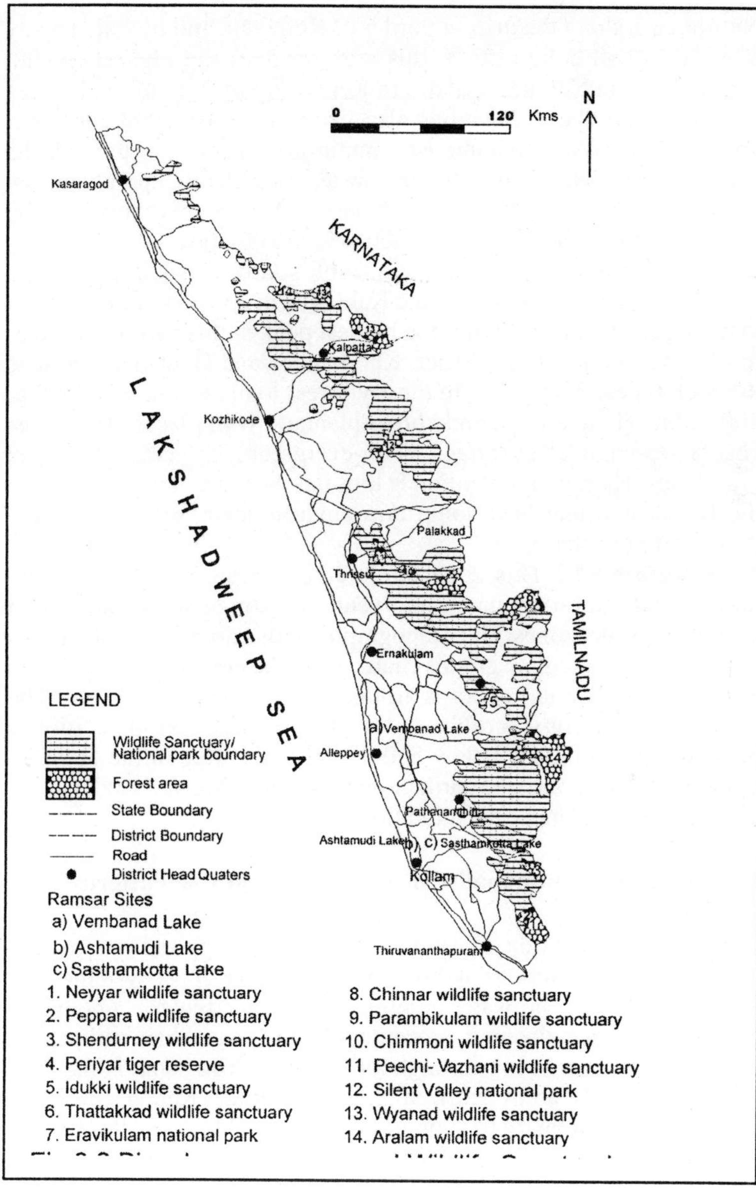

Fig. 3.2 : Biosphere Reserves and Wildlife Sanctuaries in the Western Ghats and Ramsar Sites in Kerala.

Naturarea 2 along the district border of Kottayam and Idukki. Mostly lateritic in soil composition, this area receives the highest rainfall (>5,000 mm in Neriamangalam) in Kerala. Apart from tea and coffee plantations, rubber plantations also occur. A considerable area has been deforested to accommodate immigrants from other parts of the state. Human intervention is high in terms of landscape alteration. Coconut, tapioca, pepper, arecanut and fruit trees cover much of the agricultural land. Rubber plantations are also found.

Naturarea 4 : This zone lies to the south of Naturarea 2, and covers the area extended from the Kaki hills in Pathanamthitta district to the Agasthamalai (1869m- the highest peak in this part of the state) in Thiruvananthapuram district. Khondalites and Gneisses dominate the rock types. Soil types are mostly forest loam. Rainfall is 2000 to 3000 mm. There are tea and coffee plantations in places. The forest vegetation includes evergreen, semi-evergreen, temperate shola and grasslands. Human intervention is low, but there are reservoirs within the forest area that have long term environmental effects we shall describe in chapter 5.

Naturarea 5 : This zone borders Naturarea 3. Relatively low rainfall and intensive human intervention distinguish this area from the adjoining ecozones. Hilly topography with a dominance of various rock types of crystalline groups characterise the area. Rainfall is 2000–3000 mm. Well water depth is around 5–6m. Soils are lateritic. The area has been considerably deforested to accommodate rubber plantations in altitudes below 500m, and tea, coffee in the 500m to 1,000m altitude areas and cardamom plantations above 1,000 metres. Human induced landscape alteration is high.

Diversity and Commonality in Kerala's Land Use Patterns

The above description of landscape ecology supports our overall observations that Kerala can boast a wide variety of ecosystems with a number of land and water combinations. Nevertheless, a common set of crops – rice, coconut, arecanut, tapioca, vegetables and pepper – appears to be growing throughout the state. The same is the case with rubber in the areas below 500 meters altitude. Kerala's diversity along with the capacity to support certain valuable crops throughout the state is a major strength of its natural resource base.

Micro Level Zones

The ecological variation described in the previous section and earlier throughout chapter 2 brings out the scope for utilising Kerala's production base on a macro scale. A peculiar topographic feature of Kerala, however, is the large amount of micro-level variation that also occurs.

The micro level variation emerges from landscape diversity because of the great local variations in geomorphology and soil types. Kerala's landscapes correspond to geographical hierarchy theory as established by studies in other parts of the world (O'Neill et al. 1989). This hierarchy can be established by carving out the landscape according to the drainage basins that are governed by the hydrological processes. A main drainage basin can be divided into sub watersheds, mili watersheds, micro watersheds etc. Kerala has 44 major river basins, 151 sub watersheds (100–500 km^2), 960 mili watersheds (10–100 km^2) and more than 2000 micro watersheds (less than 10 km^2) (Chattopadhyay 2002)

Taking note of landform conditions and geomorphic processes operating in a given area and the specific actions that are involved therein we can identify the micro level zones and their specific characteristics. These microzone characteristics influence the production base.

The topography of Kerala is mostly undulating. This means that each small area can be characterised by valley, slope and ridge formations. This arrangement is known as a 'topo sequence'; the smallest unit of a landscape. The valleys can be broad, narrow, flat bottomed or sloping. Similarly the slope joining the valley and ridge can have a steep or gentle degree of incline. The ridge can be elongated, narrow or wide. Whatever be the case, the basic element of any landscape unit is this topo sequence. Similar micro level units can be identified in the coastal plain and other low lying areas. Ridges (sandy ridges) and runnels (low lying areas between two sandy ridges) are the normal topo sequences in the coastal plain.

Each of these micro level units or microzones, also known as terrain components (Chattopadhyay and Chattopadhyay 1995) has a different production potential. The nutrient laden sediment and water from the ridges and slopes move down to the valleys, thus the valleys are replenished and maintain their fertility. Due to Kerala's high water availability, topography, and soil characteristics the valleys are normally given for planting seasonal crops like rice, while tree crops, which have deep root penetration are planted on the slopes and ridges.

The Panchayat Resource Mapping programme (PRM) introduced in Kerala in 1990 (CESS 1991), attempted to delineate these micro level variations to enable the panachayats to plan for more appropriate land, water and biomass management. Through mapping 25 representative panchayats distributed throughout the state, 23 terrain components were identified for Kerala. An example comes from Onchiyam panchayat, which extends from the coast to the midland, about 3.5 km inland in Kozhikode district of northern Kerala. This panchayat covers 9.06 sq km. Table 3.3 provides a list of the units identified in Onchiyam. Combining terrain conditions and soil types, analysts could recognise eight zones. This is shown on Figure 3.3 which gives a map of the Onchiyam Panchayat ecozones.

Table 3.3 : Micro Ecozones in Onchiyam Panchayat, Kerala

Micro-zone *Sl. No*	*Unit Descri-ption*	*Area in Sq. Km.*	*Soil*	*Water availa-bility*	*Land-use avail-ability*	*Problems*
1.	Ridge/hill crest with laterite	1.36	Gravelly loam	Low	Coconut, mixed tree crops, laterite quarry	Shallow soil, water shortage
2.	Steep and moderately steep slope with laterite	0.75	Gravelly clay loam	Moderate	Coconut, settlement mixed tree crops	Gully erosion, water shortage
3.	Gentle to moderate side slope with laterite	2.38	Loam to gravelly clay loam	Moderate to high	Coconut, mixed tree crops, tapioca, arecanut, banana, vegetable	Sheet erosion settlement.
4.	Valley floor with ravine alluvium	1.27	Silty clay loam and loam	High	Coconut and vegetable , settlement	Water logging and land reclamation
5.	Coastal plain with coastal alluvium	1.68	Sandy loam	High	Coconut and settlement	Water logging and salinity
6.	Sandy ridges and runnels	1.03	Sandy	High	Settlement and coconut in runnel	Excessive drainage, salinity and water logging
7.	Beach	0.20	Sand	High	Open space	Sea erosion
8.	Tidal marsh	0.39	Sandy clay		Coconut	Reclamation

Source : CESS. 2000. Panchayat Resource Mapping: Final Report, p. 25

Each of these ecozones has distinct characteristics. Problems of sea erosion, gully erosion, sheet erosion, water logging, salinity and

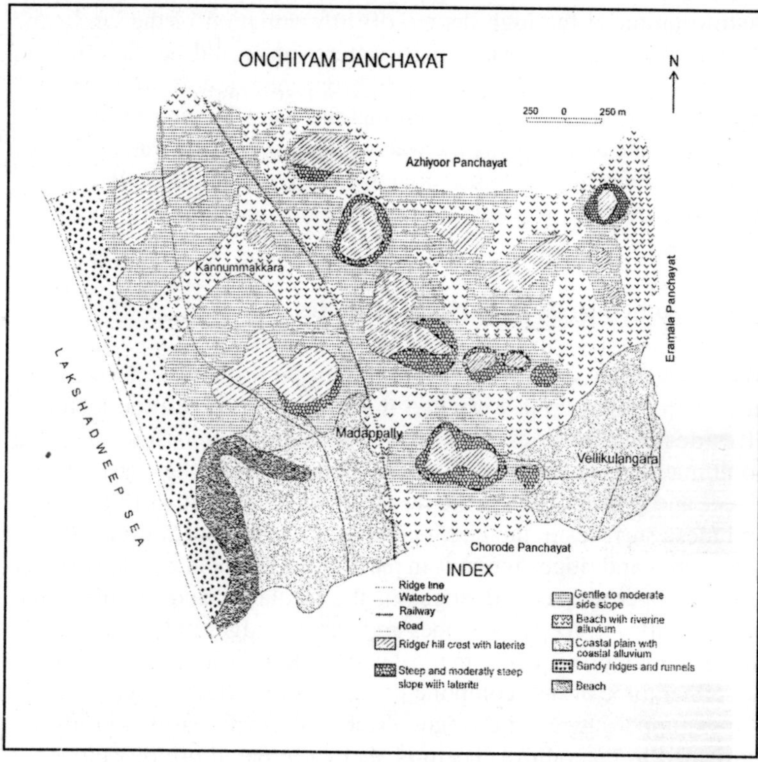

Fig. 3.3 : Ecozones of Onchiyam Grama Panchayat

reclamation are found to be unique to particular zones. So zone specific intervention could be worked out for better environmental management. The land use types had overlapping characteristics, but the range of crop variation within short distances is significant. All the paddy fields in this panchayat have been made on land reclaimed by draining and filling tidal marshes. Reclaiming the land in this way leads to water logging in the unit of the valley floor with riverine deposits. This association was seen more clearly when micro-zones 4 and 5 were identified. Water logging prevents the soil from recuperating, interferes with transportation links, and facilitates mosquito reproduction in some areas, leading to the threat of malaria, filaria, and other health problems as well as the problems of sleeping properly when multiple mosquito bites are occurring. The problem of water shortage and water logging within the same panchayat is a manifestation of the high degree of differentiation of the landscape system. As many as seven micro watersheds could be identified in this single panchayat. Because the different zones are so close spatially, the interventions required for combating water logging in one zone can contribute to increasing the water availability in the water shortage zones by specific watershed management actions. The Panchayat Resource Mapping programme is described in more detail in chapter 8.

The Topo Sequence at Various Altitudes

Macro and micro zones interact to produce a further set of variations in that the micro zones arrange themselves differently at different altitudes as one goes east from the Kerala coastline. Typical combinations are ridge and runnel arrangements in the coastal plain; levees and back swamps in the fluvial plains; valleys, mesa side slopes and mesa surfaces in the lateritic hard crust area; flat bottom valleys, side slopes and ridges/mounds in parts of the midlands; valleys, lower slopes, upper slopes and ridges in the midlands and foothills; and narrow valleys, steep side slopes and narrow ridges in the mountains. The general topo sequence in most parts of Kerala, as indicated earlier, has either three or four components. These are valleys or plains, lower slopes, upper slopes and ridges. Rice and other seasonal crops are cultivated in the valleys. There is also widespread use of valleys for settlements and tree crops. The lower slopes are planted mostly in tapioca, arecanut, and coconut. They occur as ribbons around the valleys; the upper slopes typically have settlements, coconut and fruit trees. Ridges are given for settlements, roads and other structures. Coconut trees are also planted on the ridges.

With an increase in elevation the micro-zone pattern changes. The ridges and upper slopes are covered with plantations of rubber and cashew while the lower slopes have settlements and coconut trees. At even higher altitudes, within the forest area grasslands mainly occupy the ridges. The change of crops from valley to ridge is a pointer to changes in water availability and soil composition within the topo sequence. Spanning the entire state, these micro level variations in ecological characteristics create enormous opportunities for local level resource management.

4 Causes of Concern : Land Use Change

Spices, Traders, and Colonialists

The previous two chapters have described the main features of Kerala's intricate ecological variations and have highlighted the great diversity of climate, soil, water and biomass combinations found there. The many micro-ecozones and their variations provide the basis for Kerala's high population density and for the wealth produced by its people. Over many centuries, Kerala's position as a major trading center was facilitated by the rich variety of its micro-ecozones. Kerala's fabled spice trade – itself based in part on the rich micro-ecozone variability – goes back possibly to the Sumer (Babylonia) of 3000 BC. Historically verified contacts between parts of the Roman Empire and the Kerala coast give circumstantial support to the widespread folk belief in a missionary visit by the Apostle St. Thomas in 52 AD where he is said to have founded the Palayur church in the central Kerala town of Chavakkad, one of seven original Christian churches in southern India. Nearby are the remains of an ancient Jewish synagogue, and in Kodungallur (Cranganore) one finds the Cheraman Juma Masjid. Built in 629 AD, it is considered the oldest mosque in India, and its date of construction only 7 years after the founding of Islam suggests the degree of international contact with Kerala and the products of its ecozones during the European middle ages.

In 1498 the Portuguese sailor Vasco da Gama arrived in Kozhikode (Calicut) seeking the spices of the Indies. He took back a large shipment of the black pepper so valued by European cooks and doctors as well as the bright printed cloth Europeans came to refer to almost immediately as "calico." Dutch, French and British invaders followed.

Under British colonial rule from 1792 to 1947 a large expansion of production took place when Kerala's ecozones were exploited for coffee, tea, increased spice production, and the export of teak. In the nineteenth and twentieth centuries Kerala's population increased rapidly while the exploitation of its ecological resources intensified. The great potential of Kerala's ecozones to produce wealth for humans has begun to unravel under this relentless assault of population growth and modern extraction technology. In this chapter and the chapter that follows, we shall investigate the main causes for concern about Kerala's environment. These concerns are widespread both among scientists and among ordinary people. We shall first consider the environmental dangers through an examination of changes in land use.

Landuse in Theory and Practice

Analyses of landuse and land cover change have drawn global attention in recent years, particularly in the context of growing concern about environmental degradation, climate change, and the quest for sustainable development.

Landuse is primarily an expression of people's use of natural resources. As an interface of atmosphere, hydrosphere and lithosphere; land carries the imprints of all human activities. The Food and Agricultural Organization of the United Nations (FAO 1995) identified several functions of land (Box 4.1).

Landuse patterns change at varying rates over time and scale. Long-term changes without any intervention are parts of a natural process and result in various biomass climaxes. Accelerated changes result from outside forces that have become concerns at the global, regional and local levels.

Boserup (1965) drew the attention of the scientific community to the possible link between population growth and landuse change. Marvin Harris's intensification-depletion model that we introduced in chapter 1 bears the influence of Boserup's work. The central point of her argument was that population growth leads to technological change, thereby increasing agricultural productivity through landuse intensification. Her model of agricultural change was debated from the 1980s through the 1990s (Wiegers et al. 1999). However, this debate hardly addressed the role of ecological differences within a locality or, more specifically, the presence of different landuse systems or production zones at the local level. Landuse change and associated environmental problems such as deforestation, soil erosion, flooding and desertification are often seen as global in scale despite the fact that their proximate causes and effects are inherently local (Turner 1990; Buttel and Taylor 1994). More often than not the interconnections between local, regional and global processes are

Box 4.1 : Functions of Land

* It is the basis for many life support systems through the production of biomass that provides food, fodder, fibre, fuel timber and other biotic materials for human use, either directly or through animal husbandry including agricultural and inland and coastal fishery (the production function).
* Land is the basis of terrestrial and gene reserves for plants, animals and micro-organisms, above and below ground (the biotic environmental function).
* Land and its use are a source and sink of greenhouse gases and form a co-determinant of the global energy balance: reflection absorption and transformation of radioactive energy of the sun and of the global hydrological cycle (the climate regulative function).
* Land regulates the storage and flow of surface and ground water resources, and influences their quality (the hydrological function).
* Land is the storehouse of raw materials and minerals for human use (the storage function).
* Land has a receptive filtering, buffering and transforming function of hazardous compounds (the waste and pollution control functions).
* Land provides the physical basis for human settlements, industrial plants and social activities such as sports and recreation (the living space function).
* Land is a medium to store and protect the evidence of the cultural history of mankind and a source of information on past climate conditions and past land uses (the archive of heritage function).
* Land provides space for the transport of people inputs and procedure, and for the movement of plants and animals between discrete areas of natural ecosystems (the connective space function).

Source : FAO (1995).

overlooked. Social variables at the micro level are hardly considered at the regional or global levels.

An understanding of the dynamics of regional landuse changes requires not only focusing on explanations for local landuse changes, but also considering the aggregate effects of these local changes at the regional level (Turner 1999). Broader political and economic changes also influence landuse at the local level.

Two of the prime movers of landuse changes are multiple usability of land and resource use competition. Land degradation takes place when land use stress exceeds the assimilative or carrying capacity of the system. To meet a growing population's demand for food and shelter, agricultural and urban encroachments on natural areas can occur. An increase in the land-person ratio and consequent consumption increases lead to intensification of landuse. The spread of one landuse type at the expense of another initiates the processes of environmental changes, which have both on-site and off-site manifestations. However, it is important to note that there are limitations to using crude measurements of natural area loss as an indicator of environmental change and, by implication, ecological sustainability on the local scale (Walker and Solecki 1999). Patterns of conversion, heterogeneity of the ecosystem, habitat quality and mode of landuse can be used in combination to determine environmental impacts or degree of ecological transformation.

Changing from one land use to another automatically initiates the on-site changes through cover modification. The functional alteration of the ecosystem may be manifested with a time lag, and across boundaries through changes in water quality, water flow, and levels of pollution which all may be evident at various spatial-temporal levels.

In view of the wide variation in physical settings and developmental patterns in Kerala, aggregate data at the state level do not adequately capture the dynamic processes involved in the state's landuse transformation (George and Chattopadhyay 2001). Map based analyses, field studies and area specific case studies are more useful. In this chapter we analyze landuse patterns and changes in Kerala at both the macro (state) level and the micro level. We focus on (1) forest ecosystems, (2) agriculture, and (3) urban landscapes. Changes in wetlands, water quality and environmental problems like flooding, soil erosion, landslides and coastal erosion are dealt with separately in chapter 5.

Landuse Change in Kerala : An Overview

Government departments all over India follow a common classification scheme for gathering landuse statistics. The classification system is called the Nine Fold Classification. According to this classification, Kerala has 27.85 per cent of its total area under forest. This figure has remained more or less unchanged since 1960–61. The net sown area in Kerala increased from 49.87 per cent in 1960–61 to 57.89 per cent in 1992–93. There has been substantial growth in the area under "land under non-agricultural uses." This figure rose from 5.31 per cent in 1960–61 to 7.80 per cent in 1992–93. Table 4.1 provides the changes recorded over the 30 years from 1961 to 1991.

Forest. Before the organization of the state in 1956, the forests in Kerala were owned both by the government and by individuals. Once the state was formed, all forestlands were vested in the government.

The area under forest is administered by the Department of Forests. As shown on table 4.1 in 1960–61, the forest area was 10,640 square kilometres and constituted 27.4 per cent of the total geographical area of the state. In the period from 1986–87 and 1992–93, the forest area was recorded at 27.9 per cent.

Land Put to Non-Agricultural Uses. This includes buildings, courtyards, industry, and roads. Table 4.1 shows as increase from 5.3 per cent in 190-61 to 7.8 per cent in 1992-93. The most rapid part of this increase occurred between 1960 and 1971 with a tapering off thereafter.

Barren and Uncultivable Land. This includes rocky and sandy wastelands, marginal lands and degraded areas. The area in this category declined from 4.0 per cent in 1960–61 to 1.4 per cent in 1992–93. While the rocky wastes could not be put to any use other than rock quarrying, the other types could be put to constructive uses. The barren hard crust lateritic wastelands in north Kerala are utilised for building purposes. Sandy patches are used for settlement and industry. The reduction of area in this category indicates increasing utilisation of marginal lands.

Table 4.1 : Distribution of Area in Hectares and Per Cents Under Different Landuse Categories in Kerala in 1960–61 and 1992–93

Landuse Class	*1960–61*		*1992–93*		*1960–1993*
	Area in Ha	*% of Total*	*Area in Ha*	*% of Total*	*Per cent Change*
Forest	*1,064,000	27.4	*1,082,000	27.9	-
Land put to nonagricultural uses	206,000	5.3	303,000	7.8	+ 47.80
Barren and uncultivable land	155,000	4.0	55,000	1.4	- 63.6
Permanent pastures and grazing land	47,000	1.2	2,000	0.1	- 96.0
Land under miscellaneous tree crops	206,000	5.3	34,000	0.9	-83.0
Cultivable waste	144,000	3.7	91,000	2.3	- 36.8
Current fallow	67,000	1.7	42,000	1.1	- 37.3
Fallow other than current fallow	62,000	1.6	27,000	0.7	- 56.5
Net area sown	1,934,000	49.8	2,249,000	57.9	+ 16.9
State Totals	**3,885,000**	**100.00**	**3,885,000**	**100.00**	-

Source : Department of Economics and Statistics, Government of Kerala.
* There was some area readjustment in the forest over the years.

Note : 100 hectares = 1 square kilometre. Figures in various tables are in one or the other of these measurement units.

Permanent Pastures and Grazing Land. Land in this category has declined substantially. In 1960–61 the average area per panchayat in this category (not shown in table 4.1) was 45 hectares. By 1992–93 some panchayats were down to just 2 hectares. Reduction of this area indicates a loss of village commons, which had supported livestock and supplied wood to the marginal sections of society. Most of these areas have been converted to agricultural use.

Land under Miscellaneous Tree Crops. This category also experienced a reduction in area. It has come down from 5.3 per cent to 0.9 per cent. A variety of tree crops were formerly grown in dry land areas. With the expansion of agriculture these areas declined indicating that village level biodiversity is being lost. When converting to agriculture, the range of crops is reduced. Loss of this type of land reduces the amount of free firewood available to the poorer sections of society.

Cultivable Waste. This category includes the type of land that can easily be brought to cultivation but is kept as wasteland. There are certain reductions in areas under this category. It would be expected that with an increase in population pressure and demand for land, the areas of cultivable waste would be substantially reduced. This was not the case however. One reason might be that the land is only cultivable for a certain type of crop that the cultivators do not favour.

Current Fallow. This category includes seed beds or fields which could not be put to cultivation for reasons not related to agriculture. Over the seasons and years, spatial distribution of fallow varies. The same lands are not kept as fallows year after year. Over the 30 year period the area in this category has declined by 37.3 per cent.

Other Fallow. After a few years of being kept fallow an area will be placed in this category. The areas are suitable for agriculture but are not put to use for various reasons that may or may not be directly related to agriculture. This land type registered a reduction from 1.65 per cent in 1960–61 to 0.6 per cent in 1970–71. After 1970–71 the rate leveled off at 0 .7 per cent and the average area for each panchayat was 27 hectares.

Net Sown Area: In Kerala, this accounted for 57.9 per cent in 1992–93, well above the all-India average of 46.9 per cent. Net sown area in Kerala includes wetland agriculture, tree crops like coconut, and agricultural plantations. Growth in net sown area has been at the expense of almost every other category of land.

With the progress in remote sensing, in 1989 an all-India level landuse/land cover mapping was done for the districts at a scale of 1:250,000. The landuse data generated from this exercise are presented in table 4.2. It may be observed that agricultural land accounts for 70.41 per cent of the total geographical area. Rural settlements throughout the state are also accounted for within this category. Around 2.80 per cent of land is under water, which includes freshwater lakes, reservoirs, brackish water lakes, backwaters and estuaries.

These brief background details are necessary to understand the landuse transformation in the state of Kerala.

Table 4.2 : Areas under Various Landuses or Land Covers in Kerala in 1989

Land Use Class	*Area in km²*	*Per cent*
1. Built up land	237	0.61
2. Agricultural land	27,354	70.41
3. Forestland including degraded forest	9,437	24.29
4. Water bodies	1,088	2.80
5.Wasteland or degraded land	715	1.84
6.Other (e.g. grass and grazing lands)	19	0.05
Total	38,850	100.00

Source : Kerala State Landuse Board. 1995. Land resources of Kerala, Thiruvananthapuram. p. 177

Changes in the Forest Areas

Forests in Kerala are mostly confined to the Western Ghats region of the state. Mangrove forests occur in isolated patches along the coast. The discussion here will be confined to the Western Ghats part. Kerala has experienced substantial change in the forest cover. In the absence of recurrent surveys of the area classified as forest, the official estimates continue to indicate no change in the area (Table 4.1). Analyses of topographical maps since 1905, combined with Landsat images taken in 1973 and 1983 indicate decline in forest vegetation cover over time. It is estimated that forest cover came down from 44.4 per cent in 1905 to 14.7 per cent in 1983 (Chattopadhyay 1985; Chattopadhyay et al. 1986). Table 4.3 provides information on forest and non-forest area in the state at different time points along with the annual rate of deforestation.

Table 4.3 : Forest Area, Non Forest Area and Rate of Deforestation in Kerala – 1905 to 1983

Year	*Area as Percent of Total Geographical Area of the State*		*Average Annual Rate of Depletion*
	Forest	*Non- Forest*	
1905*	44.07	56.93	na
1965*	27.71	72.29	0.27%
1973*	17.06	82.94	1.33%
1983**	14.17	85.83	0.29%

Sources : * Chattopadhyay (1985) ** Chattopadhyay et al. (1986)

There are variations in the estimated rate of deforestation in the Western Ghats in Kerala. Prasad et al (1998) suggest an annual rate of deforestation on the order of 0.5 per cent during the period 1920–1973 and 0.9 per cent for the period from 1961–1989. Notwithstanding these variations, it may be inferred that the natural vegetation cover of the state has changed significantly. Comparison of IRS (Indian Remote Sensing Satellite) images from 1995 with those of 2000 covering the southern part of the Western Ghats around the Agasthamalai area in Thiruvananthapuram district indicated a 19.5 per cent loss of vegetation cover having a crown density of 20 per cent or more. Crown density indicates the quality of the vegetation type. Forests with intact natural ecosystems will be much less than this figure. One field-based estimate for the district of Thiruvananthapuram in 1984 was that natural vegetation cover with intact natural ecosystem was 8.71 per cent of total geographical area of the district (Chattopadhyay 1985) against an area of 22.81 per cent recorded by the government forest administration department for the same district.

Forests have been diverted for plantation agriculture, road building, river valley projects, forest plantations, settlements, rehabilitation of people displaced by hydel power reservoirs and similar other non-forestry purposes. The process of deforestation dates back to the first half of the 20th century, to accommodate plantation crops such as tea, coffee, cardamom and economically valuable trees such as teak and rubber. The first teak plantations in the Malabar district of the then Madras Presidency were in 1842 in the Nilambur area. Commercial rubber was introduced in Kerala in 1902 when the Periyar syndicate in Travancore was formed and cultivation of the *Hevea Brasiliensis* rubber plants started (George et al. 1988).

Diversion of forests for other activities was intensified after the formation of Kerala State in 1956. Even the official (GOK 1995) estimate indicates that around 3,363 sq km of forest area have been divested for plantations (1,538 sq km), reservoirs/river valley projects (296), plantations under the Forest Development Corporation (378), the state farming corporation and for various other uses by other departments (1,151).

Exact figures on encroachments are not available. However, even in 1984 around 238 sq km of encroachment was recorded (GOK 1984). Puri et al. (1983) reported 946 sq km of previously forested area to be under agricultural activities. Cardamom plantations, which are mostly within the forest areas have extended to over 442 sq km in 2002–03.

The spatial pattern of deforestation indicates that in the higher altitudes tea, coffee, and cardamom replaced forests, but in lower altitudes, it is rubber and tree crops that had the dominant role. In fact, rubber was introduced in many places where the forest cover was already degraded.

Loss of Biodiversity. We noted in chapter 1 the international importance of maintaining the earth's now threatened biodiversity. Tropical forests are considered to be very high in biodiversity. A rough estimate of the extent of biodiversity loss and species reduction due to deforestation can be gathered from a study by Pandurangan (1995) in the Idukki reservoir area. This reservoir covers an area of around 62 sq km. The project site and township spreads over another 40 sq km. The study reported that as many as 44 species were lost due to agriculture, 18 species died out from exploitation of forests for timber and firewood, 17 are gone for extension of townships, 14 via grassland diversion, 11 as a consequence of dam construction and 11 from overg-razing. The total loss of 115 species in such a small zone suggests that much larger numbers could be threatened by deforestation in larger areas.

Causes of Deforestation in Kerala

The major causes of deforestation that manifest themselves statewide, can be divided into those before and those after the formation of Kerala state in 1956. The major causes prior to 1956 are as follows :

(1) Government initiatives to open the forest areas for plantations in the 2nd half of the 19th century.
(2) The 'Grow More Food' campaign in the mid 1940s that opened up forest land for the cultivation of food crops.
(3) Pre-1956 internal migration and government resettlement programs that supplemented spontaneous migration.

After 1956 the major deforestation processes resulted from irrigation and industrialization that Kerala promoted along with the rest of India. Prominent among these were :

(1) Irrigation and hydel (hydroelectric) power projects or multipurpose river valley projects.
(2) Industry related activities such as forest plantations, farming corporations (oil palm plantations, horticulture development, etc.) and fodder development.

These five issues are briefly discussed here to provide some overall understanding of the process. Plantations of tea and coffee were introduced in the high altitudes of Kerala replacing natural forests like other parts of the country. Teak was also planted in the fôrest areas. Apart from furniture making this hard wood was used for ship building and railway sleepers. The Nilambur valley in central Kerala (part of Madras Presidency during the pre-independence era) is well known for teak plantations. Even railway lines were laid to facilitate the transfer of teak logs. The Travancore native state was not under direct colonial rule; however it took favourable policy decisions to introduce tea, coffee, and subsequently rubber. Rubber plantations occupied foothill areas to begin with and started expanding both towards the high altitudes in the east and the low altitudes in the west. Although the majority of rubber plantations are in small holdings, there is evidence of forest clearances, particularly shrubs and bushes to accommodate rubber plantations.

The country was in the grip of famine after the 2nd World War. Food grains were in short supply. To meet the shortfall there was an all out effort in the 1940s to increase food production under a programme called the "Grow More Food Campaign." More and more government lands were brought under agriculture. People were officially encouraged to clear forest vegetation to raise food crops. As a result, people from other parts of the state particularly the midlands moved to the highlands and initiated food crop cultivation.

Non-marriage internal migration is an important issue in Kerala. People migrated from south Kerala to north Kerala and also from lowlands and midlands to the highlands. This movement occurred in two stages. The first stage commenced in the 1920s and the second phase was in the 1930s and 1940s (Tharakan 1976; Joseph 1998). Initially the movement was from the Travancore principality to Malabar, which was part of Madras Presidency and subsequently it spread to other areas. The destinations were mostly Wayanad in north Kerala and Idukki in south Kerala. Economically backward farmers who migrated mostly in the first phase, were given forest areas for agriculture and for settlements. From 1940 onwards big farmers having experience in the cultivation of plantation crops migrated in large numbers to open new estates. All these movements finally resulted in deforestation. The relation between net migration and deforestation is found to be positive and significant (CESS 1997:84). Micro level studies in villages in the Idukki district of Kerala indicate that until

1960 the migrants acquired land only through deforestation for cultivation purposes and after 1960 most of the settlers acquired land through purchase. Only a small number continued deforestation, which was stopped by 1980 (Jeena 1997; George and Chattopadhyay 2001).

Changes in Agriculture

Agriculture occupies large areas of the earth's land surface. Therefore whatever is done within agriculture can often have major environmental effects (Lele 1991). Changes are manifested in the diversion of prime agricultural land, crop selection, seed variation, fertiliser use, irrigation and other factors. Pretty (1995) pointed out that pesticides have replaced biological, cultural and mechanical methods for controlling pests, weeds and diseases. Inorganic fertilisers have been substituted for livestock manure, compost and nitrogen fixing crops; information for managerial decisions comes from input suppliers, researchers and agricultural extension agents rather than from local sources; machines have replaced labour, and fossil fuels have been substituted for local energy supplies. External inputs as the means to increase food production have made the whole agricultural system more vulnerable to sudden failures.

Agriculture in Kerala has changed both in its area coverage and in the intensity with which it impacts on the land. These two factors interact synergistically. Concomitant with intensification mechanisms such as irrigation, machines and pesticides, cultivation has spilled over to increasingly marginal areas with scant regard to land suitability, and crop substitution takes place in response to market forces with little concern for long term ecological consequences. At the other end of the land use hierarchy, prime food producing lands have been diverted from agriculture to housing, roads and industry.

A study by the Kerala State Planning Board (GOK 1999) identified three major agricultural trends: (1) Shifts in the cropping patterns, (2) a declining share of agriculture in the state's domestic product despite an increasing dependence of the population on agriculture, and (3) low productivity per land unit.

The shift in the cropping pattern requires an in depth discussion. A brief note on the two other items is provided first. The contribution from agriculture and allied sectors to state domestic products has come down from 38 per cent in the early eighties to 30 per cent in the late nineties. Mainly the tertiary sector is growing. This could have been a

Box 4.2 : Likely Impacts of Conversion of Prime Agricultural Land

Ecological Impact: Intrinsic productivity declines; hydrological cycles are disrupted; water cushioning areas are reduced; residual chemicals for agricultural fields pollute water bodies and thereby toxic chemicals enter into food chains.

Local Economic Impact: Commercial crops are less labour intensive; surplus labour generated due to mechanisation is not always absorbed in other sectors due to the low level of industrialisation; little chance of alternate employment.

Food Security Impact: Nonfood crops replace food crops resulting in food shortages; rice fields in fallow periods no longer produce some types of wild vegetables and fish which had been harvested by the marginal community without any cost.

Social Impacts: Nonlandowners, access to farms is reduced; the marginal community had used these areas for raising livestock and poultry, which are now restricted; further marginalisation takes place.

Sources : Compiled by the authors from various widely recognised studies.

welcome scenario provided it had absorbed the surplus labour from the agricultural sector, which is not evident at this point. Instead, the population depending on the primary sectors has increased over the years.

The second issue is low productivity of major crops, but high gross income. Lands in Kerala are highly productive. General biomass production per hectare is around 12 tons (Central Plantation Corp Research Institute, Govt. of India, Kasargod, Kerala, personal communication). Taking coconuts as an example, the number of nuts per tree has come down, but gross output per unit of land is still high. This has been attained by planting more trees and through multi tier cropping. However the long-term ecological viability of high crop density is not yet clear.

Given this background, the changing agricultural scenario is discussed here under various sub-heads.

Shifts in the Cropping Pattern: From Rice and Tapioca to Coconuts and Rubber

At the formation of Kerala State in 1956, 24 crops were raised. Cocoa was added to the list in 1980–81. By 1996, the number of major crops was the same 25, but a shift had taken place in their coverage. The area in food crops was declining, while plantation and commercial agriculture had gained considerably. The years 1975–76 and 1980–81 marked two important turning points in the cropping pattern. The area under tapioca peaked in 1975–1976 and then declined.

Groundnuts and sesamum followed the same trend. The area under rice was at its maximum in 1980–81. The same was the case with cashew nuts. Considering the area under total food crop the years 1975–76 mark the turning point after which a decline set in.

The total cropped area increased by 37.2 per cent from 1957 to 1996. The increase was 32.6 per cent during the period from 1957–58 to 1975–76 and 4.6 per cent from 1975–76 to 1996–97. In the first phase pulses, sugarcane, tea, and cereals other than rice recorded decline in area. The decline continued for these three crops from 1975–76 to 1996–97 in addition to five other crops: rice, tapioca. arecanut, cashewnut and cardamom (Table 4.4). Four of these – rice, tapioca, other cereals, and pulses – are food crops. These four food crops together accounted for 46 per cent of total cropped area in 1957–58, but this dropped to 42.5 per cent in 1975–76 and barely made 20 per cent in 1996–1997. Rice and tapioca experienced the maximum reduction. Until 1975–76 rice was the first ranked crop in the state, followed by coconut, tapioca and rubber. By 1996–97 rice ranked third after coconut and rubber. Rice accounted for 35 per cent of the total cropped area (TCA) in 1956–57, but came down to 14 per cent; a decline of 44 per cent in a span of 40 years. Areas under coconut and rubber have increased by 117 per cent and 350 per cent respectively. Tapioca experienced a decline of 33 per cent during the 40-year period from 1957 to 1997. Some minor food crops like sweet potato almost disappeared from the agricultural scene in Kerala. Figure 4.1 indicates temporal variations in food and non food crop coverage.

There is a clear shift from food crops, mainly rice and tapioca in favour of tree crops such as rubber and coconut. Export crops like

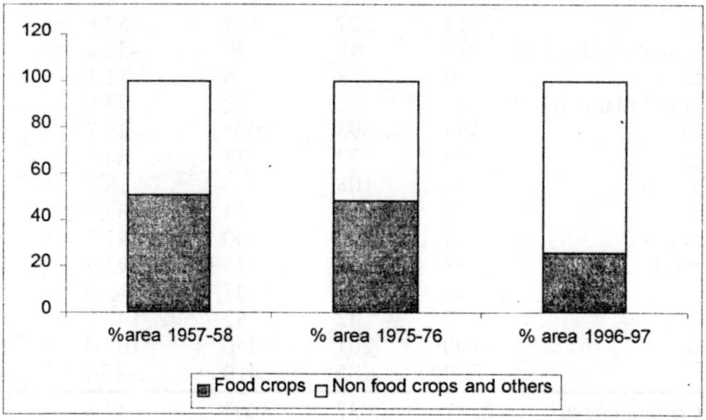

Fig. 4.1 : Temporal Variations of area under food and non-food crops

pepper, ginger and coffee have also gained in area. The maximum gain in terms of area was for pepper (173,000 ha).

In order to understand the change in the cropping pattern an in-depth analysis of rice and rubber has been attempted. Rice lost the maximum area while rubber gained the maximum area. Rice is a lowland crop planted in valleys and plains. Rubber is cultivated in the highlands and midlands regions where laterite and forest loam soils are found. The overall process of landuse change is as follows :

Coconut replaces rice in the lowlands and upland valleys
Tapioca replaces rice in the valleys
Rubber replaces coconut in the upland valleys
Rubber replaces tapioca and coconut in the midlands and highlands
Rubber replaces wasteland and natural vegetation in the foothills

In recent years rubber seems to be replacing rice in the plains and in the midland valleys. An analysis of the expansion of rubber and conversion of paddy will bring out the main thrust of the transformation that Kerala is experiencing today. Because these two crops represent two different agro-ecosystems they can be used to represent all or most of the issues to be addressed under the ecological transformation of the state.

Table 4.4 : Changes in Cropping Patterns in Kerala, 1957–1997

Crop	*Area in Thousands of Hectares*			*Change in Percent*	
	1957–58	*1975–76*	*1996–97*	*1957–58 to 1975–76*	*1975–76 to 1996–97*
Rice	767	876	431	14.2	–50.8
Tapioca	214	327	142	52.8	–56.6
Other cereals and pulses	28	45	29	–22.4	–35.5
Sugarcane	9	8	6	–11.1	–25.0
Banana and plantain	42	52	72	23.8	38.5
Coconut	463	693	1,005	49.7	45.0
Areacanut	50	77	73	54.0	-5.2
Pepper	91	108	173	18.7	60.2
Ginger	9	12	14	33.5	16.7
Cashew nut	44	109	100	147.7	–8.3
Cardamom	28	54	43	92.9	–20.4
Tea	40	38	37	–5.0	–2.4
Coffee	17	42	82	147.0	95.2
Rubber	100	207	450	107.0	117.4
Other	309	285	410	–7.76	43.86
Total Cropped Area	2211	2933	3067	32.6	4.6

Source : Department of Economics and Statistics, GOK: Agricultural statistics – various issues.

Conversion of Rice Lands

As indicated in Table 4.4 the area under rice cultivation has declined from 876,000 hectares (30% of TCA) in 1975–76 to 431,000 hectares (14% of TCA) in 1996–97. The contribution of rice crops to the state's income fell from 19 per cent in 1989 to 10 per cent in 1995 (George and Chattopadhyay 2001). This downward trend continued even after 1995, as increases in yields could not compensate for the decrease in area. Total rice production in the state came down further increasing Kerala's dependence on other states for staple foods. The environmental consequence of this conversion will be dealt with below.

The Kerala Statistical Institute (1994) conducted a survey to assess the changes in the rice growing areas of the state. They found that in 1992–93 only 58.1 per cent of rice fields were under paddy cultivation while 3.5 per cent were kept fallow. The remaining 38.4 per cent were being utilised as follows: 8.5 per cent for annual and seasonal crops, 23.8 per cent for permanent crops, and 6.1 per cent had been converted to non-agricultural use.

Disaggregating their data at the district level we can observe that the traditional rice growing areas like the districts of Palakkad and Alappuzha have 78.3 per cent and 65.1 per cent of rice land under paddy cultivation. Predominately rural districts like Wayanad and Idukki within the mountainous Western Ghats region have 77.1 per cent and 63.4 per cent of rice lands still under paddy cultivation. Ernakulam and Thrissur districts have more than 9 per cent of former paddy fields converted for non-agricultural use against the state average of 6.1 per cent. The causes underlying this high rate of conversion in these two districts are different. Industrialization and urbanisation are the main driving forces in Ernakulam. In Thrissur, clay mining for bricks, and tile manufacturing have contributed the most. Earlier in Thrissur the practice was to remove top soils from the paddy fields for a certain depth to reach the clay horizon. After the mining, workers spread the top soils back over the field so as to facilitate agriculture. But with more and more demand for bricks and tiles to construct houses, the mining is done for four to five years after removal of the topsoil. This intensive mining far exceeds the capacity of sedimentation in the monsoon flood season. The rains create artificial ponds and then these lands can seldom be brought back for growing rice.

Of the rice lands converted to non-agricultural use 44 per cent was devoted to buildings and courtyards, 31 per cent was given for brick making and tiles; roads and railways accounted for 15 per cent and another 10 per cent was for canals and water tanks (KSI 1994). Except for the canals and tanks, all other activities are directly or indirectly related to the expansion of human settlements.

To capture the spatial dimension around Thiruvananthapuram an analysis was done using topographical maps from 1966–67 and IRS (Indian Remote Sensing) images from 1991. The area administrated by the Thiruvananthapuram Development Authority and nearby surroundings had around 11.6 per cent as rice fields in 1966, which came down to 7.55 per cent in 1991. Of the total original rice land around 33 per cent had been diverted for human settlements that included mixed tree crops, 2 per cent had gone for agglomerated or clustered settlements with maximum built up areas and the remaining 65 per cent was still designated for agricultural use, but not necessarily for paddy. Vegetables and tapioca are also grown on the paddy fields.

Conversion of paddy fields and other lowlands to built up areas follows a sequence of gradual steps :

> Rice lands/wetlands→ coconut groves→ settlements with mixed tree crops→ agglomerated settlements (too crowded for tree crops).

Rice lands are low-lying areas and are usually flooded during monsoon months. In the laterite terrain they constitute the valley floor and receive rainwater flowing down from the surrounding slopes and ridges. Due to wetness and waterlogging the land cannot be readily converted to housing. The land level must first be raised and loose material consolidated. In the majority of cases the semi-natural process of consolidation and a raising of the level are followed by converting rice lands to coconut groves. This process follows two patterns (George and Chattopadhyay 2001). The first involves planting coconut saplings on the bunds along the periphery of the rice fields. Once the trees are well established, farmers widen the bunds and plant additional rows of coconut saplings. The process continues until the entire land is converted into a raised coconut garden. In the second method, mounds are created within the paddy fields at regular intervals and coconut saplings are planted. Rice is grown in the intervening areas. Once the coconut plant grows to a certain stage more mounds are created until the whole plot of land is converted to a coconut garden.

Coconut saplings are planted in such a way that houses can be constructed with minimal removal of coconut plants. Near urban centres such as Cochin, Ernakulam, and Thiruvananthapuram, piling is done to consolidate the loose soil of the wetlands for straight conversion to settlement areas. But these activities are limited to big government buildings or multistoried companies, of which there are very few in Kerala.

Causes of Conversion

Why is there a shift in cropping patterns? A survey conducted by Gopinathan and Sundaresan (1990) in Thiruvananthapuram in the south and Malappuram in north Kerala, looked at selected households to determine why people converted rice land for other uses. It was reported that between 1979 and 1989, 50 per cent of farmers in Thiruvananthapuram and 81.5 per cent of farmers in Malappuram district had discontinued rice cultivation. Tapioca, another major food crop in the state, was discontinued by 46.4 per cent of farmers in Thiruvananthapuram and 8.5 per cent of farmers in Malappuram. It was found that coconut was introduced in place of rice by 52 per cent and 83 per cent of farmers in Thiruvananthapuram and Malappuram respectively. Rubber was introduced by 30 per cent of farmers in Thiruvananthapuram and 8.5 per cent in Malappuram.

The findings of the Kerala Statistical Institute in 1994 about the underlying reasons for shifting out of rice cultivation showed that 71.2 per cent of farmers interviewed considered the low profitability of rice cultivation, followed by 18 per cent reporting a lack of irrigation. However, the responses were not the same for the entire period of the survey from 1968 to 1992. Those who discontinued rice cultivation prior to 1968 pointed out a lack of irrigation facilities (55.3%) followed by inadequate returns from rice cultivation (41.4%). During the subsequent periods, 1968–1982 and 1983–1992, uneconomic paddy cultivation was given as the main reason by 75.8 per cent of the respondents.

During the period from 1983 to 1992, building construction was given by 10.5 per cent of the respondents. Earlier studies by Panikar (1980), George (1982), Unni (1983), Sivanandan (1985) and Kannan and Pushpangadan (1988) indicated that the increased cost of cultivation and the consequent reduced profitability had influenced farmers' decisions to shift cropping patterns. During the 1960s and 1970s rice prices were fluctuating while coconut prices were increasing steadily. A reduction in the relative profitability of rice compared to coconut was being experienced by farmers even in the 1960s.

After 1970 the cost of producing rice tended to rise faster than the returns. The ratio of costs to returns is called the *parity index*. It reflects both wages and the cost of other inputs such as seeds and fertiliser. When the parity index is 100, the rice farmer breaks even; when it is above 100 the farmer profits and when it is below 100 the farmer sustains a loss. In 1971 the index was 93. In 1974 it rose to 109 but dropped to 83 in 1983. From 1971 to 1986 it averaged 96, meaning that farmers were not able to recover their costs of production most of the time (Franke 1996:133–34). In 1987 the index hit 100 for the last time and has dropped precipitously from 84 in 1990 to 77 in 1995 and collapsing to 48 in 2001 and 49 in 2002 (GOK 2002:Table 12.4). As the parity index has held below 100, the area planted under rice has declined steadily. Rice requires over 160 person days to cultivate one hectare, compared to 75 person days for coconut. Apart from the cost of labour, other input costs such as fertiliser, pesticides and costs of cultivation – especially plowing – were also increasing. This situation is compounded by higher fertiliser requirements to keep up the productivity of rice per unit of land.

A village study by Mahesh (1999) covered four categories of households: farming cultivator, non-farming cultivator, rural labourers, and others. The study village has 280 hectares of land categorised as rice fields. The farming community owned 51 per cent of this land, of which 40 per cent had been converted out of rice cultivation. The non-farming community possessed 40 per cent of the land, of which as much as 76 per cent had been converted. The rural labourers and other households, who owned 9 per cent of rice lands had diverted their entire areas.

People convert paddy fields both for agricultural and non-agricultural purposes. Except where seasonal and annual crops are introduced all other types of conversions are irreversible. Therefore these conversions have long-term impacts on the environment.

The conversion of rice land is not only a recent phenomenon. Around 11 per cent of land was shifted before 1968, another 11 per cent in the period from 1968–1982, and from 1982 to 1992 another 8 per cent was diverted. Conversion thus set in long before the 1990s but the purposes in earlier years could have been different. When joint families broke down in the 1960s and individual housing units were being sought after, but could not be accommodated in the raised garden lands, households in some areas had no option but to convert the adjoining wetlands to settlements. Property division is disadvantageous for rice cultivation. Although population growth in

Kerala is less than 1 per cent, the land requirements for settlements are still increasing.

Even large tracts of existing rice land are not put to use for cultivation. Non-availability of labour is a problem cited by many. Due to the spread of education, young people look for jobs outside of agriculture. Having only a small plot in their possession, the rural labour groups give first priority to construction. To sustain a family of 5 (the average size in Kerala) the land requirement is around 0.6 hectares in south Kerala given the present level of productivity. In Thiruvananthapuram city, paddy fields have decreased by 33 per cent since 1968. Almost all the paddy fields have been converted to settlements. Areas under agglomerated settlement have increased by 155 per cent.

Expansion of Rubber

Rubber as a commercial crop was introduced in Kerala in 1902 when the Periyar Syndicate in Travancore started cultivation of the *Hevea Brasiliensis* variety (George et al. 1988). As we noted earlier in our discussion of table 4.4, the area devoted to rubber production in the state grew from 100,000 hectares in 1957–58 to 450,000 hectares in 1996–97; a growth of 350 per cent in a span of 40 years. As shown on table 4.4 rubber was the fourth ranking crop behind rice, coconut and tapioca from 1956–57 until 1975–76. By 1996–97, rubber had become the second ranked crop behind coconut. Kerala accounts for 86 per cent of the total area for rubber in India and 94 per cent of the total national production.

Rubber is primarily a midlands and foothills crop. Production falls dramatically when it is planted above 500m. Wayanad district, located 700m above sea level, recorded a productivity of 281 kg/ha against 831 kg/ha and 768 kg/ha in the traditional rubber growing areas of Kollam and Kottayam (Chattopadhyay 1996b).

Originally rubber was introduced into areas with degraded forests. From there, it spread all over. It replaced natural vegetation (below 300m), tapioca, cashew, fruit trees and coconut. Comparisons of crop data over the years indicate that since 1975–76 areas used for tapioca declined sharply in areas where rubber production showed progressive increase. Figure 4.2 shows temporal variation of areas given for raising crops like rubber, tapioca, cashew, and for wasteland. Wherever possible, wastelands are planted with rubber. Cashew was also replaced by rubber, and since 1980–81, the area under cashew has decreased significantly. However, in drought affected areas such as Kannur and Kasargod districts of north Kerala where hard crust laterites limit normal agriculture; initial attempts to convert areas from cashew to rubber were not successful. Cashew, being drought resistant, is well adapted to the dry lateritic midland areas of Kerala. Rubber needs moisture and cannot withstand edaphic drought (prolonged dry soil). Therefore, many places in north Kerala have been converted from rubber back to cashew.

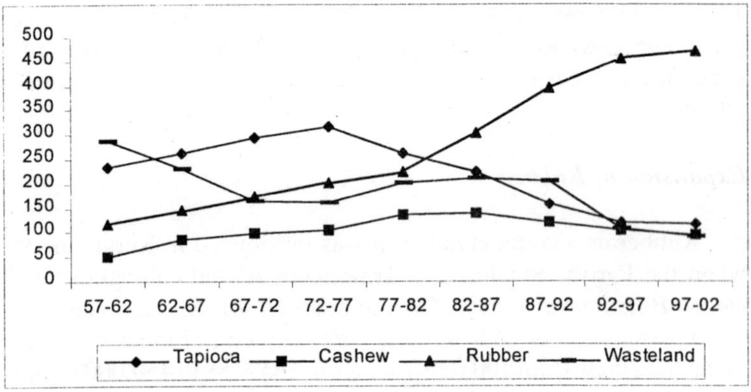

Fig. 4.2 : Temporal Variations of Area under Selected Crops and Wasteland

A sample area covering 500 sq km in the Thodupuzha-Painavu stretch of forest in Idukki district was studied to understand the consequences of the expansion of rubber plantations. Surveys of India maps for the time periods 1911–12 and 1976–77 were used. During 1911–12 forest covered 87 per cent of the total area. Rubber was grown on only 0.3 per cent (1.5 sq km). The rest was given to settlements, tree crops, paddy, and other plantation crops. Five rubber estates—Kaliyar, Mundunad, Martoma, Malankara, and Vellianatom— were below 100m altitude. Other significant cultivations were the cardamom plantation in the forest, dryland paddy cultivation in terraces, and wetland paddy in the valleys.

By 1976–77 the situation had changed dramatically. Forest cover was only 41 per cent and the rubber area had increased to 4 per cent (20 sq km). The same numbers of rubber estates were in operation but with bigger areas, spreading especially along the down slopes. Small holding plantations had increased significantly, extending to even higher altitudes. Other plantations that grew during this period were teak, eucalyptus, and tea, all of which are under the Kerala Government Forest Department.

Another dimension of the changes in landuse sometimes related to rubber involves topography. A traverse was taken from the Malankara rubber estate to Palkulam Medu in Idukki district. The stretch of land lying below 100m now alternates between rubber and settlements with some mixed open scrub areas. Rice cultivation is restricted to narrow valleys along a rivulet. The pattern continues up to the present-day forest boundary along which a patch of teak plantation is evident.

The transect analysis shows that rubber plantations are mostly restricted to within the 300m altitude in this area. One or two patches have sprung up in the higher altitudes. The area that was earlier earmarked as open scrub has been encroached upon by tree crops that were subsequently replaced by rubber plantations. On the steep slopes and ridges, rubber is not planted due to shallow soil and poor water availability. Isolated hills with thin soil cover continue where there is the open scrub. Grasslands occupy the higher ridges and cardamom is common on the plateau above 700m.

In the beginning of the 20th century, big estates of over 600 ha replaced the natural forests, scrublands or natural plantations. Large tracts of land are no longer available. In Kerala, 84 per cent of the total rubber area is in small holdings of less than 2 ha. Small holdings typically held tapioca, fruit trees, coconut and mixed tree crops. The ecological impacts of these two types of changes are not necessarily the same. The large plantations reduced forest cover and biodiversity. On the small plots agro-diversity and homestead-diversity declined. Homestead biodiversity had been used to provide fuel and fodder for the livestock population. For both types of biomass there are problems of availability in the state.

Landuse Change at the Micro Level

Just as Kerala has micro-ecozones that we described in chapter 3, land use can bring about micro-level changes. These micro-level changes are among the most important of Kerala's current environmental challenges. We shall present the dynamics of landuse change at the micro level via two specific examples, one from a rural area and one from an urban area.

Landuse Change in Udumbanoor Panchayat, Idukki District

Idukki district falls within the Western Ghats. Seventy-nine per cent of the land is highly undulating. Around 54 per cent of the total area is under forest or forest plantations. Udumbanoor panchayat has had considerable in-migration from other parts of the state. An agricultural plantation was set up in this panchayat in 1911. At that time the area was mapped under the first topographical survey in the state. Survey data are available for the year 1976 and in 1991 this village was described in detail through the Panchayat Resource Mapping Programme that we describe in chapter 8. In 1911, 42 per cent of the land in this panchayat was already in non-forest uses. The area under forest was around 58 per cent in 1911 (Table 4.5). Seasonal crops covered about 7 per cent of the area with settlement with tree crops and agricultural plantations accounting for the rest. By 1976, the area under natural forest had come down to 33 per cent, and teak plantations covered another 18 per cent. After Independence, forest areas in India were declared reserved lands and encroachments were officially prohibited apart from allocations of land for river valley projects or similar activities by the government. However, the forest department started raising plantations of various types of trees.

Table 4.5 : Variations in Landuse in Udumbanoor Panchayat, 1911 to 1991: Areas as Per cent of Total

Landuse Category	*Per cent of Total*			*Per cent Change*		
	1911	*1976*	*1991*	*1911–76*	*1976–91*	*1911–91*
Forest	57.75	32.52	32.50	-25.23	-0.02	-25.25
Forest Plantation		18.25	18.25	18.25	0	18.25
Agricultural Plantations with settlements and mixed tree crops	35.44	41.03	41.87	5.59	0.84	6.43
Seasonal crops	6.81	8.20	7.38	1.39	-0.82	0.57
Total	100	100	100			

Sources : Topographical maps. 1911, 1976 and PRM 1991

Given the total area under natural forests and managed forests (forest plantations), the loss of forest area to non-forestry purposes at Udumbanoor in the span of 65 years is only 7 per cent. It can be inferred that prior to 1911, easily accessible areas were already diverted from forest to agricultural and settlement purposes. The untouched forests lie mostly in rugged terrain. The period between 1911 and 1976 witnessed encroachment in selected areas with less intensity. The forest plantations became a buffer between the natural forest and the agricultural or settlement areas. At the same time, the forests were limited to the rugged slopes where they are more difficult to encroach upon. The areas accommodating agriculture and settlement increased from 35 per cent to 41 per cent between 1911 and 1976, but only had a 1 per cent increase between 1976 and 1991. Areas under seasonal crops (mainly rice) had increased until 1976. After 1976, this area declined slightly.

Although it might appear from Table 4.5 that there was not much change in the landuse patterns between 1976 and 1991, there were ground level changes in the composition of crops, intensity of settlements, and homestead gardens. Reductions in the seasonal crop area, albeit small, parallel the trend experienced elsewhere in the state with declining rice area. Even the remote rural areas are not free from this problem of conversion. Field investigations show that in garden areas (agricultural plantations with settlements), tapioca was previously the crop of choice. Now the same land is used for coconuts and rubber. Density of settlement areas is also increasing to compensate for a 31 per cent rise in population during the period 1971 to 1991.

In this context, landuse and landcover change can be viewed in two stages. In stage one, cultivated land expands at the cost of the forest area. In Udumbanoor panchayat, this mostly occurred prior to 1911. In the second stage, rubber was introduced and many other crops were replaced. In a study in Idukki district Jeena (1997) identified two stages similar to these. The first stage during the 1940s coincided with the government's campaign to grow more food. The second stage that introduced rubber in the 1960s was based on high profitability and incentives for commercial plantation crops. Table 4.6 illustrates this clearly. The timing of deforestation may differ from one place to another, but the basic trend remains the same. Analysis of the landuse change (CESS 1997) for the entire Idukki district brought out a similar trend. Natural vegetation cover for the entire district declined from 82 per cent between 1913 and 1928, to 65 per cent between 1954 and 1970, and to 54 per cent in 1992. Simultaneously, agricultural plantations rose from 16.3 per cent to 33 per cent to 43 per cent respectively.

Table 4.6 : Land Acquisition by Deforestation or by Purchase by Settlers in Idukki District, 1950 to 1980

Year of Migration	*Per cent of Land Acquired by Deforestation*	*Per cent of Land Acquired by Purchase*
Prior to 1950	100	0
1950–1960	100	0
1960–1970	36.4	63.4
1970–1980	19.1	80.9
After 1980	0	100

Source : Jeena, T. S. 1997. Internal Rural Migration and the Impact of Deforestation on Cropping Patterns in Kerala. New Delhi: M. Phil thesis, Jawaharlal Nehru University. P. 58

Landuse Changes in Urban Areas

Kerala presents an urbanisation pattern very different from the rest of India. The rural/urban difference is significantly lower in Kerala and there is a coexistence of rural and urban functions across the state (Chattopadhyay 1995). We noted in chapter 2 that Kerala's dispersed settlement pattern in effect reduces rural-urban distinctions. While urban functions spread along the ridges following the transport networks, the valleys are dominated by agricultural activities.

Landuse in and around urban centres is changing rapidly. The state capital, Thiruvanthapuram provides an example. A set of data was available covering the corporation area for 1961 and 1976. Wetlands and rice lands covered 24 per cent in 1961. This number dropped to 9.6 per cent in 1976 and to 5 per cent in 1991 (Chattopadhyay and Jayaprasad 1991). This decline occurred despite restrictions imposed by the government of Kerala against the conversion of wetlands, which had been identified as greenbelts within the city limits (George and Chattopadhyay 2001). Within India it is peculiar to Kerala that even in the city the predominant settlement pattern is linear, single or double storied, with a courtyard, located along the ridges and upper slopes. Settlements have spilled over into the valleys and wetlands over the years. The cost of land in the valleys is lower than that on the ridges and slopes. People desiring to live in or near the city but who have limited financial resources, opt for the lowlands.

Thiruvananthapuram corporation and its surroundings covered an area of 320 sq km, and were considered for a Geographic Information System based analysis (CESS 2000). Topographical maps for the year 1966 with a scale of 1:50,000, and IRS Image Geocoded data on the same scale for the year 1991 were used. Landuse maps for the two time points were prepared and an overlay operation was conducted. The results are shown in Table 4.7.

Table 4.7 : Landuse Change, Thiruvananthapuram City – 1966 to 1991

Landuse category	*Coverage in 1966*		*Coverage in 1991*		*Change (1966–91)*	
Rice land	37.13	(11.6%)	24.16	(7.6)	-12.97	(-35%)
Settlement with mixed tree crops	270.41	(84.5%)	270.54	(84.5%)	+ 0.13	
Agglomerated settlement	8.31	(2.6%)	21.15	(6.6%)	+12.84	(+160%)
Water body	4.33	(1.3%)	4.33	(1.3%)	Nil	
Total	320.18	(100%)	320.18	(100%)		

Source : CESS 2001. GIS Analysis for Urban Space Management.

In 1966, 11.6 per cent of total area was paddy fields. This declined to 7.6 per cent in 1991. The loss of rice land is about 35 per cent. Detailed data analysis indicated that out of this 35 per cent as much as 33 per cent was converted to settlements with mixed tree crops – while the remaining 2 per cent went into agglomerated settlements (CESS 2001:18). Agglomerated settlements represent the clustered areas or main built up areas where inter-settlement-space (non-built up or open space or homestead area between two buildings or dwellings) is minimal. These are mostly found radiating from the core of the urban centre along the transport lines. Even in the centre of Thiruvananthapuram urban area rice lands had previously been found. The category of settlement with mixed tree crop that covered slightly above 84 per cent in both 1966 and 1991 had taken area from rice lands. Mixed settlements had intensified in some areas to become agglomerated settlements. So the main built up areas forming the core of the city had increased from 8 per cent to 21 per cent; an increase of 155 per cent in 25 years.

The rate of conversion is higher in the lands adjacent to the urban areas than in more distant lands. The reduction of rice lands and low lands in the coastal plains is less. This is attributable to the fact that some of the wetlands in the coastal plains have been acquired by the government and therefore are temporarily protected. Also, this area experienced an intensive conversion spell in the early days so that the remaining wetlands need considerable effort and finance to be converted.

The expansion of agglomerated settlements indicates that open spaces within settlements are decreasing. Green space between individual settlements is a unique feature of Kerala within the Indian context. Reduction of this space represents a fundamental change in the settlement system. Although the increase in population density is extremely low in this area, and for that matter the entire state, settlement density is found to be increasing at a high rate. As we noted earlier in this chapter in connection with the demand for rural settlement space, this could be attributed to the demographic structure and the change in the joint family system. Young people, constituting a large portion of Kerala's population are getting married and moving to homes separated from their parental families. High population density coupled with the decline in the joint family system has resulted in a high demand for settlement space. Families in urban areas that do not want to lose location advantages, but who have limited financial resources, opt for settling in the lowland areas. Almost all lowland areas have been recently settled.

With the rise of agglomerated settlement areas, total surfaced areas are increasing while vegetation cover proportionately decreases. Apart from losing the wetlands, which has serious ecological impacts, increases in surfaced areas result in the creation of "heat islands", which may have an impact on the micro climate.

Cropping Intensification

Crop intensity expresses the number of times a land unit is put to use in one agricultural year or calendar year. If a plot of land is cultivated once in a year, it is considered to be of 100 per cent intensity. With today's technology a single plot of land can raise up to three crops annually. Therefore, crop intensity can be expressed by a number that varies from 100 per cent to 300 per cent. This is applicable for seasonal crops, like paddy, vegetable, pulses, etc. In the case of tree crops, or perennial crops where multi-tier cropping is practiced, the system of coverage assessment is different. The Department of Economics and Statistics, Government of Kerala has evolved a system of counting trees and converting the number into area by applying certain factors determined for each tree crop separately. For example, 80 coconut trees can be raised on one acre of land. The number of coconut trees divided by 80 will give the area under coconut cultivation. It is possible that there are some other tree crops in the same plot of land. Applying similar factors evolved for other tree crops, the area covered by tree crops individually has been worked out. By adding up the area under all tree crops raised in a single field, the total cropped area is estimated. Therefore, an increase in crop intensity for tree crops indicates an increase in density of trees or an increase in biomass density.

Agricultural statistics in Kerala have certain limitations. The figures for total cropped area presented in government reports include both seasonal and tree crops. As a result, the intensity figure does not result in an accurate picture. Therefore, apart from considering overall intensity, the intensity for rice has to be worked out separately.

The total cropped area in the state increased from 2,349 thousand hectares in 1960–61 to 3,067 thousand hectares in 1996–97 and the corresponding net sown area in the years of 1960–61 and 1996–97 were 1,924 thousand hectares and 2,249 thousand hectares. The cropping intensity, which is a ratio between total cropped area and net area sown increased from 122 per cent in 1960–61 to 136 per cent in 1996–97. The increase in cropping intensity is directly related to the increase in total cropped area as the net area sown cannot be further stretched. Keeping aside the three plantation crops of rubber, tea and coffee, which have limited scope for intercropping, we calculate that the intensity figure covering the rest of the crops ranged from 124 in 1960–61 to 149 in 1996–97.

As indicated earlier the crop intensity for rice has been calculated separately to highlight pressures on the lowlands. However, before embarking into the details it may be pertinent to discuss certain facts concerning rice cultivation in Kerala. Rice is grown here in all three seasons. Locally these seasons are known as *virippu*, *mundakan*, and *punja*, equivalent to the all-India nomenclature of *kharif*, *rabi* and *boro*.

Mundakan refers to the winter or small harvest. In Kerala planting for this harvest takes place just after the *Onam* or New Year celebration in late August or early September. In the end of October and November the North-East or "retreating" monsoon deposits water across much of the state filling irrigation canals that allow for the crop. The mundakan harvest takes place in January in most areas. From February through May Kerala experiences the long dry season, but on or about 1 June, the Southwest or major monsoon hits the state with substantial rains that last through June, July, and early August. The plantings done in early June lead to the virippu, or major harvest. The punja harvest was previously a vegetable or legume harvest done on wet rice lands during the post-monsoon season. In some parts of Kerala such as Kuttanad and in kole lands paddy fields are usually submerged during the monsoon. Cultivation is only possible by dewatering. These lands are referred to as punja lands and the cultivation is known as punja cultivation.

Although the punja land is primarily a water logged area that used to be cultivated during post-monsoon dry months after draining out excess water, more recently it has been brought into cultivation in other seasons also with the strengthening of embankments all around the field and the capacity for draining at any time of year. In the absence of a detailed plot by plot enumeration of rice cultivation it is difficult to assess net area sown under rice. However, it is possible to estimate it using available data. Rice areas are confined to particular types of land that are well identified in Kerala. With the introduction of short duration high yielding variety (HYV) seeds nowadays there is a tendency to grow paddy in all three seasons, however, spatial coverage is always limited within particular land types. In normal cases, rice cultivation covers the maximum area in the main (virippu) monsoon season, but in the waterlogged lands, where draining of water is required to implant rice saplings the maximum coverage under this crop is recorded during post-monsoon (punja). In some parts of Kerala, where irrigation facilities are well developed, the maximum rice crop coverage could be marked in the winter season (mundakan). Given this background it can be reasonably assumed that the maximum area covered by rice cultivation in any of the three seasons is the net sown area under paddy. The gross area under rice has been worked out by adding figures pertaining to all three seasons. There is a problem in calculating the crop intensity for rice because a ready figure of net sown area for rice is not available. We therefore use the planted area for whichever season was greatest in order to make the calculation. To illustrate let P1, P2 and P3 be the areas covered by rice in three seasons respectively. The gross rice area (Gp) = P1 + P2 + P3. The intensity of rice = Gp/ P1 or P2 or P3 which ever is the highest. The intensity of paddy cultivation has been worked out for all the years, for which these disaggregated data are available.

In 1975–76, when paddy cultivation was at its peak the intensity of paddy cropping was 223 per cent. In 1986–87, the intensity figure remained the same although the total area under paddy decreased by 25 per cent. This highlights that rice fields are used throughout the year i.e. in all three seasons. Land continuously under paddy is prone to losing certain essential micro nutrients, which cannot be compensated for by the addition of higher doses of fertiliser such as N (nitrogen), P (phosphorous), and K (potassium).

Crop Production, Fallow and Fertiliser

Associated with high cropping intensity is the intensive use of fertilisers. The land does not get sufficient time to recover by lying fallow in a two-crop culture. Raising the same crop continuously year after year robs the soil of nutrients. This is partially made up by the application of fertiliser.

Uninterrupted paddy cultivation and overuse of fertiliser is leading to a decline in land productivity in Kerala. A key measure of productivity is the ratio of crop production to fertiliser consumption (CP : FC), which indicates how land responds to fertiliser applications over time. If the integrity of the soil is preserved under cultivation, the CP variable will increase in response to an increase in fertiliser use. Declining ratios point to fertiliser use that has increased to the point of diminishing returns or to insufficient replacement of organic matter through fertiliser inputs leading to soil depletion (Munasinghe and McNeely 1995). USDA (1989) data indicate that most of the countries in the world are experiencing declines in the crop production to fertiliser ratio, with certain variations. Japan is one of the few countries which has maintained the ratio over the years.

In Kerala we attempted to work out the ratio (Chattapadhyay and Sasidharan 1994). Fertiliser consumption and yields of paddy over the years were plotted on a graph. From the trend line it was possible to compute the rate of increase of paddy yield corresponding to fertiliser use. During the period 1976–1980, when fertiliser consumption increased from 30 kg/ha to 40 kg/ha – an increase of 10 kg/ha, the paddy yield increased by 50 kg/ha or five times as much. This gives a ratio of 1:5 for fertiliser to yield. Then from 1984–86, fertiliser consumption increased from 50 kg/ha to 60 kg/ha but the yield increased by only 35 additional kg/ha. This gives a ratio of 1:3.5. The decline in the ratio from 1:5 in 1976–80 to 1:3.5 in 1984–86 indicates a decline in soil quality. The relation of soil quality to landuse change will be discussed in the section on changes in soil quality in chapter 5.

5 Causes of Concern: Changes in Landscape Ecology

Climate, Wetlands, Soil, and Water

This chapter presents the changes in the landscape ecological system of Kerala. Our analysis covers micro climate, wetlands, soil and water. We discuss the trends in terms of the quantity and the changing quality of the resource base.

Changes in the Micro Climate

Climate change is currently a hotly debated topic in the scientific community. As we noted in chapter 1, at the global level warming seems to be occurring as a result of elevated concentrations of CO_2 and certain other "greenhouse gases." The human contribution to climate change is now well recognised. Carbon dioxide (CO_2) released into the atmosphere by burning fossil fuels is the prime source of global warming. This change is related to micro climates, the atmospheric characteristics prevailing within small spaces, usually in the layer near the ground bearing the impacts of human-made changes of vegetation cover on humidity, temperature, winds, and rainfall and the impacts of built-up areas on wind speed, air turbulence, increased evaporation, drainage and runoff. The Intergovernmental Panel on Climate Change (IPCC) was created in 1988 by the World Meteorological Organisation and the UN Environment Programme (UNEP). The IPCC assessments that were released in 1990, 1995, and 2001 dealt with available scientific information on climate change, its potential impact on society and possible responses (Dunn and Flavin 2002). The Kyoto protocol in 1997, although contentious in the beginning, appears to have been brought into force quickly. This was done despite the United States' unwillingness to abide by the Kyoto emissions targets.

While CO_2 is mostly related to industrialization, changes in landuse and land cover can also influence micro-climate conditions. Carbon emissions from industry are not yet a problem in Kerala. However, Kerala experiences some micro-climate changes that are most likely due to landuse and land cover alterations. One such change is deforestation.

Whether deforestation influences rainfall patterns or not, has been widely debated, and so far no conclusions have become accepted by a consensus of scientists. As early as 1888, the influence of deforestation on rainfall was discussed in the *India Meteorological Memoirs* (Blanford 1888). Researchers have presented both positive and negative evidence on the consequences of forest clearing on rainfall (Meher-Homji 1980). After analysis of rainfall trends and the number of rainy days at 28 stations in the Western Ghats from 1906 to 1975, Meher-Homji concluded that there is evidence to link deforestation to diminished rainfall and fewer rainy days. The Munnar plateau in Kerala, which recorded 22 per cent deforestation during this period, showed a declining trend for 11 out of 12 rainfall and rainy days-related variables considered in the study.

A similar study in the Andaman and Nicobar Islands brought out a link between the decreasing levels of rainfall and rainy days and deforestation (Biswas1980). A detailed analysis of rainfall from 1901 to 1980 covering 75 stations in Kerala was attempted by Soman et al. (1988). That study found trends that could be indirectly related to landuse and land cover change. It was reported that rainfall in south Kerala decreased significantly between 1941 and 1980, compared to the preceding 40 years between 1901 and 1940. Extreme rainfall, seasonal rainfall and annual rainfall all showed a decrease of between 15 and 20 per cent in the hilly tracts covering Ernakulam, Kottayam and Idukki districts. This area had lost around 56 per cent of its forest during the same period. The concentration of rubber plantations was also high here. Why should rubber plantations correlate with lower rainfall? Research suggests that rubber plantations have a reduced evapo-transpiration rate compared to the forest ecosystem. This can be deduced from the fact that the total amount of water consumed by rubber plantations was estimated to be 900 to 1000 mm per annum, whereas the annual water consumption rate of a typical tropical rain forest of the same area was estimated to be 1500 mm (Satheesan et al. 1993:72). Field experiments brought out that rubber plants with old foliage recorded an evapo-transpiration rate of 1 to 2 mm per day whereas agricultural crops such as vegetables, tapioca, banana and coconut had average daily evapo-transpiration rates varying between

3.0 and 5.8 mm per day (Satheesan et al. 1993:72; Varadan et al. 1990: Table 6). Expansion of rubber plantations therefore might have been reducing the moisture content of the air leading to a reduction in rainfall.

The physical mechanisms by which vegetation loss alters the climate conditions of a region are not yet well understood. The trend however is unambiguous : the rainfall is decreasing in Kerala. It should be noted that north Kerala only experienced this trend during the northeast monsoon. Table 2.3 gave typical monthly rainfall patterns. In chapter 1, we noted that temperature increases in Kerala are also associated with declines in photosynthesis and with lower crop yields.

Changes in the Coastal Ecosystem

As we stated in earlier chapters, Kerala is divided into three physiographical zones : lowlands, midlands and highlands. The lowlands (below 8m) are considered the coastal zone. Three major towns in the state, Thiruvananthapuram, Cochin-Ernakulam and Kozhikode (Calicut), are located in the coastal zone. Other large towns including Kollam (Quilon), Alappuzha (Alleppey), and Telicherry-Kannur (Cannanore), are also along the coast. The population density in the coastal zone is over 2,000 persons per sq km, the highest of the three zones.

The coastal area of Kerala is characterised by beaches, mudflats, mangroves, backwaters, rock cliffs, and lateritic and sandy plains. The 570 km long coast line is subjected to both accretion and erosion. There are zones of intense erosion and also extensive accretion distributed throughout the state. Out of 570 km of coastline around 304 km exerience accretion while the remaining 266 km are affected by erosion of various degrees of intensity (Baba et al. 1997). The Kerala coastline can be divided into three distinct zones (Nair 1987): gently sloping permeable coastline dominated by sand and clay and covering the central part of Kerala, patches of sedimentary cliffy semi-permeable coastline, and rocky (crystalline) coastline also occurring in patches. Erosion and accretion both occur in all these units. The conditions under which erosion takes place in the coastal tract are (Thomas and Samsuddin 1989) :

- seasonal erosion associated with heavy monsoon rains
- erosion induced by human activities such as beach sand mining, construction of jetties, and breakwaters

- end erosion due to rock wall, groins and other protection measures
- erosion at inlets
- mudbank induced erosion
- erosion of laterite cliffs
- erosion beside the rocky promontories

A combination of these factors is involved in any given locality. The sand and clay dominated coastline experiences erosion as well as accretion. In Aryad panchayat of Alappuzha district, a stretch of land more than 150m wide was built up between 1928 and 1992. In Chellanam panchayat in Ernakulam district, the coastline receded by 130m during that same period (Chattopadhyay 1996a).

About 40 km north of the capital at Thiruvananthapuram, the sedimentary cliffs at Varkala are subject to heavy erosion. Crystalline rocks form the promontory, as can also be seen at Kovalam on the coastline 17 km south of Thiruvananthapuram where the area experiences heavy erosion along the sides of the promontory. A number of pocket beaches have formed within the rocky coastal stretches throughout the state.

Mangroves and Backwaters

Mangroves usually grow in places where fresh water mixes with saline water. The delta fronts are natural habitats for mangroves. Mangroves in Kerala can usually be found in small isolated patches. Some of the larger concentrations are found in the Valapattanam mouth in Kannur district, Kadaluni-Beypore in Kozhikode district, and Kumarakom in Kottayam district. The mangroves are generally under considerable stress. The state has lost a sizable area of mangroves over the years. It is estimated that there were 700 km^2 of mangroves in the near past but at present it has been reduced to less than 50 km^2 (Mohanan 1997).

The Kerala backwater system is unique within India. It provides life-supporting resources like fish and lime shells and it facilitates transport. The chain of backwaters and connecting canals has provided a waterway that runs from south to north in the state. The backwaters have been shrinking. They have been experiencing depth loss due to excessive siltation, and clogging from silvania in many places. Apart from the physical changes, the alteration of the chemistry of the water has significantly reduced its productivity. The high population density along the backwater system has compounded the pollution problem through the disposal of household waste, sewage and industrial waste. Rosamma Stephen (1991) reported a decrease in zooplankton density, from 4,574/m^3 in 1968 to 2,348/m^3 in 1971 to 742/m^3 in 1991 in the Vembanad estuary, which is the largest backwater system on the west coast of India.

The mudflats are mostly associated with estuarine mouths, where rivers bring in silt and freshwater. Dams have arrested rivers in their upper reaches. The sediment supply has become irregular. Most of the mudflats have been converted to prawn farms or coconut gardens. The extensive mudflat in the Thottapally area has been converted mostly to coconut gardens. In some patches where the soil is degraded and has become excessively alkaline due to submerged decomposed vegetation, coconut trees cannot grow. These patches are now left as unproductive marshes.

Sand Mining, Sea Walls and Coastal Water Pollution

The Kerala coast is well known for black sand deposits. These are strategically important for radioactive minerals. These minerals are extracted through beach sand mining. This sand removal accelerates coastal erosion.

To combat erosion, sea walls have been constructed along major parts of the coastline. However, this has not proved effective. Due to underscoring—when the sea waves push under the base of the sea wall—the sea wall sinks in a number of stretches and requires frequent strengthening. At both ends of the sea wall the coastline is heavily eroded.

The coastal waters in Kerala also have problems. It has been estimated that more than 500 thousand cubic meters of effluents, containing a variety of pollutants, are being dumped into the rivers of the state daily from major industries. These pollutants eventually find their way into the coastal waters by way of the backwaters.

Under the aegis of the Department of Ocean Development of the Government of India the Center for Earth Science Studies regularly monitors the pollution levels in Kerala's coastal waters. Research results available so far indicate that the coastal waters at various locations are becoming increasingly polluted. High concentrations of dissolved and particulate cadmium (0.18–0.29 mg/l=milligrams per litre), and mercury (35–50 mg/l) were reported in the water samples collected from the Cochin river mouth and the Veli outlet at Thiruvananthapuram (Ramalingam et al. 1994). The coastal canals are equally affected. A study of Connolly canal, which passes through Calicut, by Nirmala et al. (1994) indicated that the dissolved oxygen value of the canal water was 2.93 mg/l. The nearby wells were affected by this canal, as evidenced by the fact that there was a DO value of 2.4 mg/l in the well water. Good quality water will have a DO value above 5mg/l. The rotting of wooden logs is perhaps the main reason for the high pollution level. Coastal waters are more affected than river waters as they are the receptacles and suffer stagnation at times. Economic activities such as fishing, coir making, and agriculture are now under stress. Human populations along the coast face problems with drinking water, health, and hygiene. All of these problems are interrelated.

Changes in the Wetland Ecosystem

Wetlands are considered to be among the most productive ecosystems. Previously, wetlands were considered to be wastelands and the only way to develop them was either to fill them for settlements or to drain them for agriculture. It is now understood that wetlands perform important ecological functions including flood cushioning, recharging of ground water, controlling surface water filtration, providing nurseries for fish, offering sanctuary for birds, and maintaining biodiversity in general. As we noted in chaptere 2 filling or draining these lands for human activities became a cause of global concern, which led to the Ramsar convention. Wetlands in various countries are designated as 'Ramsar sites'. Funds from international agencles are available to protect and manage Ramsar wetlands. India has designated 20 Ramsar wetland sites through the Department of Environment. Two sites in Kerala are included on the list: Vembanad Lake in central coastal Kerala and the Ashtamudi estuary. Recently the freshwater lake Sasthamkotta has been included in the list of Ramsar sites (See Figure 1.2 for the locations).

There are various estimates of the extent of wetlands in Kerala. Both brackish water and fresh water wetlands are found here. In some estimates parts of rice lands are included. Man-made reservoirs which cover around 296 sq km dispersed throughout the state are also considered wetlands.

The Kerala government estimates that brackish water wetlands cover 2,426 sq km in the state (GOK 1992). Wetland mapping conducted by the Government of India Space Application Centre, for all the states in India based on satellite imagery estimates Kerala's wetlands to be around 2,349 sq km.

The brackish water wetlands located in the coastal plain where population density is high, are subjected to extreme stress. Transformation of wetlands into other types of environment is a common phenomenon here. Shrinkage of the surface area, siltation and a decline in water quality are also common problems encountered in the wetlands.

Lake Vembanad : From Major Wetland to Polluted Estuary

A detailed survey of the extent and consequences of wetlands transformation is yet to be done. However, some studies on wetlands provide a basis on which the situation can be gauged. Vembanad estuary – usually known as Lake Vembanad – is the largest brackish water body in the state. Its surface area fell from 315 sq km in 1912 to 179 sq km in 1983 (Nair et al. 1987). The area having Kerala's highest population density surrounds this body of water. The Cochin urban area is located at the outlet of this water body to the sea.

A sizable portion at the southern part of the lake was drained in the 1940s as part of the "grow more food campaign." Concrete embankments were constructed to divide the area into several cultivable sections for paddy. Popularly known as the Block Reclamation area, its sections are identified with English letters like R block, D block etc. The outlet of the Vembanad estuary to the sea is occasionally dredged to maintain navigability. Seven major rivers with a combined catchment area of 15,000 sq km, debouch into this lake. Huge quantities of silt brought down by these rivers are deposited here. Although time series data are not available, it is reported by the local people, particularly by the fishermen, that the lake is becoming shallower. The need for dredging is an indication of siltation. Reclamation—filling of the lake to create dry land—continues all around this lake to various degrees.

A barrage was constructed across Vembanad Lake during the 1970s converting its southern part to a freshwater body. The main objective was to facilitate rice cultivation in non-monsoon months. Part of a brackish water body was also converted to fresh water. This part with an area of 81 sq km now receives silt deposits from a

5,838 sq km catchment area of four rivers along with the pollution from fertilisers and pesticides used on agricultural fields, sewage from settlements without sanitation facilities and household waste water from settlements all around. In the absence of regular flushing, the area has effectively been converted into a pollution sink. From a land utilisation point of view there are six types of activity bordering the southern part of the lake: urban settlements, dense settlements with mixed tree crops, settlements with tree crops, block reclamation areas (mainly rice, coconut and settlement), rice lands and a rice and prawn alternation locally known as *pokkali* cultivation. This water is also subjected to coconut husk retting which adds to the pollution.

Analysis of water quality brings out close relationships among variations in the water quality parameters and the activity zones (Sooraj et al. 2000). Coconut husk retting which is widespread in coastal waters contributes to increased BOD (Biological Oxygen Demand) and reduced DO (Dissolved Oxygen). Nitrate-nitrogen (NO_3-N) and nitrite-nitrogen (NO_2-N), which reflect fertiliser pollution, showed concentrations of more than 1.00 micro mol/litre (micro molecules per litre, a measurement of the molecular concentration of the water) and more than 3.00 micro mol/litre respectively towards the southern and eastern parts of the lake bordering predominantly rice fields. The higher the concentration the lower is the water quality.

Figure 5.1 gives a map of the Lake Vembanad area and surroundings. The pollution is increasing, as regular flushing of the lake has been restricted due to the Thaneermukkom barrage – shown in the photo of Figure 5.2 – which prevents tide water incursions from the northern part of the lake. This created a further problem of heavy infestation of salvinia or African payal. Salvinia is an exotic hydrophyte (water-loving plant) that thrives in fresh water. Proliferation of this species was first reported in the 1950s (Gopalan et al. 1987). Many of the water bodies, ponds, and channels in the Kuttanad area of Alappuzha and in Kottayam district are choked with this weed. Construction of the Thaneermukkom barrage in 1974 further facilitated salvinia (floating moss) growth as saline water flow was restricted. The salt water had prevented growth of salvinia. The Thottapally spillway in the southwestern corner of the Kuttanad area along National Highway 47 (see Figure 5.1) had been installed to restrict salt water inflow to the Kuttanad area of Alappuzha district to facilitate rice cultivation and to drain out flood water during monsoon months. The spillway however also turned out to provide a congenial

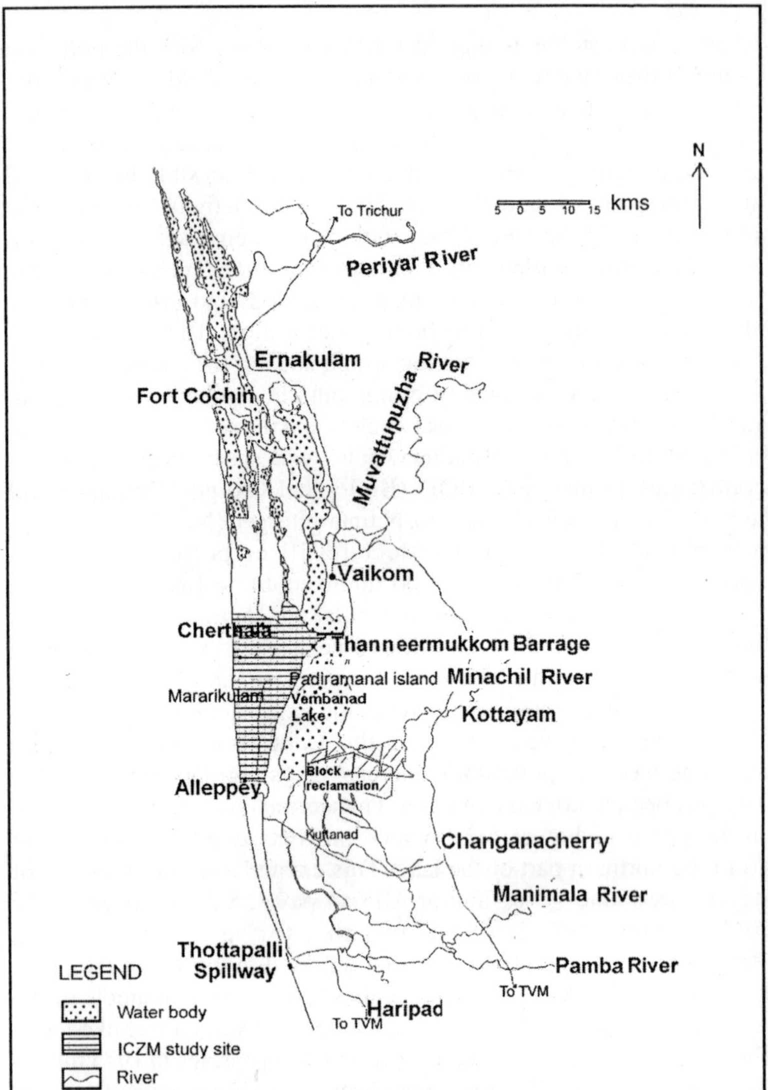

Fig. 5.1 : Vembanad Lake and its Surroundings

Fig. 5.2 : The Thanneermukkom Barrage. Photo by Richard W. Franke

environment for the growth of salvinia. During the monsoon months when the water level rises and most of the rice fields are waterlogged the African payal moves with the flow of water and in the process it colonises many of the low-lying paddy fields. This hampers the movement of boats through the water courses. The northern part of the lake is under stress from urbanisation, industrialization and port and harbour related activities. The domestic sewage produced in the Cochin urban area is estimated to be 3.5 million litres per day. Approximately 90 per cent of this sewage is untreated and is discharged into the surrounding water bodies, including Lake Vembanad (Kumaran 1994).

Pollution of Other Water Bodies in Kerala

Apart from Vembanad lake other water bodies are also affected. The Asthamudi estuary adjoining Kollam town is experiencing changes in area, depth and water quality. Population growth and drastic changes in landuse have contributed to these developments. Around 3 sq km have been filled up by dumping residual materials from the Kundra clay mine located in the northeastern part of the estuary in Kundra panchayat. There are pockets of increased enrichment of nutrients and low dissolved oxygen (DO) levels. Around 6.5 sq km of paddy fields within the catchment area of this estuary have turned into marshes or waterlogged areas. The organic matter and calcium carbonate content (contributing to hardness) in the sediment sample showed an increasing trend over recent years (Muralikrishna 2000). This signifies that the pollution load in the estuary is increasing.

The wetlands ecosystem throughout the state is subjected to various degrees of stress apart from the explicitly urban-industrial pollution. Various agricultural practices are also contributing to the changing quality. Once certain thresholds of fertiliser use have been reached, the soil and plants cannot absorb it, and excess amounts called "residuals" accumulate or are moved around by the motion of water in and near the fields. The residual nitrate in the rice fields is sometimes washed down to the lakes and backwaters where it causes water quality problems. The algal blooms observed in rice fields are possibly associated with an increase in nitrate level (Mary Kutty et al. 1996). The amount and types of algae are widely used as indicators of pollution levels in Europe and North America (James and Evison 1970; Friedrich et al. 1992). Some algae are reported to give oily, salty, sweet or bitter tastes to water while some others like the *Microcystis* species produce toxins that kill cattle and cause disease in human beings (Maya et al. 2001). There is also evidence that post harvest soils carry residual nitrates and phosphorous (George and Chattopadhyay 2001). Both of these elements finally find their way to adjacent water bodies either through surface runoff or through sub surface water. High nitrate concentrations are associated with an unsatisfactory microbiological quality of water and high concentrations of phosphorous are largely responsible for eutrophication (Chapman and Kimstach 1997).

In the freshwater wetlands of Sasthamkotta lake near Kollam municipality and Vellayani lake in the vicinity of Thiruvananthapuram city, studies by CESS (1998; 2000) indicated changes in water quality due to various human interventions. Prakasam and Joseph (1991) reported that in Sasthamkotta lake during the rainy season nitrates, phosphates and silicates registered slight increases, which is probably due to agricultural runoff. Given the setting of the lake the agricultural runoff is from the tapioca (cassava) and other tree crops in the lateritic slopes. This indicates that fertilisers have been applied even for cassava, a crop that does not respond to fertiliser, and for tree crops. The water of Vellayani lake had a high total coliform count (20–645 per 100 ml). This is attributable to the lack of adequate sanitation facilities in surrounding areas. High coliform counts mean a high likelihood of the spread of infectious diseases that cause fever and diarrhea. Even people living in Kerala's traditionally somewhat separated house compounds can become vulnerable in an area where a lake facilitates the spread of infected water. Local people living around this lake have reported deterioration of water quality over the years.

Soil Erosion

Soil erosion occurs naturally as part of the land building process. As a natural process soil erosion depends on rainfall, soil characteristics and topography. Pedologists identify three soil horizons. These horizons develop through a process called leaching. The uppermost layer is called the A horizon. The C horizon represents the weathered parent material over which the B and A horizons develop. The "A" horizon is the most productive level of the soil. Once the level A is lost, high doses of fertilisers are needed to keep the "B" level productive. A sizeable area of Kerala's hill slopes are almost devoid of A horizon soil and cultivation now occurs in the B level.

Under natural vegetation cover in the Western Ghats the rate of soil erosion is 0.062 tons/ha/year (Chinnamani 1965). This is well below the average natural rate of 8 tons/ha/year computed under natural vegetation cover in various countries. Soil erosion becomes a problem when the cover type is altered. The erosion rates in the Western Ghats varied from 0.224 tons/ha/year on a tea plantation with 75 per cent canopy, (25 per cent of the soil is exposed to direct sunlight throughout the day) to 4.662 tons/ha/year on a plantation with a 15 per cent canopy.

Soil erosion takes place in three forms: sheet erosion, rill erosion (less than 30 cm deep) and gully erosion (more than 30 cm deep). Gully formation gives the highest rates of erosion. We compared gully formations in four settings: natural vegetation and rubber plantation, tea plantations, tea plantations and rock outcrops, and mixed crops. The density of the gullies was the lowest under natural vegetation and rubber (154 metres/ha). It was the highest (340 metres/ha) on the tea plantations (Chattopadhyay 1985). The erosion rate is directly proportional to the gully density.

Sedimentation is a further problem. Sediments deposited in reservoirs constitute accelerated erosion in the catchments. This problem is encountered throughout the world. In India, Nizamsager in Andhra Pradesh, is perhaps the most silted in reservoir. It has lost 52 per cent of its storage capacity. Kerala has 30 important reservoirs and three lakes; their catchment areas are subjected to various degrees of erosion.

Using the internationally accepted Universal Soil Loss Equation (USLE) James et al. (undated) estimated that the catchment areas of the Mangalam and Peruvannamuzhi reservoirs were experiencing annual soil loss at an average of 23 tons/ha/year and 66 tons/ha/year respectively. The average annual sediment contribution per hectare of catchment area was estimated at 11.26 cubic metres at Mangalam and 286 cubic metres at Peruvannamuzhi. Surveys by KERI in 1977 and 1982 had reported an annual sedimentation rate of 25.39 cubic metres at Malampuzha – the site of a large dam and popular tourist area at an altitude of 100 to 200 metres on the upper reaches of the Bharatapuzha river and 90 cubic metres at the Peechi reservoir. It was found that although soil erodibility is only 70 per cent in Peruvannamuzhi of that in Mangalam, for all the contributing factors such as length of slope, rainfall erosivity and cover management, Peruvannamuzhi recorded higher levels. The Peruvannamuzhi catchment also had some areas under tapioca cultivation. Tapioca cultivation may perhaps be the explanatory variable since it contributes to soil erosion because the roots have to be pulled up to harvest them and the soil comes loose. This typically occurs in the dry season when the soil is subject to blowing away as dust.

Landslides

Landslides are massive movements of soil, rock, debris, and mud down a slope. They are generally classified under the category of natural hazards. Geological processes are involved in creating most landslides. The landslides in Kerala are different. They mostly occur during heavy monsoon rains. Tropical areas have thick soils even on the steep slopes. The humid tropics are characterised by chemical weathering, under which hard crust cap rocks are formed. This is known as lateritisation. The soil is rich in clay. During continuous rainfall in the monsoon months, the soil gets saturated. The contact zone between the soil and bedrock, along which sub-surface water flows, is lubricated with excess water, facilitating movement of the soil. Most landslides occur in hollows, i.e. areas of slope convergence.

In general, slopes steeper than 18 degrees are unstable. Kerala has a sizeable area of slopes above this level. Plantation agriculture has spread over all types of slopes. Landscaping, particularly for plantations and agriculture also facilitate high water infiltration. This makes the area more vulnerable to landslides. Terracing is a common measure for soil conservation but may actually be hazardous if inadequate attention is given to drainage and the water holding capacity of the soil.

Landuse alteration throughout the foothills and up the mountain slopes has contributed to the hazards. Kerala has recorded deaths due to landslides. One of the most devastating landslides occurred on 9 November, 2001 at Amboori in Thiruvananthapuram district which caused 39 deaths. Figure 5.3 shows the consequences of the Amboori landslide. Geological investigation highlighted the existence of a multiple rock fracture zone filled with regolith in saddles. Hydrostatic pressure developed due to water oversaturation that could have produced this slide with or without microseismicity (Soman et al, 2002). Large scale landuse and landcover change had drastically modified the slope wash process, which directly affects hill slope hydrology. When these alterations take place along geological weak planes the risk of landslides increases manifold. As society has very little control over geological processes, amelioration is possible only in the realm of land management.

Fig. 5.3 : The Amboori Landslide. Photo by G. Sankar

Changes in Soil Quality

Vegetation plays an important role in soil formation. Landuse alteration also changes soil quality. Leaching which is partly responsible for soil development is associated with root systems, organic matter, and the composition of the vegetation. Replacing natural vegetation with introduced species induces changes in soil properties. It has been reported that the silt to clay ratio is higher in natural forests than in plantation areas such as eucalyptus. The studies of Dabral et al. (1965), Mathur et al. (1976), Chinnamani (1965) and CESS (2000) pointed out deteriorating soil fertility due to changes from natural cover to plantations.

Soil fertility, soil structure, soil erosion and water availability to the soil are partly related to the organic matter present in the soil. Therefore, the quantum of organic carbon available in a stretch of soil is a measure of its quality. Non organic carbon derives from minerals, salts, and the like and does not carry the same properties as organic carbon.

The source of organic carbon in soil is primarily the vegetative matter: plant litter, roots and other plant parts. The concentration of organic carbon in the soil depends directly upon the amount of available organic matter, which is related to the biomass generated by the ensuing land use type.

Balagopalan (1995) and Karthikakutty Amma et al. (1995 and 1996) found that organic carbon and total nitrogen are lower on rubber plantations than with natural vegetation. Degraded forest and teak plantations had even lower values. Similar observations have been reported in Hainan in China. The organic matter content on rubber plantations lies mostly in the range of 10 to 20 grams per kilogram of topsoil (g/kg) against 158 g/kg in evergreen forests. In all cases forest soils contained more nitrogen than soils in rubber plantations. This was attributed to the high litter production (leaves, twigs, etc.) under natural forest conditions, which varied from 5.09 tonnes per hectare per year (t/ha/year) to 10.24 t/ha/year in Uttar Kannada, Karnataka to 7.9 t/ha/year to 12.5 t/ha/year in Kallar, Kerala (Bhatt 1990; Shaji Abraham 1994).

Vegetation types and soil nutrient levels are definitely related. The surface runoff that comes out of a forest area carries nutrients along with the silt. When this water floods over agricultural fields, the soil regains fertility. This is especially important in the case of irrigated rice agriculture. The rice plant has an unusual cell structure in its roots that facilitates taking up oxygen and nutrients directly from water. Flooding the fields with water in the early stages of plant growth can greatly increase the output. The warm water carries blue green algae that help to fix nitrogen. Decomposed organic materials provide natural fertilisers. The gentle movement of the water across the field aerates the soil and the plants, further assisting in the delivery of oxygen. However, when the vegetation cover is changed in the upland areas where the irrigation water originates, the runoff flowing through the downstream-irrigated fields may not be as nutrient rich. Similar, though less dramatic effects can occur with other crops including coconuts.

Infiltration, Surface Runoff, and Stream Flow

Infiltration and surface runoff depend on rainfall, slope, soil type, landuse management, and environmental protection practices. If infiltration is high, the surface runoff will be less, but a nearby stream will have a higher baseflow as the sub-surface water will eventually feed the stream. The degree of soil erosion also depends on surface runoff and infiltration. Infiltration is a complex process influenced by a large number of factors. Organic matter in the soil and the surface litter (decaying leaves, etc.) play important roles in infiltration. Litter layers usually act as a sponge for water at all times of the year and tend to counteract the changes in the moisture content of the soil that occur because of the alternation of rainy and dry seasons. In natural forests, where litter production is high, infiltration rates will automatically be high. Kerala is dominated by latosals (lateritic soils), which cover nearly 60 per cent of the geographical area of the state. Experimental studies have shown that the distribution of gravel and organic carbon in the soil partly explains the differential rate of infiltration in hill slopes of 15 per cent to 20 per cent incline and having an elevation difference of 3 meters or more (Varadan and Raghunath 1985). The basic infiltration rate varied from 6 cm/hr (centimetres per hour) in soils planted in coffee to 35 cm/hr where rubber was the crop (Sushanth and Jayakumar undated). The high infiltration under rubber can be partly attributed to the generally good management practices followed by rubber cultivators. Infiltration in the forest area is significantly higher due to the presence of a thick litter layer on the forest floor (Lee 1980).

Satheesan et al. (1993:72) conducted infiltration experiments covering rubber, teak, coffee, and cardamom plantations in selected watersheds in Kerala. They reported that the ratio of direct runoff to total precipitation remained very low throughout the year, ranging from 2.8 per cent under tea and coffee to 5.1 per cent under cardamom. For rubber the ratio was 3 per cent. Similar findings had been reported earlier by Sushanth and Jayakumar (undated : 5). They also reported that direct runoff under coconut was very high at 11.25 per cent. Table 5.1 provides information on basic infiltration and direct runoff of selected landuse types.

Table 5.1 : Infiltration and Runoff Under Different Landuse Types in Kerala

Landuse Type	*Average Slope of the Plot in Per cent*	*Basic Infiltration Rate in Centimetres per Hour*	*Direct Runoff as Per cent of Total Rainfall*
Tea	25	24	1.5
Coffee	24	6	3.5
Rubber	40	35	2.5
Cardamom	46	25	6.5
Coconut	57	15	11.25

Source : Centre for Water Resources Development and Management, Undated.

Baseflow varied from 66 per cent under rubber to 95 per cent under coffee. The direct runoff data from table 5.1 give an idea of the impact of landuse and land cover on stream flow. With a good canopy cover, as is the case with coffee and tea, baseflow potential is high. Becaúse coconut plantations do not have a good canopy cover, their baseflow is poor, and the direct runoff is high. The steepness or gradient of the coconut plot as indicated in table 5.1 is another factor to consider. Direct runoff is always higher in steep slope areas. In the case of rubber, the canopy is an important factor, but if the rubber is planted with proper terracing, a high infiltration rate can be achieved. Rubber uses less water than natural forest cover and many other plants. Water use in rubber plantations is estimated to be 500 to 600 mm lower than in the typical tropical rainforest ecosystem. The rate of evaporation is also lower in rubber plantations. As indicated earlier in this chapter lower rates of evapo-transpiration can influence local rainfall and thus the micro climate. Therefore rubber monoculture has the potential to contribute to local climate change although effects on a regional scale are difficult to demonstrate.

Alteration of land cover from forest to another type causes changes in the overall ecosystem. During the initial years of alteration the soil has enough organic carbon from its natural forest period. The soil's hydrological functions are barely perturbed at first, but over the years the disturbances become evident. Many of the streams in the hill areas run dry for part of the year. Springs are lost or became seasonal. When noted one by one, these changes are often ignored. If they are considered cumulatively, they turn out to hamper the functioning of the ecosystem. People who depend directly on these environmental resources lose out. They are often among the poorest—part of a worldwide phenomenon to which we refer in chapter 11.

Landscape Factors Influencing Drainage Discharge

Drainage discharge varies a great deal seasonally from one place to another depending on rainfall. Since Kerala's rivers are rainfed, bankfull stream conditions are found only during the monsoons. In the dry seasons it is baseflow that maintains the river. The amount of water that will come from a catchment or river basin depends on atmospheric factors, catchment parameters, vegetation, and land cover types (Lee 1980).

A study conducted in the Wayanad Plateau (Chattopadhyay and Chattopadhyay in Press) compared eight sub-catchments. Forests and plantations were considered together as not all catchments had forests, and some plantations (mainly tea and coffee) provide good canopy cover. Plantations and forests together present the total vegetation cover. Correlation analysis indicated that the total discharge is highly correlated (+0.88) to the area of the catchment. The hill areas correlated with discharge positively at 0.33. Stream length was found to have a negative correlation.

A multiple regression analysis showed that 74 per cent of the variance in discharge can be explained by a combination of the catchment area, stream length, average slope in per cent, hilly area in per cent to total area, and per cent of area under forests and plantations. As was indicated in the previous section, plantations with good canopy cover accelerate infiltration and contribute to baseflow. The slope is often modified through terrain and contour bunding such as terracing. This will have a positive impact on infiltration, but if vegetation cover is reduced, the positive impact of slope management may be nullified.

Changes in River Water Quality

Water flowing out of a drainage basin can be used as an indicator of the condition of the basin. The quality of water and the silt load both reflect the health of the basin which is composed of rock types, terrain conditions, soils, vegetation, and the effect of human activities. Alteration of biogeochemical and nutrient cycles can be detected by river studies (Meybeck and Helmer 1989).

As we reported in chapter 2, with its 44 rivers and a total usable yield of 42,772 million cubic meters of water Kerala is well endowed. Around 850 sq km are taken up by rivers (GOK 1995). Kerala receives an annual rainfall of over 3,500 mm. The major rivers in Kerala are mostly perennial. People in the rural areas, and even in the small urban centres previously relied heavily on river water for domestic and agricultural purposes. Wells, which are major sources of water for domestic use today, still rely on rivers for recharging. When the river water is no longer potable, the rural community becomes hard pressed. The poorer groups that often cannot afford to dig a well see their access to water threatened. The problem occurs not only in the interior. The coastal and backwater areas where the rivers debouch are equally affected. In fact, loss of productivity in coastal waters has become a major global concern (Meybeck 1998).

Studies on water quality in south Kerala rivers (Chattopadhyay et al. 1996; Soman et al.1997) point out that river water quality is affected by landuse. The dissolved oxygen (DO) value was low in samples from areas downstream from rubber plantations. The stream Killi Ar, which originates in the foothills and passes through Thiruvananthapuram city before joining the Karaman river, was found to be extremely polluted through most of its course. Pollution levels get very high in the urban areas where sewage disposal is a major contributor. In the upper catchment, the dense rubber canopy extends out over the stream and prevents sunlight penetration to the surface of the stream. Studies of the distribution of photosynthetic active reduction (PAR) in rubber plantations show that 80 per cent of the PAR is accounted for by the upper canopy within 5m of the top of the canopy (Satheesan et al. 1984). Penetration of sunlight is poor, which constrains photosynthesis resulting in poor aeration and consequently lower DO values. The Vamanapuram river, which is comparatively less disturbed, also showed lower values in the samples taken in proximity to the rubber plantations when compared to the area under forest. River water is also affected by higher counts of coliform (e-coli) which causes diarrhea especially downstream where higher population densities are found (Soman 2002).

A segment of the Chalakudy river was studied to understand the relationship between landuse and land cover and water quality (Chattopadhyay et al. 2001). Water samples were collected in four different seasons under various land use types and hydro-geochemical analyses were done. The degree of acidity or alkalinity of the water (pH values) was found to vary substantially across the seasons along with rainfall and also with different landuse practices. Urban areas as zones of intense human activity contribute significantly to water quality alterations, sometimes polluting the water beyond the self purifying capacity of the river. The stretch of river adjoining the urban settlement and the uncultivated paddy fields came out as highly polluted. For example nitrate and phosphate levels were 7.30 mg/litre and 0.580 mg/litre respectively. In the forest area, nitrate values varied from 0.05 to 0.12 mg/litre and phosphates from 0.005 to 0.009/litre. The disposal of untreated urban waste in the stream might have contributed to the high levels of pollution. The areas dominated by settlement with mixed tree crops and the river segment flowing through uncultivated paddy lands are also polluted. High concentrations of nitrates and phosphates are also attributed to high fertiliser pollution runoff.

Fertiliser and pesticides are sources for water pollution in the areas dominated by agriculture. Where landuse type is constant—settlements with mixed tree crops—the differences in pollution levels may be attributable to the differences in landuse intensity and fertiliser application.

The forest area was not generally affected but the reservoir was marginally polluted with medium levels of phosphates and high levels of nitrates. Bacteriological analysis showed that the total coli form count and fecal coli forms were high in all the samples. The physical and chemical characteristics of the water and their associated land-use types are given in Table 5.2. Figure 5.4 shows variations of selected water quality parameters under different landuse categories in the Chalakudy river basin.

Until now, an abundant water supply has been a cornerstone of the quality of life in Kerala. Studies in various parts of the state indicate that declining water quality is a problem requiring immediate attention. The rivers are now stressed and their resilience limits will soon be exceeded. Even the wells in Kerala are affected. Pollution sources are classified as point (localised) and non-point. Non-point sources are mostly agriculture and dispersed settlements. Industries are point sources. Management of non-point pollution sources is now required through decentralised local initiatives. If water security is threatened, symptoms of which are already beginning to manifest themselves, there will be a high price to pay.

Table 5.2 : Landuse Type and Physical and Chemical Characteristics of the Chalakudy Basin, Kerala in 2001

Landuse Type	*Physical and Chemical Characteristics of the Water*
Forest	Very low hardness, nitrate, nitrite and phosphate, low conductivity (as water is not saline conductivity increases with an increase in the salinity of water), low suspended solids
Plantation	Very low hardness, nitrate, nitrite and phosphate, low conductivity, low suspended solids compared to forest
Settlement with mixed tree crops	Except one station, all other stations in this category show high levels of nitrate, the excluded station had high phosphates
Uncultivated paddy	High nitrate, low hardness
Urban settlement	High hardness and conductivity, very high nitrate, nitrite and phosphate
Paddy	High values of nitrate and phosphate, low suspended solids
Coconut cultivation	Low hardness, moderate calcium and magnesium, moderate dissolved solids
Clay mine	Low hardness, moderate magnesium and nitrate
Prawn farm	Very high values of hardness and conductivity, very high calcium and magnesium, total dissolved solids very high

Source : Chattopadhyay et al. 2002

Changes in the River Regime

Human activity can cause widespread changes in rivers. These include changes in runoff and sediment transport leading to adjustments in stream channel dimensions. Wolman (1967) first drew the attention of geomorphologists to this line of investigation and subsequently,

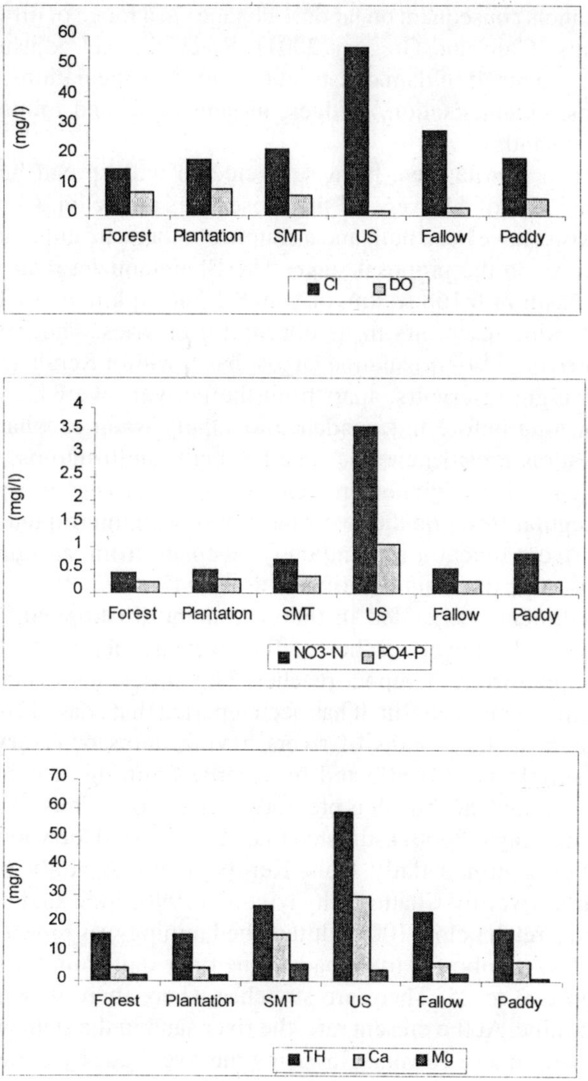

Fig. 5.4 : Variation of Selected Water Quality Parameters
SMT= Settlement with Mixed Tree Crops
US = Urban Settlement
See Table 5.4 for additional information

large numbers of studies have investigated morphological adjustments of river channels consequent on landuse changes in a range of different environments (Chin and Gregory 2001). River channel adjustment takes place as a result of dam construction and river regulations, land use changes, channelisation, bridges, urbanisation, and mining of riverbeds for sand.

Rivers in Kerala have been subjected to widespread human interventions. Out of 44 rivers, 11 have reservoirs upstream. As many as 30 reservoirs have been built and a couple of others are either under construction or in the proposal stage. The Bharatapuzha river, with the largest basin of 6,168 sq km, of which 1,786 sq km are in Tamil Nadu, hosts nine reservoirs in its different tributaries. The Periyar, the longest river (244 km) and the largest basin within Kerala (5,284 sq km) has eight reservoirs, apart from the Periyar lake which was constructed long before Independence to supply water to what was then the Madras Presidency. The state has eight multi-purpose river valley projects, 10 irrigation projects and 11 hydroelectric power projects. Regular flows on the rivers have thus been interrupted. As a result, the rivers function intermittently. Seepage from surrounding areas has become the main contributor to river flow.

The sediments generated in the catchment are trapped in the reservoirs and the lower reaches of the river are deprived of any sediment supply from the upper reaches. This upsets the balance of the total sediment of the basin. It has been reported that coastal erosion is more rapid at the mouths of rivers having dams or reservoirs. Sediment supply is also affected by riverbed mining, a common practice throughout the state. It is presently a matter of serious concern. Padmalal and Maya (2000) estimated that about 20,000 lorry loads of sand are being mined daily from Kerala rivers. A study of the Vamanapuran river by Chattopadhyay et al. (1996) indicated that at 27 sites in a stretch below 100m altitude bed mining was prevalent in this river. The number of truckloads mined per day from each site varied from two to 25. There are stretches where there is no more sand left to mine. At the present rate, the river sand in the state will be exhausted within 5 to 10 years, lowering the river beds by more than 5 meters from the present level (Soman 2002).

Lowering of the riverbeds has several implications. It changes the thalweg profile (the line along the lowest elevation in a valley) and the long profile. The rivers adjust themselves with new base levels and move towards equilibrium; then cross profiles develop accordingly, responding to seasonal fluctuations of water levels. Once the riverbed is lowered, the bank slumps and the adjoining agricultural fields become unstable. During the monsoon months, heavy water flow erodes these areas.

In some areas property owners with land adjoining the bed mining sites have gone to the courts to obtained judicial stays against sand mining (Chattopadhyay et al. 1996). In some cases, the sand miners place sediment traps in the main rivers, and obstruct regular flow. This causes disputes between people in the upper reaches and the lower reaches.

Another impact of lowering the riverbed is the drop in water tables in adjoining wells. The groundwater level is locally controlled by the water level in the rivers. Once the water levels in the rivers go down, the water levels in the wells will go down. Many of the wells need deepening as a result. There is also the possibility that some wells will become non-perennial due to the excessive lowering of nearby riverbeds. Some instances have already been reported. In the Vamanapuram basin, hard rock is exposed in the riverbed around Pullampara, Aruvipuram, Meenmutti, and adjoining areas. These areas have been marked as grey zones (areas of lowered water tables) by the Central Ground Water Board. The direct relationship between river sand mining and grey zone declaration is yet to be established, however the spatial association is quite alarming.

River channels in Kerala are lined in most cases, particularly when they pass through the urban centers. Side walls built of rock along the riverbank are a common sight. In many cases, embankments have also been constructed along the river, and settlements have sprung up right along the riverbanks, leaving virtually no space for the rivers to spill into during monsoon rains. This results in breaching of the weaker areas, increased flood heights in whatever open area is available, and pushing the water level upstream.

Brick clay mining in the riverbanks is another issue. The natural levees are removed making the surrounding areas more vulnerable to flooding during the monsoons. Riverbank vegetation is also cleared. This results in riverbank failure (slumping), and is especially common in the monsoon months.

Management of the rivers needs a watershed approach as the source of waters and sediments are hydrologically determined. Which watershed could supply how much and which ones would be affected, could be identified at the lowest level and intervention programs could be chalked out. Aggregate data do not help in this context. Rivers as resources warrant ecosystem approaches in management, not commodity approaches of utility value. The dimension of ecosystem service is too large to understand and apparently simultaneously too important to ignore. During the People's Campaign that we describe in chapter 9 steps were taken to delineate watersheds and develop a watershed based approach to development planning.

Changes in the Groundwater Regime

Groundwater is an important component of the hydrological cycle. Part of the rainwater infiltrates the sediments and recharges the groundwater. It is a very slow process. To meet the growing demands for irrigated agriculture throughout the world groundwater is used. As we noted in chapter 1, overpumping has become a major global concern.

Groundwater is an important resource for irrigated agriculture in India. The areas beyond the irrigation canals rely solely on groundwater for raising crops in the non-monsoon seasons. When the rate of pumping exceeds the rate of replenishment (recharging by rainfall), the water level goes down. In Punjab, Haryana, Bihar and West Bengal, where intensive agriculture is practiced, declines in the water tables have been reported. Water deficient states like Gujarat are also reporting water table overdraws. Overpumping in and around mega cities is a further concern. In Kolkata since the 1960s each house in the new suburbs was built with its own tubewell. The general groundwater table dropped 6 meters. In some parts of central and south central Kolkata, the groundwater dropped 10 to 15 meters (Biswas 1995; Mitra 1995). Large areas of Punjab have already experienced water table drops. The net draft for irrigation is equivalent to 658 mm of rainfall. Groundwater withdrawal is now equal to 99.4 per cent of groundwater recharge leaving little or no amount of water as storage (Singh 1997). This will eventually result in falling groundwater levels and gradual depletion of the groundwater.

In Kerala, the historically lavish water endowment has so far meant that the problem is not yet acute, but symptoms of overdrawing are beginning to appear. Presently 5.8 per cent of Kerala's rainfall goes to recharge groundwater tables. Aggregate data for Kerala indicate that the state is only using up to 15 per cent of its annual rainfall recharge. This ranges from 4.93 per cent in Wayanad district to 32.88 per cent in Kasargod district. Disaggregated data available from the Central Groundwater Board for 1992 indicate that four Blocks, Chirayinkil in Thiruvananthapuram district, Kottarakara in Kollam district, Mala in Thrissur district, and Kasargod in Kasargod district are using over 50 per cent of their annual recharge. Kasargod block is the only block out of the four not dominated by agriculture. River sand mining and clay mining are also widespread in these areas. As indicated earlier, river sand mining causes a fall in the water tables of the surrounding wells. Clay mining also causes a decline in the water table. When clay is mined, a sizeable area is removed that served to hold water, and the ditch that is created by the mining creates a new base level for the surrounding wells. This invariably leads to a decline in the water table. Field investigations clearly indicate that a decline in groundwater recharging capacity is due to clay mining (Reshmi 1996).

The case of Kasargod is slightly different. Here, there is extensive wasteland composed of hard crust laterite. Although laterite is a good recharging medium, the hard crust generates high surface runoff. There is no irrigation project in this area. Well water is used for drinking and irrigation. Water is being drawn to utilise former wastelands for tree crops.

Considering the developments until 1992, the groundwater department identified these four blocks that would reach a critical stage in the next five years. The groundwater estimation committee categorises the areas as white, grey and black. Explanations for these categories are provided in Table 5.3.

Table 5.3 : Types of Groundwater Development in Kerala

Type	*Stage of Development = Annual Net Draft as Percent of Utilisable Ground Water*	*Restriction Level*
White	Less than 65	No restriction for further development
Grey	65 to 85	Restricted future development
Dark	More than 85	No further development

Source : Ground Water Estimation Committee, Ground Water Department, Government of Kerala, 1992.

It was estimated that Chirayinkil block in southern Kerala would soon reach the dark stage when no further development is recommended and the other three areas will be in the grey stage of restricted future development. Analysis of 1999 data by the Central Ground Water Board (CGWB) showed drastic changes. In 1989 all 152 blocks in Kerala were safe (white category), but by 1992 three blocks were approaching the grey category while one block (Chirayinkil) was heading for the dark category. By 1999 as many as 15 blocks reached the grey or dark categories. These 15 blocks were reclassified as overexploited (3 blocks), critical (6 blocks) and semi critical (6 blocks) (Tamta et al. 2004:194). These blocks are distributed throughout the state from north to south mostly along the midlands. The CGWB reported in 1999 that overexploitation of groundwater reserves for domestic and agricultural purposes contributed to the lowering of the water table in patches of the midlands and hill regions of Kerala. Although data are not available for the most recent years it can be gauged from the trend that the situation has likely been deteriorating further.

Monitoring the water level in 41 wells in the Vamanapuram basin continuously from January 1994 until January 1995 (Chattopadhyay et al. 1996) revealed that in 18 wells the water level in January of 1995 was lower than that of 1994. The maximum decline was 2.37 metres. However, 21 wells showed increases with a maximum of 2.11 metres. Two wells showed no change. If we extrapolate these results to the entire basin, hypothetically it experienced a drop of 25 cms or one-fourth of a metre. Data for one year are not sufficient for a definitive measurement since water levels fluctuate depending on a number of parameters like site, lithology, distance from a river or other water body, rainfall and antecedent moisture in the soil. Nevertheless, the tendency of the area towards a decline in the water table, which was marked as critical by CGWB, is a matter of concern.

The Decline in Water Harvesting

Alongwith overdrawing, riverbed mining, and clay mining, the lapse of traditional water harvesting practices also contributes to the reduction in ground water recharge. A state level survey showed that there were 995 ponds and tanks throughout Kerala with a dry season storage capacity of more than 1500 cubic metres each (CWRDM 1989). More than 50 per cent of these ponds and tanks are silted and not in use. A survey by the graduate students of the Geography Department, University College, Thiruvananthapuram in 1997, found that out of the 80 ponds within the Thiruvananthapuram city corporation area, 12 ponds had been filled to make land for housing, marketplace and playgrounds. Sewage was dumped in four ponds. Out of these 80 ponds 41 ponds were silted and needed immediate maintenance. Repairs were initiated in four ponds.

Table 5.4 : Types of Ponds, Thiruvananthapuram in 1997

Characteristic	*Number of ponds (Total N=80)*
Ownership	
Owned by the urban corporation	56
Owned privately	24
Present use	
Ponds reclaimed for housing	10
Ponds reclaimed for playgrounds	2
Ponds used for soaking medicinal plants	1
Ponds used for washing and bathing	14
Ponds used for agriculture and related activities	30
Ponds used for sewage dump	4
Ponds left unused	19

Source : Post Graduate Student Project, Geography Department, University College, Thiruvananthapuram 1997.

Table 5.4 provides details about the status of the ponds. Many of the valley head ponds locally known as "talakulam" (see chapter 2) have been filled in or left unused throughout the state.

Groundwater Quality

Overpumping of water can cause quality deterioration in the coastal areas, and in areas adjoining the estuaries, with brackish water flow. When the groundwater level goes down in a particular place a trough is created, altering the water to flow in from adjoining areas. In Kolkata, the salinity level of the water that was pumped out from certain depths increased (Basu and Main 2001). The problem of arsenic in West Bengal, Bangladesh, and parts of Madhya Pradesh is well documented (Mandal et al. 1996; Chakraborty et al. 1999). Leaching from waste dumps, industrial discharge, and sewage can also lead to groundwater contamination.

Kerala has not yet experienced large-scale groundwater contamination like other states. But it is now confronted with two sets of problems. Firstly, the coastal areas cannot use well water due to saline incursion. Overpumping around the estuaries has led to rising salinity levels in well water in some areas. Secondly, high coli form counts have been reported in wells in various parts of the state. Single pit latrines are widely used in Kerala. Often, the bottoms of these pits are not sealed. Wells, often located down slope, are contaminated by leaching from the pit latrines.

Analysis of the water of the wells from the Chalakudy basin indicated the presence of fecal coli form primarily due to the location of the wells near a sewage pit (Chattopadhyay et al. 2001). Random sample analysis of well water in south Kerala has shown high e-coli counts (Soman 2002). Domestic water should be totally free from fecal coli form contamination.

When taken together quality deterioration and reduction in quantity reduce the water security of the state. A majority of the people in Kerala depend on well water for domestic use. The density of wells in Kerala is more than 150 per sq km. Every household has a well, and in some areas like Alappuzha district, each household has more than one well. The state is not in a position to supply piped water to all households in the foreseeable future. People have to depend on well water. If the water dries up or the quality deteriorates and the water is not potable; we could witness the undoing of many of Kerala's health achievements. In chapter 10 we describe some local democratic initiatives for alternative water development practices.

Performance of Irrigation Projects

For many years irrigation projects have been the major infrastructural device for agricultural development in Kerala. The rivers are tapped at a reservoir in the upper reaches, and water is diverted through canals to irrigate *ayacut* or command areas where the water is distributed to the fields. This practice is prevalent throughout the world. India, China, and the United States have arrested major rivers for irrigation and drinking water. There are now more than 3,000 major dams in the world.

During the first Five Year Plan in India (1952 to 1957), agriculture was given top priority. Large scale river valley projects were introduced. Kerala state also started setting up irrigation projects immediately after its formation in 1956. Kerala presently has 17 irrigation projects some of which are yet to serve the expected ayacut area originally proposed. Kallada, Chittoorpuzha and Kanjirapuzha, started in 1961 but could be completed only by the end of the 1990s. Nine other projects that were started between 1975 and 1981 have not made any progress at all.

Large cost overruns and failure to meet targets even after project completion were two common problems encountered not only in Kerala but throughout India. The much talked about Kallada project had a revised cost estimate 34 times higher than the original. The original estimate was 133 million Indian rupees in 1961, and was revised to 4,578 million (4.6 billion) Indian rupees in 1992–93 (GOK 1993). The Beypuzha (Chaliyar) project began in 1981 and is yet to take off. The cost escalated from 106 million in 1981 to 3,780 million in 1992–93.

A detailed study of the performance of the Pamba irrigation project was conducted by Chattopadhyay (2001). The Pamba irrigation project started in 1964 and was completed by 1993–94. Its ayacut area spreads over Kottayam, Pathanamthitta and Alappuzha districts. The net ayacut area to be covered was estimated to be 21,135 ha. Water did not reach the western parts where the main benefits were expected. The major findings are :

- The soil irrigability potential was not considered while laying out the canals;
- In many areas canals cut across the valley caused waterlogging upstream;
- Many areas received water only two or three times in the past 11 years;
- The mundakan (rabi) crop in the sandy soil area was affected due to inadequate water supply;
- Some rice fields that were supposed to benefit from irrigation had been converted to plantations; on some rubber was planted and some rice lands were kept fallow;
- Some areas benefited from groundwater recharging;
- Sugar cane cultivation in Pathanamthitta district benefited to some extent;
- The high productivity of rice that was expected due to irrigation was not reached;
- Opinion Surveys showed that the canal had not benefited a majority of the farmers in these areas.

The Pamba is supposed to be a surplus river basin with rich water potential. If adequate water could not be supplied there, the potential of other irrigation projects can well be doubted. Widespread water logging was reported from the Periyar valley ayacut area near Aluva (Alwaye) in Ernakulam district. Some canals in the Pazhassi irrigation project in Kannur district could not carry water due to the high porosity of the substratum in certain areas.

Performance of the irrigation systems is well below expectations. The undulating nature of Kerala's topography is not suitable for canal irrigation. Areas under seasonal crops are dwindling. Tree crops occupy mainly the upper slopes and canal irrigation of tree crops, which is now frequently talked about, is not practical. Considering all these aspects, the wisdom of investing more money in canal irrigation systems can be questioned. Except for the Palakkad area, canal irrigation contributed very little to enhancing agricultural productivity. In many places, particularly in Kallada, the irrigation canal drained the wells. With all of the investments made, canal irrigation serves less than one third of the net irrigated area of the state.

Canal irrigation turns out to be extremely costly both financially and environmentally. The all-India estimate showed that the cost to irrigate one hectare of land by a canal system is Rs 45,000, whereas by using minor irrigation schemes it costs between Rs 10,000 and Rs. 15,000. The cost is well below Rs 10,000 using pond irrigation.

Water is a decentralised resource. Through a canal irrigation scheme, a decentralised resource becomes centralised and is then redistributed. Minor irrigation schemes are not only decentralised, but they tap available resources in cost effective and environmentally sustainable ways.

Reservoir Sedimentation

Another common problem associated with river valley projects is river sedimentation. This problem is well recognised and sedimentation surveys have been done for a number of reservoirs in India. Some of the findings for selected reservoirs are given in Table 5.5. These reservoirs are located in various states with different rainfall, soil, and topographic combinations.

Table 5.5 : Reservoir Sedimentation in Selected Reservoirs in India

Name of Reservoir	Catchment in sq km	Age in Years*	Original Reservoir Capacity (million cubic metres)	Present Reservoir Capacity	Total Loss as Per cent of Original Capacity	Annual average loss in Per cent
Bhakra (Punjab)	56,980	21	9,868	9,160	7.32	.35
Maithan (Bihar)	10,960	18	1,581	1,436	9.20	.50
Mayuraklshi (West Bengal)	1,860	16	608	561	7.80	.50
Nizamsagar (Andhra Pradesh)	21,694	37	841	403	52.00	1.40
Hirakud (Orissa)	83,395	22	8,105	6,900	14.85	.68
Lower Bahvani (Tamil Nadu)	4,200	12	931	910	2.26	.19
Tunghabhadra (Karnataka)	28,180	32	3,751	3,167	15.58	.049

Source : James et al. undated.

* As on the date of completion of the survey. This date is not uniform for all the reservoirs.

The average annual loss in these reservoirs varies from 0.19 per cent in Lower Bhavani to 1.40 per cent in Nizamsagar (column 7). The annual rate of sedimentation from the catchment unit is two to three times what had been assumed during project design. Reservoir depletion begins at the moment of implementation.

The Kerala Engineering Research Institute (KERI) conducted sedimentation surveys in the Malampuzha and Peechi reservoirs (KERI 1977 and 1982). The Malampuzha reservoir in the Bharatapuzha basin was one of the early irrigation projects in Palakkad district and achieved its goals to a certain extent. The sedimentation rate was found to be 25.39 cubic metres per hectare in Malampuzha and 90 m^3/ha in the Peechi reservoir. Based on reservoir sedimentation studies in 1985 and 1986, James et al. (nd) estimated that the annual sedimentation rate in the Mangalam reservoir was 11.26 m^3/ha and for Peruvannamuzhi it was 286 m^3/ha. The Mangalam reservoir lost 8.8 per cent of its original capacity over 29 years while the loss in the Peruvannamuzhi reservoir was 21.6 per cent over 13 years. The average annual loss was thus 0.30 per cent and 1.7 per cent for Mangalam and Peruvannamuzhi respectively.

If we consider the Mangalam and Peruvannmuzhi rates as extremes, and if we assume a hypothetical situation in which all the reservoirs will be functional until they are completely filled, (which is not the case) the life of the reservoirs in Kerala would theoretically vary from 60 years to a little over 300 years. As all the reservoirs are constructed with a projected time span of 100 years, it can thus be presumed that only some of the reservoirs will function to their full theoretical lifespan.

Based on the Universal Soil Loss Equation (USLE) James et al. (nd) estimated that average annual soil loss for the Mangalam Reservoir catchment was 23 tons/ha and that for Peruvannamuzhi it was 66 tons/ha. The differences in the rates were attributed principally to type of land cover and to management practices. Given the extent of landuse change taking place in Kerala and given the interventions in the forest area, the life spans of all the reservoirs will be shorter than the rates projected when the reservoirs were designed.

Fragmentation of Landholdings

The size of landholdings plays an important role in agricultural productivity. Although the highest rate of productivity in India was achieved by small landholders in West Bengal (Government of West

Bengal 2004:74–75) very small holdings can pose a land management problem. Operational holdings in Kerala are extremely fragmented.

Kerala's internationally famous land reform introduced in 1957 and finally implemented in 1971 had a positive impact on many poor people's access to resources. Through a ceiling provision and a transfer of ownership process, 1.29 million tenants got titles to 1.97 million acres of rice land, about 1.5 acres, or 0.6 ha each. Another 340,000 households received house compound land of about 0.08 acres or 0.03 ha. (Franke 1996:126 for details and references). Every working farmer now owns a piece of land, but the holdings are too small in most cases to provide adequate returns to farming and the small plots are fragmented in ways that reduce the effectiveness of technological and environmental interventions.

The Kerala Government Directorate of Economics and Statistics estimated that in 1985–86 91.53 per cent of holdings were less than 1 ha, against the all-India figure of 58.10 per cent. Holdings of less than 1 ha covered 46.02 per cent of the total area in Kerala, whereas at the all-India level only 13.20 per cent was in this category. Subsequent data for Kerala for 1990–91 showed that the number of holdings below 1 ha had increased from 91.53 per cent in 1985–86 to 92.60 per cent in 1990–91 (table 5.6). By 1990–91, holdings of less than 1 ha accounted for 48.8 per cent of the area.

Table 5.6 : Per cents of Operational Holdings by Size in Kerala, 1985–86 and 1990–91

Size	*Per cent of holdings 1985–86*	*Per cent of holdings 1990–91*	*Per cent of Area 1985–86*	*Per cent of Area 1990–91*
Marginal (less than 1 ha)	91.53	92.60	46.02	48.80
Small (1–2 ha)	5.76	5.20	21.55	21.20
Semi-Medium (2–4 ha)	2.12	1.80	15.34	14.00
Medium (4–10 ha)	0.51	0.30	7.39	6.30
Large (more than 10 ha)	0.08	0.10	9.70	9.70

Source : Agricultural census 1990–91, Department of Economics and Statistics, Government of Kerala.

A more detailed breakdown indicates that in 1990–91, operational holdings of 0.02 ha to 0.50 ha accounted for 72.23 per cent of the total holdings in Kerala. For Malappuram and Alappuzha districts over 79 per cent operational holdings were less than one-half hectare. Idukki was the district with the lowest figure at 47.62 per cent and in

Wayanad it was 54.96 per cent. Fragmentation of holdings is not limited to any particular district and is related to population distribution.

Kerala's extreme landholding fragmentation threatens to undermine the beneficial effects of the land reform in distributing wealth producing assets. Due to the undulating topography, which covers about 80 per cent of the landmass in the state, each plot receives silt and water from the adjoining plot on the slope above. The sub-surface water of a downslope field is recharged through infiltration from the upslope plot. Interventions on individual plots may not succeed. When the fragmentation level and the number of holdings is so high it is more efficient for a group of farmers to operate a single topo-sequence (ridge to valley segment; see chapter 3). Resource management at the micro watershed level following the topo-sequence group approach could possibly be a second phase of the land reform.

Similar considerations may apply to the house compound plots on which people grow coconuts, mangoes, papayas, teak, and other crops. When an individual owns only a small plot the preference might be to build a house. Many owners of marginal rice plots have opted to convert them to residential use. This should also be an issue in any second phase of land reform.

Floods

Flooding is a natural hazard that affects most of the world's countries. The tropical countries with seasonally concentrated high rainfall are subjected to recurrent flooding and suffer heavy economic and human losses. From 1960 to 1981, there were 96 internationally recognised natural disasters, 28 of which were floods (CSE 1990). India was the second most affected country behind Bangladesh. Various international agencies are involved with flood management and mitigation measures but they have not made much headway. Although flood control technology exists, the damage that occurs over the years seems to be increasing. High population densities in the flood plains is one of the major reasons. Alteration of land cover, deforestation and accelerated soil erosion have resulted in a shorter "residence" time of water in the upper catchments, increasing the incidence of flash floods. Available data indicate that the average area annually affected by floods in India increased from 6.86 million ha in the 1950s to 16.57 million ha in the 1980s (CSE 1990). The National Flood Commission estimated that flood prone areas increased from 25 million ha in the 1960s to 34 million ha by the end of the 1970s.

Incidence of Floods

Kerala may not figure largely in the devastating floods of India. However, it has a sizeable area susceptible to annual floods. The Central Water Commission (CWC) estimated that about 8,700 sq km are flood prone in Kerala. Historical records indicate that an extremely high flood in 1341AD in the Periyar river led to the silting up of the Cranganore port that had been a major trading center for centuries. It also opened a new outlet to the sea at Cochin (Menon 1965). Since there were low population densities at that time, no loss of life or property was recorded.

In terms of damage, the most devastating flood affecting Kerala was in 1924. The year 1961 witnessed another severe flood that caused considerable property damage. Floods were recorded in 1968, 1975, 1981, 1985, 1986, 1989, 1992, and 1994. Limited floods occurred in 1964, 1971, 1978, 1996, 1998 and 2001. The cumulative damage is staggering. In 1989, the losses in terms of money amounted to more than Rs 7,750 million. More than 2 million people were affected; about 135 thousand houses were damaged and crops on more than 65 thousand hectares were destroyed. In 1992, up to 988 villages were affected and 60 lives were lost (GOK 1997). The all-India figure indicates a loss of Rs 14,000 million annually on average, not counting the loss of human lives and the serious socio-economic disorders caused by flooding.

Although the landform of Kerala is mostly characterised by hills, undulating terrain, and valleys, about 25 per cent of the total area is prone to flood. These areas include Kuttanad around Vemband lake, the "kole" (depression) lands of Thrissur and Ernakulam districts, other low lying areas, floodplains throughout the state, and even the flat bottomed valleys in the plateaus of the Wayanad and Idukki regions. Greater Kuttanad, covering 874 sq km is the most susceptible to floods due to its peculiar landform. This area contains "saucer shaped basins" mostly 1 to 2 metres below mean sea level. Every year during the monsoon the entire landmass gets waterlogged. In 1981 Kuttanad was flooded 6 times. Thiruvananthapuram and Cochin have been experiencing repeated waterlogging during the rainy season.

Besides the greater Kuttanad region, floods during 1989 devastated the hill districts of Idukki and Wayanad. Other affected areas include Palakkad, Malappuram, and Kannur districts. All of these districts have very limited lowland areas. In 1992, which happened to be the major flood year of the 1990s, the affected areas were distributed across all districts of the state except Kannur in north Kerala (GOK 1997). The broad distribution of flooding indicates that flooding in Kerala does not depend on absolute elevation but rather on relative height with respect to the river and the sudden rushing of surface runoff (flash flooding) that is generated in the surrounding highlands and slopes. Kerala experiences riverine floods, stand floods and flash floods. Some areas are affected by sea surge, but not as extensively as the east coast of India.

Floods and Rainfall

To understand the links between flood and rainfall, we analysed precipitation data from 1960 to 2003 for the entire state of Kerala. The highest rainfall at 4009 mm was recorded in 1961 and the lowest was 1465 mm recorded in 1982, the driest year in recent history. Based on the mean annual rainfall (2776 mm) during these 44 years, we created a rainfall departure curve. Figure 5.5 shows the graph.

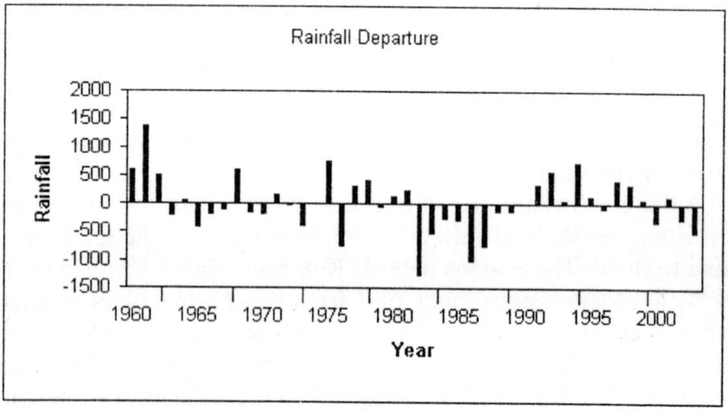

Note : 0 = 2776 mm, average rainfall between 1960 and 2003

Fig. 5.5 : Rainfall Departure Curve for Kerala

Correlating the floods for the same period with rainfall departure curve data we find that the floods of 1961, 1968, 1975, and 1981 appear to occur on a 6 to 7 year cycle. In all of those years except for 1981 the rainfall was more than 128 per cent of the average and the damages were extensive. The floods of 1971 and 1977 were limited and affected mainly the Kuttanad region. In 1964 and 1978, Thiruvananthapuram and Kollam districts were affected. 1981 witnessed considerable damage even though the corresponding rainfall was only 107 per cent of the average. The floods of 1994, 1996 and 1998 were limited in their extent. Although rainfall was below average in 1985 (91%), 1986 (76%), and 1989 (95%), flood damage was still extensive. The floods of 1992, which devastated the whole state, correspond to a rainfall well above average at 121 per cent. The rainfall departure curve points out that the peak rainfall years causing floods are separated by years of low rainfall, often much lower than average. As a result, the intervening period between two floods is characterised by a drought year. This trend was especially prevalent during the 1960s and 1970s. During the 1980s, the trend changed. After 1981, when rainfall was slightly above average, up until 1990 the rainfall was below average. In 1982 the rainfall was 55 per cent of the average. Rainfall started to increase in 1983. Another low was recorded in 1986. From 1987 to 1992, there was a steady increase followed by another period of fluctuation. In the 1990s the highest rainfall was recorded in 1994 when it was 126 per cent of the average. This fluctuating trend may be related to a broad climatic cycle, which is beyond manipulation by present day technology. However, flood incidence in the years with average and even below average rainfall, and consequent damages are all areas of concern.

A possible climatic reason for flood havoc during the years with below average rainfall may be a concentrated pattern of rainfall, i.e. a heavy downpour over a couple of days. To elucidate this point, data on monsoon rainfall were analysed for 1989, a year when major flooding occurred, although the total rainfall was less than average.

Case of a Severe Flood : 1989

Rainfall for 1989 was 5 per cent below the average. We computed average daily expected rainfall and the actual daily rainfall for a period of 63 days from June 1st, the bursting day of the monsoon, to August 2nd, the normal end of the regular monsoon rains during the major monsoon in Kerala. Except for Idukki in the Western Ghats region and Kollam in the southern part of the state, actual rainfall was below normal in all districts. Wayanad district in the Western Ghats region of north Kerala received 34 per cent below average rainfall. However in a span of 6 days from July 22 to July 27, the total rainfall was 533 mm, and on the 24th of July alone the rainfall was 169 mm. Munnar in Idukki district received 1160 mm of rainfall in one week, from July 20 to 26 in 1989. Extensive damage occurred during this flood in various districts.

A similar phenomenon was reported by Saseendran et al. (1983) for 1961. During that year, Vayithiri in Wayanad district received 2,430 mm of rainfall in a span of just 10 days, from June 29 to July 8. Gupta (1990) reported an unusually heavy rainfall at Thiruvananthapuram on October 17, 1964, when the precipitation was 401mm in a span of 10 hours. This is far above the expected daily average monsoon rainfall of 20 to 30 mm.

Unusually heavy rainfall thus occurs in Kerala from time to time. There is every reason to believe that this trend will continue in the future. The reduction in areas for the water to spread over and percolate into aggravates flooding. While the incoming rainfall that causes the flood is a natural phenomenon; water distribution, its areal spread, and its flow patterns are heavily influenced by human activity. We can summarise the human effects by considering the way human activities can alter basin ecology.

Alteration of Basin Ecology and its Impact on Flooding

Human intervention alters the basin ecology, contributing to increased flooding and changing flood patterns. The main human activities influencing basin ecology are filling or draining of wetlands, shrinking of water bodies, floodplain occupancy, embankment construction, and changes in the landuse and land cover patterns. In Kerala and in many other places these activities continue unabated. In developed countries like the United States, flood losses also continue to increase primarily due to expanded use, occupation and development of downstream floodplains.

Reclamation of the greater Kuttanad area surrounding the Vembanad lake was initiated over 100 years ago. Low lying areas adjacent to the backwaters were subjected to draining or filling. Conversion of wetlands to paddy fields and to dry lands for settlement and tree crop culture are widespread phenomena throughout Kerala. It has been estimated that only 58 per cent of the 5,736 sq km of wetlands and paddy fields have retained their wetland condition (KSI 1999). Leaving aside 4 per cent for fallow land, it can be estimated that as much as 38 per cent of the wetlands or flood-cushioning areas are no longer available to hold spill water during periods of excess rain. As a result, the height of the flood water in the rest of the open areas increases, and the flood moves upstream.

Drastic reductions and clogging of canal and drainage systems in and around urban areas like Cochin and Thiruvananthapuram have compounded flooding. The duration of "standing floods" has increased significantly. Even during normal monsoons, these urban centres experience deep and prolonged waterlogging.

Embankments have been laid down to protect reclaimed tracts, particularly in the greater Kuttanad region and in lands adjoining the backwaters. On average there are 4 km of embankment for every 1 sq km in Kuttanad. Spill areas meant to accommodate rainwater and excess river discharge during a monsoon are continually diminished. This leads to a gradual heightening of the water column in open areas. The areas beyond the reach of the flood height in lower Kuttanad are 3 to 4 times less than that in upper Kuttanad. The more elevated upper Kuttanad has a number of enclosed basins. Once flood water enters these pockets, it takes a long time for it to drain. Edatuva in upper Kuttanad was under the grip of the 1989 flood despite considerable reduction in drainage discharge. This is a typical example of flood aggravation due to human intervention.

As a result of embankment construction along the rivers, silt is deposited the river beds. Continuous silt deposition leads to raised thalwegs and increases the height of the water column in the river compared to the surrounding land. This process is much faster in tidal channels. The height of the embankments needs to be raised regularly. This makes the developed areas more vulnerable to floods. The drainage system deteriorates in tandem. Cochin city provides an example: at present, almost all the secondary channels in the central city are silted up. Water logging and drainage congestion are regular occurrences during the monsoon months. One effect of the stagnant water is to provide breeding areas for mosquitoes. In October 2002 the Cochin city government contracted for a major mosquito reduction program that is based heavily on schemes to clean up drainage canals and make them drain. This project is based on the biological mosquito control project we describe in chapter 10. The raised water columns resulting from years of silting in will make this a difficult and expensive undertaking.

About 24 per cent of the reclaimed wetlands in Kerala are designated for multiple cropping. In built up areas, considerable investment has been made in infrastructure and institutions. This enhances the economic value of the lands. As a result, even insignificant floods now cause a considerably high economic loss. Flash floods are common in the hill regions and plateaus where forest cover has been lost. During high intensity rainfall, rushing waters from the surrounding slopes sweep down the narrow valleys at high speeds causing extensive damage. The floods in the Wayanad and Idukki regions are of this kind.

Another problem is that the flood prone areas of Kerala are densely populated. Alappuzha records the highest population density among all the districts in Kerala (table 6.2). At the micro level, Cochin has the highest population density. Since we cannot expect any substantial reduction or redistribution of the population within the foreseeable future, the flooding is apt to worsen with continued floodplain occupancy and ongoing reclamation. Flood management may need to be added to landscape management, linking both to a larger development process.

Reducing the intensity of the use of flood prone areas is one alternative for overcoming the problem. Micro level intervention and site-specific measures are essential. A river, to function as a river, has to flood in response to monsoonal precipitation; otherwise it is a dead river. It becomes incumbent on the flood plain users and planners to understand these mechanisms and fashion policies to keep the rivers alive and functioning more harmoniously with human settlements.

Postscript to Chapter 5: The Great Tsunami of 26 December 2004

The earthquake that struck the north Indian Ocean near Acheh in Sumatra on 26 December, 2004 caused a giant tsunami leading to unprecedented human death and suffering. Across the region, more than 2 lakh (200,000) people lost their lives, while millions had their property destroyed and their livelihoods undermined. Economic growth was derailed and even the geography of some parts of the region changed. With a 9.3 magnitude on the Richter scale this earthquake released energy equal to about 5 megatons of explosives. The surface waves on the Indian Ocean grew to a wave length of about 800 km, the same as their velocity of 800 km/hr in the deep sea. The crests of the waves were higher by 50 cm than normal sea level while the troughs were 40 cm lower than normal (Ballukraya 2005). In India, the Andaman and Nicobar islands, and a coastal stretch of Tamil Nadu were seriously affected. Andhra Pradesh and Kerala had less powerful assaults from the waves.

The first tsunami waves of 26 December 2004 hit the east coast of India between 7:30 am and 9:00 am and the second hit was around 11:00 am. The Kerala coast – around on the western side of India – experienced the first tsunami wave between 11:00 am and 11:30 am. The spread and depth of inundation varied spatially and temporally. Based on preliminary observations we can infer that there were two hydrodynamic aspects of the tsunami: modulation of wave power by the initial thrust of energy release mediated by sea floor topography, and a rise of sea water levels along the coast (Mahadevan 2005). The spread of the tsunami wave, the level of inundation and the impact depended upon bathymetry, the coastal configuration, geology, coastal geomorphology, the presence of backwaters and the land elevation.

The Kerala coast was hit in several places but the damage was greatest in Alappad and Arattupuzha panchayats on both sides of the Kayamkulam outlet in the lowland segment about 100 km north of Thiruvananthapuram. It was reported that in Kerala there were 200 deaths of which 161 were in the Alappad-Arattupuzha area. During one phase of the tsunami the sea water receded and part of the near shore seabed was exposed. This unprecedented phenomenon attracted many people, some of whom walked down into the exposed seabed but got caught by the onrushing tsunami wave of great velocity. They were killed.

The tsunami flood height rose to more than 2 metres along the road that runs very near the coast and penetrated a kilometre inland crossing over the backwaters. According to local people a vast sheet of water spread over the land and the backwaters. For a couple of minutes there was no distinction between the backwaters and the sea. After this the water flowed back to the sea through the backwaters and the Kayamkulam outlet. The backwaters reduced somewhat the fury of the flood. In many parts of Tamil Nadu the backwash created much more damage. In Kerala around 4000 houses were totally destroyed and another 3000 houses were heavily damaged. Fig 5.6 is a photo of a tsunami damaged house.

Recorded history indicates that the Indian subcontinent experienced tsunamis following the 1881 ar island earthquake, the August 8, 1883 famous Krakatoa volca xplosion in Indonesia,

Fig. 5.6 : Damaged House Due to Tsunami. Photo by Srikumar Chattopadhyay

the June 26, 1941 earthquake in the Andaman and Nicobar Islands, and the November 27, 1945 Karachi earthquake, prior to the present tsunami (Anon 2005). History thus warns us that protecting coastlines and coastal populations against likely tsunamis is a prudent undertaking.

Tsunamis are unavoidable events since they are part of geological processes, however, it is possible to minimise the loss of human lives and the damage to property and the environment. A tsunami is essentially a coastal event, of course of rare occurrence. However other natural disasters, such as sea surges, coastal floods, tidal waves and cyclones are not uncommon to the coastal tract. The tsunami inundation of 2004 can be considered an extreme event paradoxically useful for designing action plans to minimise the effects of all types of disasters along the coast. The Kerala coastal community is basically a fishing community. Although fishing is one of the riskiest occupations in the world with frequent accidental deaths (one in every four days during the last decade) in Kerala, the fishing communities have historically taken these losses in stride as occupational hazards. The terrible tsunami of 2004 dumbfounded the fishing community despite its familiarity with the behaviour of the sea and its traditional faith in *Kadala Amma* (mother sea). People were caught unaware and are totally shaken.

A few policy questions can be raised in the aftermath of the tsunami (Kurien 2005). Could the implementation of a proper Coastal

Regulation Zone (CRZ) have limited the loss of life and property? Is it wise to locate dwelling places very near to the sea, to strip away the natural vegetative barrier along the sea front, and to leave the population without proper awareness of the possibility of such disasters? What role could local government bodies play? Preliminary research suggests that much of the damage is directly linked to mismanagement of the coastal environment. It was reported from the east coast of India that coastal vegetation acted as a bio-shield in many stretches to reduce the velocity of the onrushing sea surge during the tsunami. The affected coastal stretch of Kerala was devoid of such coastal vegetation. Coastal dunes were also not there. Moreover settlements have sprung up very near to the coastline and coconut trees have been planted for economic benefits. Coconut trees, however, are not proper biological shields against sea intrusion (See Figure 5.7). A proper environmental management programme covering land use zonation, regeneration of coastal vegetation and settlement planning could help to reduce damage considerably. In chapter 10 we shall describe the Integrated Coastal Zone Management project in a part of Alappuzha district near to the impact area of the great tsunami of 2004. This project includes many of the disaster-resistance components that might have saved lives in affected areas then and will hopefully do so all along the coast in the future.

Fig 5.7 : Beach Near Tsunami Affected Area Showing Coconut Trees Planted Too Close to the Shore and Lack of Proper Biological Barrier to Sea Intrusion. Photo by Richard W. Franke

6 Why the Environment is Changing

Global and Local Forces

Kerala's ecosystem changes as elucidated in the previous chapter result primarily from human activities. In recent environmental literature the causes of ecosystem changes are being referred to as "drivers" (Millennium Assessment 2005:114 *et seq).* Climatic variability, the erosion cycle and sea water movement are associated with the larger global framework, but land use and land cover changes as well as associated soil erosion, soil nutrient loss, and changes in water quality are largely attributable to human activity. This includes "indirect drivers" such as population growth, migration, urbanisation, along with associated processes such as road building, plantation agriculture, replacement of a land use type, digging within the river bed, applying excessive fertiliser or overpumping water. The effects of these human interventions are both simple and complex and are manifested at various levels.

Landuse and Land Cover Change

Landuse and land cover change are considered the most important direct drivers of ecosystem changes over the past 50 years (Millennium Assessment 2005:117). Human environment analysts, more often than not, confront the formidable tasks of coming to grips with complex causal webs linking social, biological, and environmental processes (Turner 1997). Many observers claim that the driving forces behind land use and land cover changes emerge from a spatial hierarchy (Blakie and Brookfield 1987; Rayner et al. 1994; Skole et al. 1994). Little is known about the hierarchical linkages among processes at different scales (Sanderson 1994; Wilbanks 1994). The factors involved in this process can be migration, an implicit variable in bid-rent formulation (Walker and Solecki 1999) or urban population. According to the Von-Thunen model (Chisholm 1962:20–34) landuse change around urban centres can be attributed to market forces. Cropping patterns in the surrounding areas evolve to cater to the felt needs in the urban market. Land prices also depend on the market. Changes in the rural-urban interface are governed by supply and demand mechanisms. Given the complexity of the processes involved and the synergistic nature of human decisions at local to regional scales, it may not be possible to settle on one cause-effect relation.

To provide a framework for understanding some of the relationships we offer the case of Idukki district. Idukki lies in the Western Ghats region of Kerala and is at the provenance of the state's major river systems. This district houses five hydroelectric projects out of nine hydroelectric projects in the state.

Idukki district experienced the highest percentage increase of population growth among all the districts of Kerala at 74.70 per cent in the period 1951–61. Several economic and political factors favoured this increase (George and Chattopadhyay 2001). The government's "Grow More Food Campaign" was one. The 'Grow More Food' campaign was a program, in which the government issued exclusive cultivation rights to migrants from other areas coming to the forest areas of Idukki district. Dry land rice cultivation was introduced initially, which was subsequently replaced by tree crops and rubber.

According to 1981 data 47 per cent of the total population in Idukki district were migrants (CESS 1997). This district had received 269 thousand persons in the decade 1971 to 1981 while only 8 thousand people moved out of this district in the same period.

The highland area recorded an overall population growth of 1,343 per cent from 1901 to 1991 against the state average of 345 per cent in the same period. Male population dominated this very high growth rate contributed by the migrants. This is perhaps reflected in the sex ratio of both Idukki and Wayanad districts, where numbers of females to thousand males were 975 and 966 respectively, whereas for Kerala as a whole it was 1036 females for 1000 males in the census year 1991.

Landuse in this district changed significantly. Data available from the maps prepared during the period 1913–1928 and for 1992 indicate that forest area came down from 65.8 per cent of the total geographical area of the district (5019 km^2) in 1913–28 to 20 per cent in 1992; agricultural plantations increased from 16.3 per cent to 43.4 per cent and degraded forest area increased from 15.8 per cent to 31.3 per cent during the same period. The forest area, which covered sizable areas in continuous stretches in 1913–28 is now fragmented in patches. Forest plantations of teak, eucalyptus and other woods occupy only 2 per cent of the total geographical area of the district. The area covered by forest plantations thus seems to be marginal, however, with the introduction of plantations the core forest area opens up for human intervention and process of biodiversity loss is initiated. The nature of landuse change is given in table 6.1.

Table 6.1 : Area Trends in Landuse in Idukki District, Kerala, 1913 to 1992

Nature of change from 1913–28 to 1992	*Area in km²*	*Per cent Change 1913 to 1992*
Decrease in forest area	2285.0	- 69
Increase in forest plantations	89.0	+100
Increase in degraded forest area	781.0	+ 99
Increase in agricultural plantations	1359.0	+166
Increase in waste land	8.0	+ 32
Increase in reservoir area	50.0	+ 61

Source: Centre for Earth Science Studies 1997:55

The factors driving land use and land cover change in Idukki district from 1850 AD to 2000 AD can be listed as follows :

(1) Expansion of tea cultivation and introduction of rubber cultivation during the end of the 19th century;
(2) Major internal migration from lowlands to highlands within the state;
(3) Installation of hydroelectric projects and further expansion of rubber cultivation;
(4) The grow-more-food campaign and expansion of agriculture in the forest area during the 1940s;
(5) Formation of Kerala state and the initiation of new hydroelectric and irrigation projects;
(6) Land reforms;
(7) Declaration of the Periyar tiger reserve; and
(8) Enactment of environmental laws.

Actions were at different levels from global to local. Expansion of rubber cultivation in tropical countries generally can be linked to the invention of the automobile and the demand for rubber tires. In the early 20th century the stimulus provided by motor transport in the developed countries led to small farmer rubber plantations in parts of Malaysia, India and Ceylon (Redclift 1987).

Most of the other factors having local manifestations are decided at the state level, which of course is linked to national level policies. Grow-more-food campaigns or river valley projects are part of the national priorities executed by the state. All the major national and state initiatives have both direct and indirect impacts. Population growth, though observed locally, is attributable to natural growth and to in-migration. Given the background from the Idukki example we now attempt a state level analysis to cover some of the human factors that facilitate environmental changes.

Population

Population growth and population density can be strong indirect drivers of land use change. Kerala accounts for 1.2 per cent of the total geographical area of India but accounted for 3.1 per cent of India's total population in 2001. According to the 2001 census Kerala's population was 31.8 million of which 15.4 million were male and 16.4 million were female. From 1901 to 2001 Kerala's population increased around 5 times. During the first forty years of the 20th century (1901–1941) Kerala's population doubled and the next doubling took 30 years (George and Chattopadhyay 2001). The growth rate was the highest (2.31%) in the period 1961 to 1971. Since then it has come down. During 1981 to 1991 the growth rate was 1.32 per cent and in 1991 to 2001 it was less than 1 per cent. The female population has been greater than the male population throughout the century: in 1901 there were 1004 females for every 1000 males. This sex ratio increased to 1036 in 1991, and to 1059 in 2001. Figure 6.1 is a graph of the population growth in Kerala from 1901 to 2001.

The population density in Kerala was 819 per sq km in 2001, an increase from 749 in 1991. The district level distribution of population density varies from 1415 per person per sq km in Alappuzha district to 215 in Idukki district as per census data of 2001. Idukki district, in other words, has the lowest population density of any district in Kerala, but the land use changes have been dramatic nonetheless as we described earlier in this chapter.

The same ranks had been maintained by these two districts in 1981 (table 6.2). In fact comparison of density distribution at the district level in 1981, 1991 and 2001 brings out a spatial trend in Kerala,

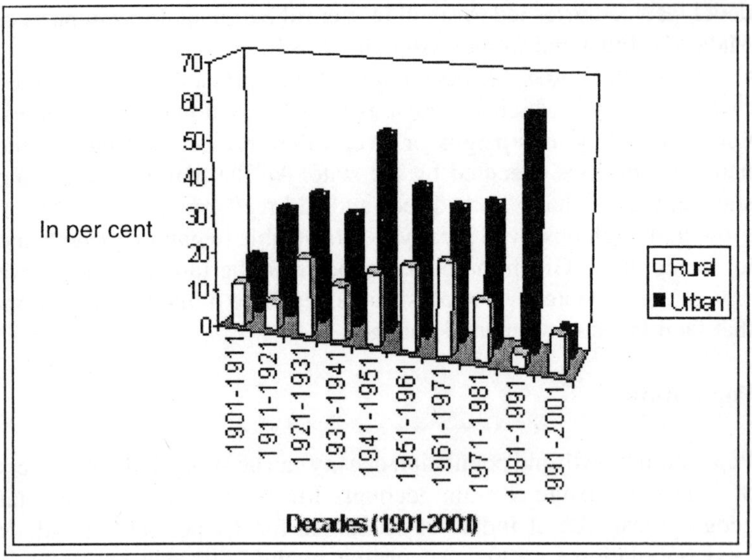

Fig. 6.1 : Growth of Population in Kerala, 1901–2001

which is quite unique. Each district had the same rank in 1991 as it had in 1981 except Kottayam, Palakkad and Kasargod. Comparing 1991 and 2001 it may be found that 10 districts maintained the same rank in both these periods. Interchange took place between Ernakulam, Kozhikode, Malappuram and Thrissur. This indicates that population growth rate in all the districts was more or less the same for the two decades from 1971 to 1991. The decadal growth rate of density for the state had come down from 19.24 per cent in 1971–1981 to 14.32 per cent in 1981–1991 to 9.35 per cent in 1991–2001. All the districts experienced a decline in growth rate. Three districts—Pathanamthitta, Alappuzha and Kottayam—all dominated by rural populations, had registered a 1981–1991 growth rate below 10 per cent. Kerala's low growth rate can be partly attributable to out-migration. Between 1971 and 1981 approximately 250,000 persons left Kerala for employment in the Gulf States of the Middle East. This trend continued during 1991–2001 with a net migration rate of 0.31 per cent of total population (George and Chattopadhyay 2001). Only Malappuram, Wayanad and Kasargod districts registered more than a 20 per cent growth rate. Among these Wayanad recorded significant decline in the 1981–91 period compared to the previous decade.

Table 6.2 : Kerala's Population by District 1981 to 2001

District	*Total Population (in 000) 2001*	*Density of Population (Persons per Sq. Km)*			*Decade growth (% increase from one decade to the next)*		
		1981	*1991*	*2001*	*1971–1981*	*1981–1991*	*1991–2001*
Thiruvananthapuram	3,235	1,184	1,344	1,476	18.08	13.5	9.8
Kollam	2,584	873	967	1,037	18.77	10.7	7.2
Pathanamthitta	1,232	426	450	467	9.45	15.6	3.8
Alappuzha	2,105	1,319	1,415	1,489	11.62	7.28	5.2
Kottayam	1,953	771	830	884	10.29	7.71	6.5
Idukki	1,129	193	215	252	26.64	11.2	1.72
Ernakulam	3,098	1,053	1,170	1,050	17.18	11.1	-10.3
Thrissur	2,975	805	903	981	14.60	12.2	8.6
Palakkad	2,617	456	532	584	21.30	16.5	9.8
Malappuram	3,630	677	872	1,022	29.43	28.9	17.2
Kozhikode	2,879	958	1,118	1,228	23.25	16.7	9.8
Wayanad	787	260	315	369	33.87	21.3	17.1
Kannur	2,412	651	757	813	24.34	16.6	7.4
Kasargod	1,203	438	538	604	27.78	22.8	12.3
State totals and averages	31,839	655	749	819	19.24	14.3	9.35

Source : Census handbook of Kerala 1971, 1981, 1991 and 2001.

Figure 6.2 is a map of the population density of Kerala by districts for 2001.

Settlement Pattern

Kerala's population is in the process of stabilisation, but the demand for housing will take some more time to stabilise. This is because, as indicated earlier in chapter 4, the joint family system has given way to nuclear families and young people after marriage prefer to construct a new house and settle separately from both sets of parents. The demand for housing automatically puts pressure on land areas for building, on clay for brick making and on sand for construction. Prime agricultural lands are encroached on and the riverbed is mined for sand. In addition, demand for water increases.

To understand the pressure on land for settlement siting, we consider first the existing settlement pattern. Certain geographic peculiarities have exerted a striking influence on the settlement system

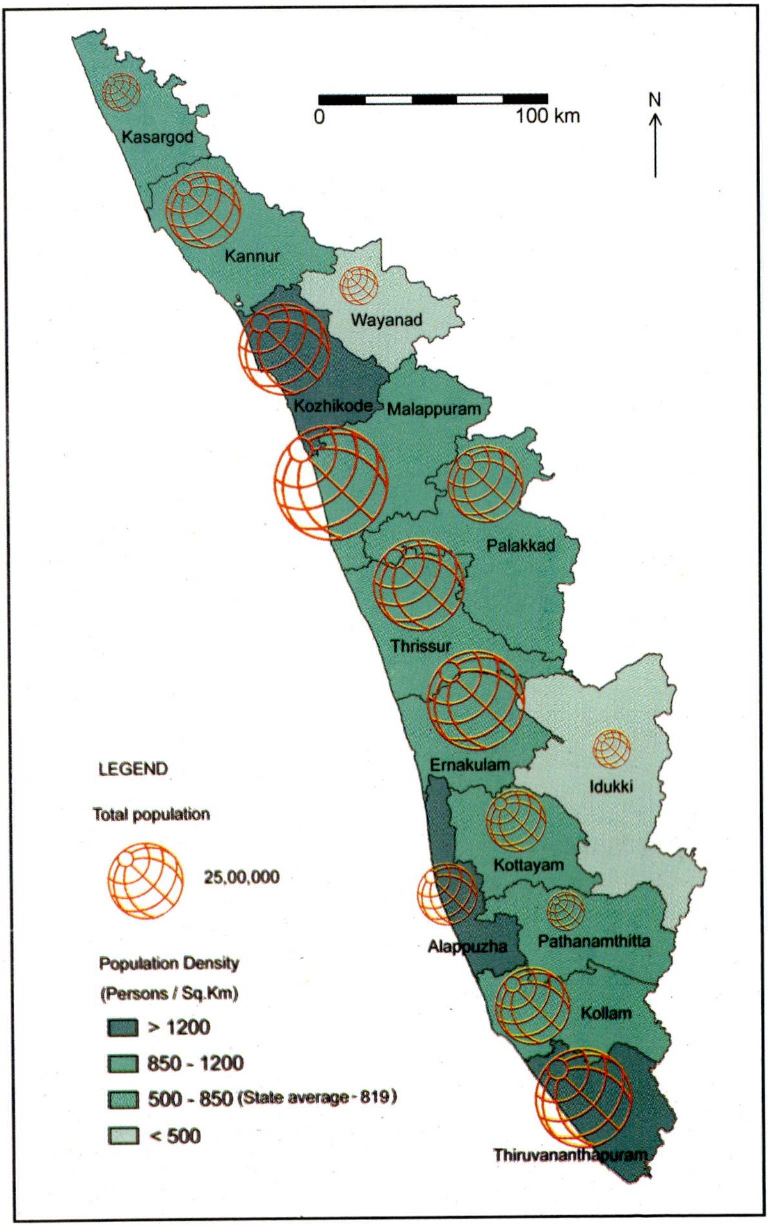

Fig. 6.2 : Distribution of Population Density in Kerala by Districts.

of Kerala (Chattopadhyay 1995). As we noted in chapter 2 the basic pattern is linear and dispersed. The landform condition does not encourage compact or nucleated settlements as in other parts of India. The relative abundance of water and its uniform distribution have also contributed to the dispersed nature of settlements in Kerala.

The settlement patterns in the state can be divided into five types :

(1) Linear along the sand ridges parallel to the coastline. Traditionally settlements in the coastal area developed over the sand ridges leaving the interdune low areas for cultivation. Coast parallel linear settlements take advantage of minimum distance to reach the sea for fishing.

(2) Linear settlements along the lateritic ridges and upper slopes, perpendicular to or radiating out from the coastal plain. The roads have developed along the ridges on both sides of which there are settlements. The valleys are traditionally avoided to keep away from flooding and to maintain food production. These linear settlements appear to spread out from the urban centres or nodes along the ridgeline making a continuous chain of settlements.

(3) Clustered or compact settlements in the Palakkad gap area. Due to low gently rolling topography, comparable to the adjoining Tamil Nadu plain, settlements here are clustered or compact like those of Tamil Nadu. Most of the settlements are located on the mounds.

(4) Clustered or compact settlements in Kuttanad. Traditionally settlements in Kuttanad were linear along the rivers, channels or canals. However with increasing population pressure, there is a tendency towards clustered settlements within the vast expanses of paddy fields.

(5) Clustered settlements in the plateaus of Wayanad, Nelliampathi, Munnar and Periyar.

These trends were initiated in the early stages of settlement history, no doubt, and have continued for a long time. As a result even in the major cities of Thiruvananthapuram and Kozhikode paddy fields are found. The case of Cochin is a little different, as it had developed in the lowland with a large sea and backwater front.

The settlements in Kerala have no distinct core, nor do the villages have any marked nodality. This picture is of ancient origin. The famous traveller Ibn Batuta (1307–1377 A.D.) recorded this situation during the fourteenth century. In the 19th century British colonial official William Logan (1995) noted that the coastal tracts of Malabar (Kerala) were so densely populated that it was difficult to say where one town ended and where the next one began.

Settlements over the coastal dunes run in an unbroken line. While the basic structure is still traceable, modifications started to meet the growing demand for habitation. The interdune lows or runnels in the coastal area, peripheral lowlands of the mounds, the prime paddy fields in the alluvial plains and intervening valleys within the lateritic terrain—all these are being taken for settlements.

Given the current preference of Malayalees to live in a single nuclear family house with a compound, the demand for habitation space is quiet high. When agricultural land and built-up land are in close proximity near the core of the settlement or on the fringe of urban areas the bid-rents for built-up land exceed those for agriculture. The peripheral agricultural lands thus constitute the advancing edge of encroachment. Once people settle, the amenities like roads, water lines, and electricity develop.

Urbanisation

Three parameters are considered in India to declare an area as urban. These are: (1) density of population, (2) size of population and (3) more than 75 per cent of male workers to be engaged in the non-primary sector. The mode of labour participation is by far the most important criterion to determine an area urban or non-urban. Apart from this, any area notified by the government as township or cantonment will automatically be considered urban.

Villages are the smallest population units in India. Identifying villages is complicated in Kerala where settlements are continuous along the ridges and upper slopes. It was indicated above that urban services spread along the ridges while the intervening valleys are given for agriculture. This juxtaposition draws the agricultural landscape into competition with nonagricultural landscapes along the contact zone. More often than not agriculture loses.

This type of interlocking has been noticed in various other Asian countries. The peripheral region along the ridges is neither rural nor urban but demonstrates features of both types. Evolution of this spatial form has been studied in Indonesia, Taiwan, South Korea (McGee

1989, 1991) and China (Lin 2001). A distinct spatial form called the zone of intense interaction emerges between two major urban centres or adjacent to urban centres in various parts of the world. The underlying processes giving rise to this spatial form include high population density, smallholding rice cultivation, close linkage with large cities, existence of cheap labour and markets, well developed transport infrastructure and the expansion of the global economy (Ginsberg 1990; McGee 1991).

Many of these postulated interaction zones are based on observations around metropolitan areas. In Kerala, where small towns dominate, these spatial forms evolved through a combination of environmental factors, particular configurations of land, the historical persistence of mixed agricultural and non-agricultural activities and the institutional base of decentralised amenities. The emergence of the interaction zone has led to various changes in landuse and land cover and thereby on overall ecosystems.

We have prefaced our discussion of urbanisation in Kerala with these background notes to elucidate the point that expressing the level of urbanisation only through the per cent of the population living in urban areas may not be an adequate indicator of the impact or the driving forces in ecological change.

According to the 2001 census (GOI 2001) the urban population of Kerala accounted for 8.3 million or 26.0 per cent of the total population of the state distributed in 17 urban agglomerations and 81 towns. All these urban centres together covered only 9.35 per cent of the total geographical area of the state. The urban population was 7.1 per cent in 1901, when the urban area was only 1.33 per cent of the total geographical area. The proportion of the population in the urban centres rose to 10.8 per cent in 1941, 18.8 per cent in 1981 and 26.4 per cent in 1991. In all census periods the urban population in the state has been lower than that of all-India.

The growth rate in urban area from 1981 to 1991 was 98 per cent (table 6.3). This is a very significant phenomenon. As indicated earlier, the area growth indicates that the interface of agricultural lands and urban land use had moved further outward making more and more agricultural area vulnerable to conversion for housing.

Table 6.3 indicates that from 1971 to 1991 the growth appeared to be higher than in other decades. This span of 20 years can be construed as the period of intense urbanisation. There appears to have been a decline in urban areas from 1951 to 1961. This was due to a reclassification of towns in 1961. This happened on account of changing definitions and also in 1951 the state was under two different administrative controls, which might have had an impact on the identification of urban centres.

The unique feature of Kerala's urbanisation is the dominance of small towns or class III towns according to census classifications. Around 32 per cent of Kerala's urban population lived in class III towns in 1981. By 2001 only14 per cent of the total urban population lived in class III towns and 69 per cent of the urban population is now accommodated in class I towns (greater than 1,00,000 population). This shift indicates a major change in the settlement pattern of the state.

The growth in urban populations and the decline in dominance of small towns and the peripheral expansions of large towns automatically translates into spatial expansion of built up areas. A districtwise discussion will clarify the point further.

Table 6.3 : Urban Population and Urban Area in Kerala by Decades—1901 to 2001

Year	*Per cent of Urban Population*	*Urban Growth Rate in Per cent Over the Decade*	*Number of Towns*	*Per cent of Total Area Under Urban Administration*	*Per cent Growth in Urban Administrative Areas*
1901	7.1	n.a.	21	1.33	n.a.
1911	7.3	1.54	27	1.51	14
1921	8.7	2.98	44	2.00	32
1931	9.6	3.46	53	2.28	14
1941	10.8	3.05	62	2.63	15
1951	13.5	5.27	94	2.73	4
1961	15.1	3.99	92	2.66	–3
1971	16.2	3.57	88	3.23	22
1981	18.8	3.75	106	4.38	36
1991	26.4	6.09	197	8.65	98
2001	26.0	7.70	98	9.35	08

Source : Computed from Census handbooks of Kerala for various years.

District Level Picture

Districtwise the urban population varies from 3.8 per cent in Wayanad, a hilly district within the Western Ghats region, to 50.5 per cent in Kannur, a district in the Malabar region of north Kerala. Dominated by a series of small towns, the Kannur urban agglomeration extends along the coast for about 30 kms. It was recognised as a class I town for the first time in 1991. It has urban concentrations at Kannur and Thalassery. Ernakulam district, the industrial hub of Kerala, was the leading urbanised district in the state until 1981. It now occupies the second position by having 47.7 per cent of its population in the urban sector in 2001. The district also ranked second in 1991 (table 6.4). Along with Kannur, Kottayam and Alappuzha districts have also experienced around 100 per cent urban growth in the decade 1981–1991. It is noteworthy that the districts of Ernakulam, Thiruvananthapuram and Kozhikode, in spite of being traditional urban growth foci, have recently experienced only average growth. This substantiates the observation that Kerala's urbanisation process is decentralised and diffuse in sharp contrast to the processes exhibited in other Indian states. Urbanised space is spreading over the coastal plain and parts of the midlands. Historically, urban centres in Kerala are well distributed to cater to local needs. As early as 1901, when there were only 21 urban centres the distribution was more or less even (Chattopadhyay 1988).

Table 6.4 : Urban Population of Kerala by District as Per cent of Total District Population – 1981 to 2001

District	*1981*	*1991*	*2001*
Kasargod	0	16.5	19.4
Kannur	23.4	51.0	50.5
Waynad	0	3.4	3.8
Kozhikode	27.2	38.4	38.3
Malappuram	7.4	9.1	9.8
Palakkad	10.1	15.8	13.6
Thrissur	21.1	26.3	28.2
Ernakulam	39.6	48.8	47.7
Idukki	4.6	4.7	5.1
Kottayam	9.5	17.6	15.4
Alappuzha	15.9	30.6	29.4
Pathanamthitta	0	13.1	10.0
Kollam	13.2	18.6	18.0
Thiruvananthapuram	25.3	34.0	33.8
All Kerala	18.8	26.4	26.0

Sources : Census of India 1981, 1991, 2001

The area under urban administration accounted for 9.35 per cent in 2001, which had been 2.66 per cent in 1961 (CESS 1997). The districtwise areas under urban administration are given in table 6.5. Apart from Ernakulam and Kannur districts, the spatial expansion of urbanisation is either stagnant or slow. Even in Thiruvananthapuram

Table 6.5 : Per cent of Area in Kerala under Urban Administration by District—1961 to 2001

Districts	*1961*	*1971*	*1981*	*1991*	*2001*
Kasargod	0.85	0.85	0.85	5.34	5.26
Kannur	4.61	5.23	11.00	21.80	21.82
Wayanad	Nil	Nil	Nil	1.60	1.91
Kozhikode	5.69	5.20	5.69	14.48	14.49
Malappuram	1.76	2.71	2.71	4.83	5.27
Palakkad	2.37	2.37	2.37	4.95	4.95
Thrissur	1.98	2.37	6.07	9.93	13.87
Ernakulam	6.50	10.77	13.41	22.49	22.50
Idukki	0.70	0.70	1.33	1.33	1.34
Kottayam	2.44	2.44	2.44	7.65	7.85
Alappuzha	7.79	7.79	7.79	20.33	20.37
Pathanamthitta	1.06	1.06	1.95	3.80	3.80
Kollam	0.75	2.14	3.40	7.59	6.32
Thiruvananthapuram	5.23	6.71	6.71	11.78	15.66
All-Kerala	2.66	3.23	4.37	8.65	9.35

Sources : Centre for Earth Science Studies (CESS), 1997; Chattopadhyay 1985 and 1988.

district the growth in urbanised areas stagnated for two decades. The case of Alappuzha is noteworthy. In 1961 it ranked first among all the districts by having 7.79 per cent of its area designated urban. There was no change until 1981 and in 1991 only an additional 12.54 per cent of total district area had come into the fold of urban administration. There was no addition of urban area during 1991–2001. Alappuzha was an important port town in the recent historical past. It played a significant role in the local economy.

Stagnation or slow growth in area combined with continuous increase in population signifies intensification of urban land use. This is precisely what was taking place in a number of traditional urban centres in the state until 1981. Urbanisation in terms of spatial expansion frog-leaped from 1981 to 1991. The growth slowed down considerably in the 1991–2001 period.

As pointed out earlier, one of the important factors that has considerable influence on urbanisation trends in Kerala is its unique rural-urban continuum. While urban settlements spread along the ridges and upper slopes, rural land use remains prevalent along the valleys. The difference between urban and rural areas is continuously getting blurred.

The ratio between population size and the composite urban index for urban and rural areas at the district level indicates that apart from the mountainous Western Ghats areas there is not much difference between rural and urban areas in obtaining services (Chattopadhyay 1995a). Data for 1981 indicate that 90 per cent of urban centres market plantation-based products. Small centres developed all along the midlands throughout the state and functioned initially as procurement centres for local products and markets for consumer goods. These small centres are connected to the large urban centres in the coastal plain either through roadways, railways or both. Some of these coastal and lowland towns also enjoy access to water transport.

The impact of urbanisation on land use is felt widely. On the one hand spatial expansion or extensification brings more and more rural areas under urban administration; while on the other hand, land use within the urban centres is undergoing changes to various degrees. Figure 6.3 is a map of the growth of the urban population in Kerala by district.

A study in Thiruvananthapuram city brought out the types of changes taking place within the urban centres (CESS 2001). The increase in agglomerated settlement area within Thiruvananthapuram city was a little over 150 per cent in the period 1966–1991. This changed the urban hydrology and created a heat island; the built-up area absorbs and radiates more heat than an open area, and reduces the water cushioning area. Thiruvananthapuram now suffers water logging during every southwest monsoon. What was once a clean city now endures problems of drainage, solid waste disposal, storm runoff and excessive heat. The built up areas are spreading along the transport lines. The main transport lines are along the ridges but connecting branch routes cut across the valleys and lowlands. The rice fields located near these routes are more susceptible to conversion to settlement areas than are those at a distance. In other words, the pattern of spatial growth had a significant impact, although the population growth as such was not so high.

Migration

Kerala is well known for migration both inside and outside India. Based on a survey in 1992–93 the Department of Economics and Statistics estimated that there were 1.2 million migrants from Kerala of which 56 per cent migrated outside the country while the remaining 44 per cent migrated to other parts of India. A sizable proportion of

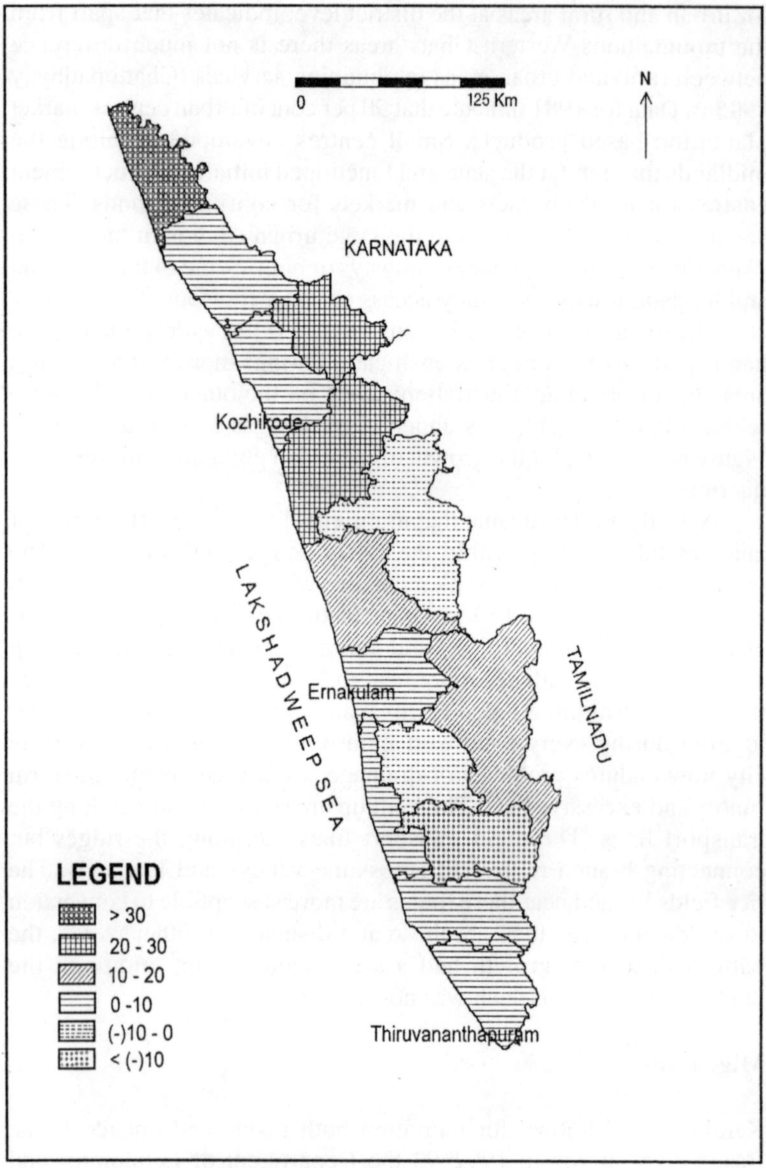

Fig. 6.3 : Districtwise Growth of Urban Population in Kerala

the remittances from these migrants, particularly from those outside India, is invested in buildings in Kerala. Well-built modern houses dot even the rural areas. This building boom has stimulated further land use changes.

Kerala is also noteworthy for rural to rural migration. A study found that 87 per cent of rural internal migrants had migrated to other rural areas within the state and about 68 per cent of migrants originating from urban areas had also moved to the rural areas while the other 32 per cent of urban migrants staying inside Kerala moved from one urban area to another (Chattopadhyay 1988). At the all-India level the migration trend is from rural to urban and from small cities to big cities or urban to urban (Bose 1978). The push-pull factors between rural and urban areas are not so important in Kerala. To examine this point more fully, we worked out the ratio between population and rural and urban amenities in each district. All the amenities in rural and urban sectors of a district were given specific weights and a total value for a particular category was worked out for rural and urban areas separately. Then the rural population in each district was divided by the amenities value for the rural sector of the same district and a value was obtained. The same was done for the urban sector. There are now two values, one each for rural and urban sectors in a district. The difference in the population amenities ratio, based on 1981 data between urban and rural areas varied from 29:31 in Ernakulam district (the lowest) to 3:12 in Idukki district (Chattopadhyay 1995). For Kerala as a whole the ratio between the amenities index and population was 1:1810 for the urban areas but 1:3070 for rural areas. In other words population pressure on amenities is less than twice that of urban areas. This indicates that the difference between rural and urban areas is low so far as availability of amenities is concerned. Another way to see the Kerala difference is to compare the access to selected amenities in rural areas with similar access for all-India. Kannan (1988:18–21) summarizes various Indian government surveys showing that in the early 1980s 98 per cent of Kerala villages had at least one all weather road, compared to 46 per cent for all India. One hundred percent of Kerala villages had a primary school within two km compared to 40 per cent for all-India; 97 per cent had a higher education facility within five km compared to 21 per cent for all-India; 78 per cent of Kerala villages had a hospital within five km compared with 35 per cent for all-India. A full 97 per cent of Kerala villages had electricity available to at least some households versus 33 per cent of villages in all-India.

Similar contrasts exist for health centres, fair price shops, veterinary dispensaries, cooperative banks, seed stores and post offices. This helps explain why the migration in Kerala is not always rural to urban. The internal migration trend in Kerala is from lowlands and midlands to highlands and from Travancore in the south to Malabar in the north. Idukki district in the south and Wayanad district in the north are the two focal points within the Western Ghats that attract migrants.

While migration to the hill region started during the latter part of the 19th century with the introduction of plantation crops, the major inflow was during 1921–31 and 1951–61. The nature of this inflow and its impact on landuse change in the hill district of Idukki was discussed at the beginning of this chapter. The migration from Travancore, the former princely kingdom in the southern part of Kerala to Malabar began in the 1930s. The migrants who settled in the hill tracts of Malabar were small farmers from Travancore and a majority belong to the Syrian Catholic Christian Community (George and Chattopadhyay 2001; Tharakan 1976). The economic depression of the 1930s contributed to out migration from Travancore. Many small farmers who got entangled in debt sold out their small holdings in Travancore and with the proceeds could acquire larger properties in Malabar. The migrants started occupying various types of land ranging from cultivable wastelands to virgin forest to cultivated land. Joseph (1998) reported that out of 341 migrants in the 1940–1980 period up to 55 per cent acquired cultivable wasteland. Figure 6.4 is a map showing the spatial trends for internal migration in Kerala.

This internal migration from the lowlands to the highlands and from Travancore to Malabar triggered large-scale changes in land use. While migrants may be industrious, they also carry their experience of resource use from their place of origin. This has more often than not interfered with the local ecology of the receiving area. Many recent soil erosion and landslide problems can be linked to the settlement of migrants and their cultivation practices. Landslides in Kerala are associated with movements of people and agricultural practices introduced by them. Slopes above 18° are normally unstable and should be kept forested. Migrants from the lowlands did not adhere to this rule. They terraced the slopes and started intensive agriculture or rubber plantations. Terracing accelerated infiltration. During heavy rains the soils became oversaturated and started moving down the slopes as debris flows. Many of the landslides in Kerala are associated

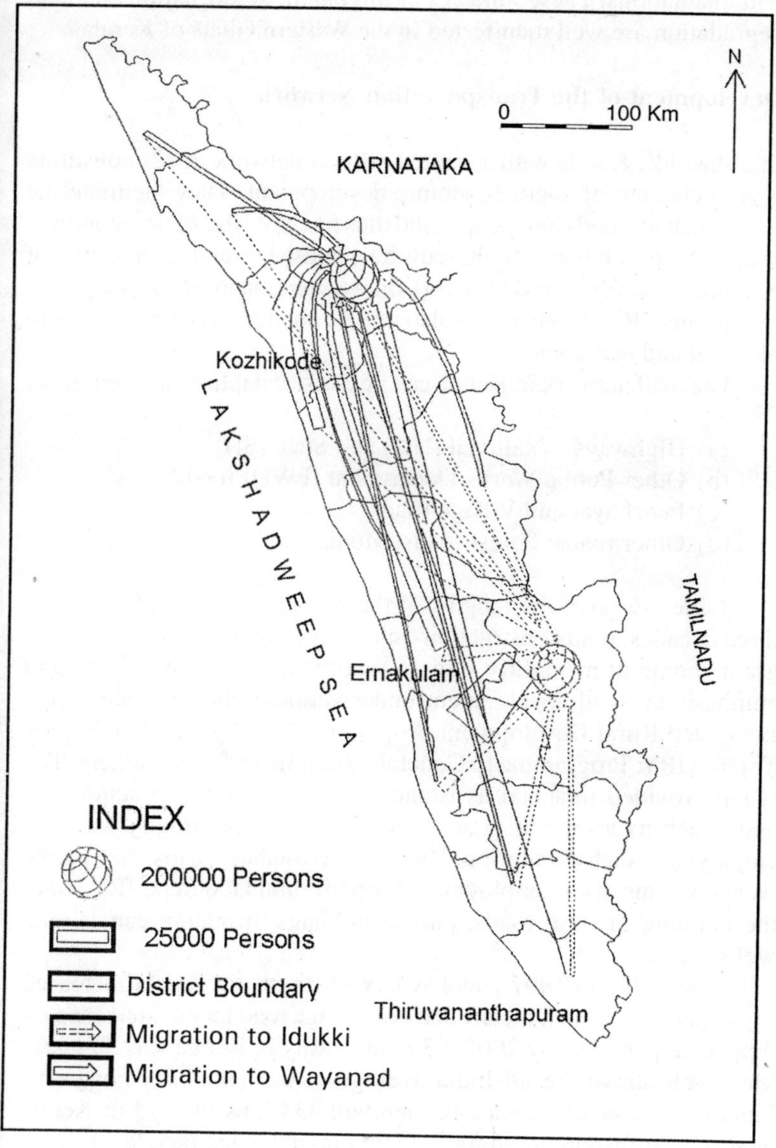

Fig. 6.4 : Spatial Trend of Internal Migration in Kerala

with plantations. These linkages of migration, deforestation and land degradation are well manifested in the Western Ghats of Kerala.

Development of the Transportation Network

In a state like Kerala with a limited railway network, roads constitute a vital element of socio-economic development. They facilitate the movement of goods and people and thus integrate the socio-economic space. As an element of the cultural landscape and as a factor of economic growth the road network draws attention in any development programme. Roads, however, also initiate changes in landscape, both physical and economic.

The road network in Kerala can be classified into four categories :

(a) Highways – National (NH) and State (SH)
(b) Other Public Works Department (PWD) roads
(c) Panchayat and Village roads
(d) Other roads: Forest roads, Municipality roads, etc.

Table 6.6 gives the lengths of these roads in Kerala. In the past three decades, road construction especially in rural areas has attained great momentum mainly due to reactivation of panchayats and emphasis on road development under various schemes such as the Integrated Rural Development Programme (IRDP), Jawahar Rozgar Yojana (JRY: Programme to Generate Rural Income), and others. The IRDP provided financial assistance for poor families to acquire an asset such as a cow or goat, or to set up an opportunity for self-employment such as tailoring. The JRY was established to create public works as a means of employment for poor rural labourers. JRY led to the building of rural roads, public buildings, irrigation canals, and wells.

From 1981 to 1997 panchayat road lengths in Kerala increased by 49 per cent, which means that on average road length increased by 3 per cent per year. By 2002–03 road density in Kerala was 3.56 km/km^2, well above the all-India average of 0.91 km/km^2. For every hundred thousand population, there are 434 kms of road in Kerala against a corresponding figure of 299 kms at the national level (State Planning Board 2004). According to a study of the Confederation of Indian Industries in 2003, the rate of infrastructure penetration in Kerala is 2.5, the same as Punjab and only behind Maharashtra (2.8) and Tamil Nadu (2.6). The adjoining state of Karnataka has the value of 2.4. The rate of infrastructure penetration is a composite index arrived at by considering transport, communication and energy related variables.

With people's desire to have a road facility at their doorsteps, construction of any number of roads would appear to be justifiable. However, experience shows that indiscriminate road construction can lead to a number of problems. It needs meticulous layout planning so as to yield optimum benefits with minimum environmental damage.

The distribution of roads indicates control of terrain conditions. A sample study at the panchayat level showed road densities of Onchiyam (Lowland), Akathethara (Midland) and Poothadi (Highland) panchayats are 4.8 km/km^2, 3.2 km/km^2 and 1.3 km/km^2 respectively (Chattopadhyay 1995a). As we saw earlier, in the coastal plain old roads ran and still run along the sand ridges, which are longitudinally placed, and parallel to the coastline. Settlements were mostly along these ridges. The low-lying areas between ridges are locales of paddy cultivation. These depressions are also the natural drainage paths of the coastal plain and are connected with a network of natural drainage channels. With an increase in population roads are constructed across the interridge areas sometimes filling up low lying areas and blocking drainage channels. The midlands and highlands of Kerala are characterised by alternating ridges and valleys, which are placed almost perpendicular to the coastal plain. Initial road developments are along these laterite ridges. The major arteries—mostly old roads—have been placed along the ridges whereas the subsequent roads are constructed across the lowlands to connect newly developed settlement sites. The result is either total or partial blockage of the natural drainage channels. Road density decreases towards the eastern highlands corresponding to the decreasing population density. However it cuts across hill slopes and virgin lands, thereby accelerating changes, which are not always favourable to development.

Roads and Deforestation

If we compare a road map with a map on deforestation (Fig. 6.5), we can see that deforestation proceeds along the roads. The roads render the previously impenetrable forest area accessible. This facilitates human intervention on a large scale. The high positive correlation (0.64) between deforestation (in km^2) and road length (in km) at the

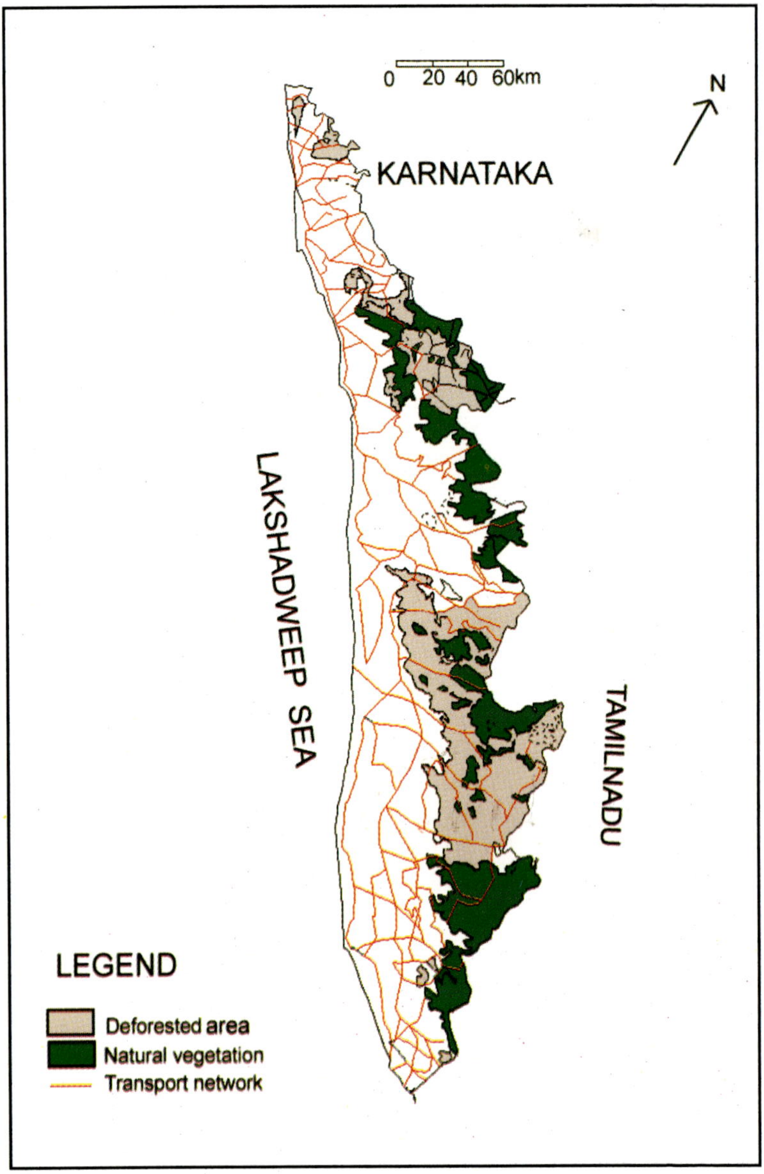

Fig. 6.5 : Deforestation and Road Network in Kerala

district level supports this observation (Chattopadhyay 1995b). The environmental harms caused by road construction may be grouped into direct and indirect.

Direct harm

1. Roads across lowlands act as dykes and restrict the free flow of water and therefore aggravate flood problems. A clear example can be cited from the Kuttanad area where the Alappuzha-Changanassery road has contributed to flooding. With hardly any wasteland available, roads are mostly constructed over cultivable productive lands.
2. Natural drainage channels are often converted into roads with a narrow and ineffective drain on one side. The natural drainage path becomes dysfunctional to varying degrees. In the coastal plains this practice is widespread.
3. In the highlands, roads cut across the slope. As a result a graded uniform slope is divided into two segments above and below the road. This directly contributes to destabilisation of slopes causing landslips in many cases.
4. Roads act as the channels of resource exploitation in the interior places. Once a road is constructed into the forest area, deforestation is aggravated.

Indirect Effects

1. Due to blockage of normal drainage in low lying areas, water stagnates. This leads to a build-up of moisture and dilutes soil nutrients. Normal soil drainage and flushing are obstructed. This ultimately reduces productivity and renders both water and soil unusable. Agricultural practices in many cases do not yield the desired results.
2. Once roads have been constructed, settlements spring up. During the last 20 years around 18 per cent of productive land has been diverted for settlements and related activities. This is detrimental to the agro-economic condition of the state.
3. Blocking of drainage due to road construction in the hill tracts contributes to high water percolation. This sometimes crosses the water-bearing limit of the soil. As a result, landslides or soil creeps take place. During the monsoons this is well evidenced in the hills tract.

4. Roads in the hill section destabilise slopes. Slope failure all along the Ghat roads is a regular phenomenon. In the Pothundi-Nelliampathi road as many as 16 slope failures were recorded in a stretch of 2 km during a single rainfall event in June 1985 (Chattopadhyay et al. 1986). All these sediments moved to the Pothundi reservoir via streams and rivers contributing to increased siltation.

It is suggested that guidelines be laid down about the maximum permissible road density for each area, considering the terrain, the population and the environmental impact. Once people become aware of the adverse environmental impact from excessive road construction, the present tendency of constructing roads in inappropriate places may stop or slow down.

Table 6.6 : Types of Roads by Length in Kerala – 1981 and 1997

Category	*Road length in Kms*		
	1981	*1997*	*Growth in Per cent*
National highways	839	1011	20
State highways	2055	4169	103
Major district roads	5650	6887	22
Other district roads	7133	8708	22
Village roads	2327	2495	7
Subtotal	18004	23270	29
Panchayat roads	73188	109058	49
Corporation and municipal roads	5135	8627	68
Others	2251	5522	145
Grand total	98596	146477	49

Sources : Government of Kerala State Planning Board Economic Review, 1981 and 1998

Rules and Acts

Government rules, laws, regulations and acts play important roles in environmental resource management. These institutional mechanisms are significant policy interventions aimed at realising the greater good for the larger society. In the process they help fashion the course of change and allow various stakeholders to satisfy their interests. A good legal and institutional arrangement fosters favourable processes to encourage parties to cooperate to achieve common goals and realistic and acceptable rules as well as to develop strategies, which comply with rather than break the rules (Thomson 2002). It is a common experience that many of the rules cannot bring about the desired results. Underlying reasons for the failures can be debated. Our interest here is to describe briefly some of the land acts that have had profound impacts as drivers of land use change.

Act No. 1 of 1894 was the Land Acquisition Act, under which the expression *land* includes benefits that arise out of land, and things attached to the earth or permanently fastened to anything attached to the earth (Joy Sankar 2003). There are several similar acts such as the land assignment act, land conservancy act, land development act, land relinquishment act, land tax act, and land utilisation rule.

The Land Reform Act

In 1959 the first Communist Ministry of Kerala passed the "Kerala Agrarian Relations Act," (KARA) which became the KARA of 1960. This law was never implemented but supplanted by a weaker Congress Party act in 1964 and finally, in 1969 by the law now considered Kerala's land reform act. The 1969 law came into force on the first day of 1971. The Kerala land reform act is based on the following principles :

(1) The broad outline of land reforms enunciated in the Indian national five year plans;
(2) The basic aims of the Kerala Agrarian Relations Act of 1960;
(3) The need for doing justice to all sections of the population;
(4) The necessity of reducing the drain on the finances of the state; and
(5) The desirability of a new law capable of smooth and easy administration

This act replaced all major land legislation then in existence in the state (Oommen 1994). Kerala's 1969 land reform contained three major components :

- A ceiling on the absolute size of holdings, with excess to be redistributed to the landless. It is estimated that 85,000 households received plots averaging 0.59 acres (0.24 ha) under this provision.

- The abolition of paddy land tenancy and thus the abolition of rental payments from actual operators to non-cultivating landlords. Stays of eviction prevented landlords from using this provision to throw tenants off the land. This provision brought ownership rights to 1,290,000 (12.9 lakhs) households that gained an average of 1.5 acres (0.62 ha) each. Former landlords received compensation from the government at 16 times the fair market price as determined by a government committee.
- The abolition of house compound land tenancy and conference of full ownership rights to 269,000 *kudikidap-pukar* or hutment dwellers (kudi = hut) who received an average of 0.08 acres (8 cents: a cent is one one-hundredth of an acre) of land on which their houses stood. Compensation was set at 25 per cent of the market value with half paid to the former landlord by the government and half to be paid by the recipient in 12 annual instalments. Unpaid instalments were to become debts to the government, not the landlord, and thus could not be used to repossess the land (Franke 1996:125–26; Franke and Chasin 1994:58).

A Land Tribunal was established to help the tenants. The land reform act provides for ceilings both on existing holdings and future acquisitions. The ceiling for a family of five members is 12 standard acres. One standard acre varies from one to four acres depending on the type of land. The ceiling limit in terms of ordinary acres is not more than 36 acres. Cultivating tenants other than those holding land in the name of a member of the armed forces, a seaman or a legal representative of these persons are entitled to fixity of tenure.

A landlord can repossess land from the tenants under the following circumstances :

(1) a trustee or owner of a place of public religious worship can repossess land for extension of the place of worship by obtaining a certificate from the District Collector;
(2) a landlord who is not in possession of any land or is in possession of land of less than 5 standard acres can repossess land not exceeding 50 cents (one half acre) for the purpose of constructing buildings for bona fide members of his family;

(3) a landlord can repossess land for personal cultivation if the tenant possesses more than the ceiling area. The tenant will, of course, be left with an inalienable extent equal to the ceiling; and

(4) a landlord who is a small land holder can resume half of the extent of the tenant's holding, but the total extent of land in the landlord's possession cannot be raised above 4 standard acres or 4 acres in extent whichever is greater.

The Kerala land reform has had a significant impact on ownership and on inequality. Every family now has a piece of land in its possession. Although it has contributed to fragmentation of land the single most important achievement is to make a land resource accessible to every household. Conferring ownership rights to hutment dwellers has benefited the poor particularly the agricultural households. The agricultural labour households improved their position as a class.

Some Other Relevant Acts

The Land Utilisation Order of 1967. This order was promulgated to control landuse practices to maintain the production of food crops. Changes in cropping patterns and conversion of land from one use purpose to another are restricted and require written permission from the Collector.

High Range Colonisation Scheme Rule of 1968. This rule is to encourage families belonging to the landless agricultural labour class to settle in the high ranges. Rules specified eligibility for assignment, extent of land to be assigned, and procedures for issuing the patta (the ownership title paper).

Wayanad Colonization Scheme Rules of 1969. This scheme was originally framed by the Government of Madras in 1943 for settling aboriginals, ex-service men, and other persons. This was basically to help settling local people.

Assignment of Government Land to the Scheduled Tribe Rules of 2001. This is an extension of the Kerala Land Assignment Act of 1960. Government land is assigned to the Scheduled Tribes for house sites, personal cultivation and "beneficial enjoyment" (a legal term). This rule became necessary in the context of the Muthanga incident that will be described in chapter 7. The assigned lands are heritable but not alienable. As the distribution is envisaged in the forest land the assignee must not cut any tree standing on the land and shall not cultivate or use the land in violation of the restrictions or instructions issued by the relevant authorities for ecological protection or plant quarantine reasons.

Environmental Offshoots of Land Reform in Kerala

Land reform in Kerala has brought many positive changes with definite social and economic benefits. In spite of certain limitations it has been more equitable than any other similar attempts made in India. Inter alia, it has promoted more participation of the people in the political and development process (Oommen 1994). However, there are some problems that have long term environmental consequences related to the land reforms. These include :

(1) Exemptions given to plantation crops like rubber, tea, coffee and cardamom have led to large scale conversion of paddy lands to crops such as rubber. Kerala had sizable private forest lands. Conversion of private forest lands to plantations is also exempted;

(2) Agricultural labourers, who became new land owners as a result of the land reform, converted their possession to build dwelling places. With possession of barely a few cents of land as a consequence of family property division, for the 2nd generation after land reform the immediate priority of many agricultural labourers was to build a shelter. If the ownership is for wetlands the former tenant has little choice but to convert the agricultural land into built up areas. A location specific study by Mahesh (1999) indicated such a trend. Analysis of field data brought out that rural labourers, a majority of whom possess less than 10 cents of land, had converted all of their wetlands;

(3) Non-tribals who migrated to the hill areas took advantage of the land reforms to become de facto owners of land supposed to be possessed by the local tribes. There has been substantial alienation of land from a set of tillers of the soil to non-tribals (Oommen 1994). Non-tribals, who had migrated from lowlands and midlands to these high hill areas, also carried their cultivation practices. As a result severe land degradation

> is observed in these hill areas. Attapadi is one such area which has suffered under these circumstances. Although the Government of Kerala passed the Kerala Scheduled Tribe Act in 1975 to restore some of these lands to the tribals, there has been little headway in its implementation.

We attempted in this chapter to discuss briefly the drivers of land use change. Some of these changes are positive and help to democratise the society, to create more equitable distribution of resources and to facilitate easy access to resources. However, in the long run some of these changes can lead to undesirable consequences. It is necessary to take note of these issues and design appropriate intervention measures, which are essential to finding the path to sustainable development. In the next several chapters we shall look at Kerala's attempts to find a path to sustainable development with social justice, democracy, and equality. Despite the unexpected problems associated with Kerala's famous land reform, the participation and empowerment it provided for ordinary villagers opens up possibilities for democratic local planning that we shall see have become hallmarks of Kerala's new generation of activists.

Part - II

Kerala's Initiatives for Sustainable Development : Employing the Power of an Awakened People

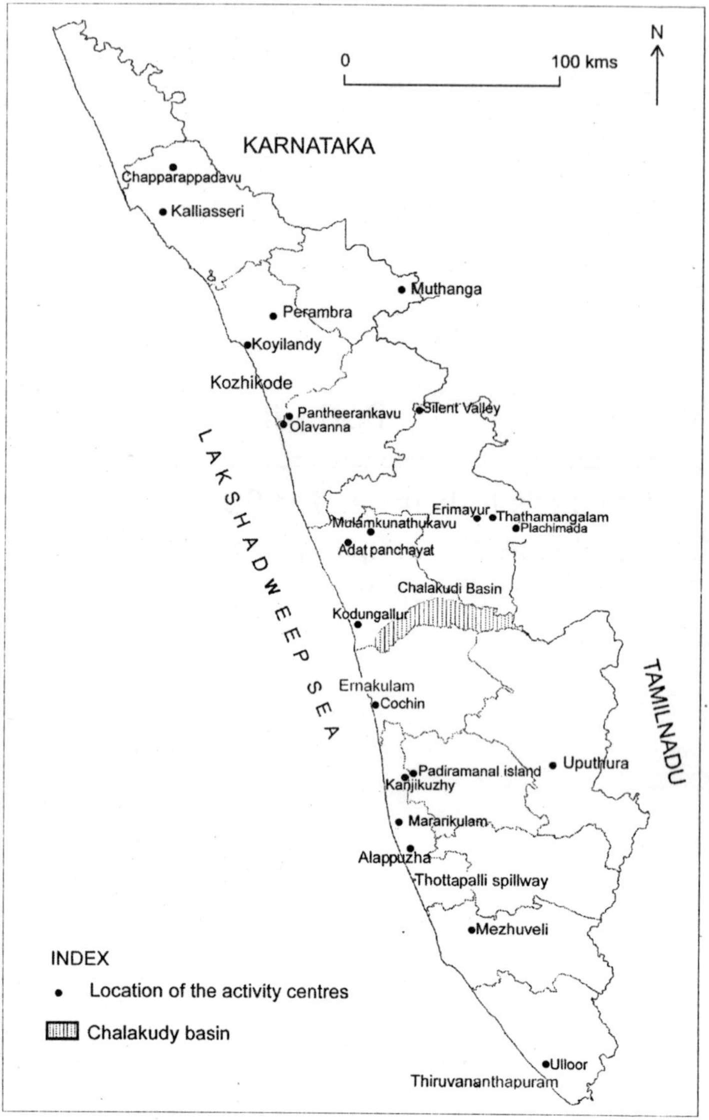

Fig. 7.0 : Location of Places Referred to in Chapters 7 to 11

Introduction

Part II of the book consists of chapters 7 to 11 in which we discuss selected democratic initiatives undertaken in Kerala at the state and local levels as well as people's struggles to save or reconstruct their environment in the face of outside powerful interests. The energy and activism in these initiatives result from decades of people's movements in Kerala. These movements have been documented and analysed by numerous observers of Kerala's history and politics. Activism is the main component of the well-known Kerala model of development. During the late 19th century this activism was expressed primarily as caste improvement associations. By the 1920s it had spread to trade union organising and in the 1930s it led to the formation of radical left political parties, mass organisations of workers and peasants, the library and reading room movement, and the development of literacy and public health campaigns.

Both before and after the victory of the first communist ministry in 1957, Kerala was a centre for land reform activity (briefly discussed in chapter 6), education reform, major achievements in reducing caste discrimination, and in the 1970s and 1980s, the enactment of agricultural labour protections and pension plans for agricultural labourers and casual (headload) workers.

As we have shown in earlier chapters, Kerala faces a plethora of serious environmental crises and challenges. Against these, Kerala has the capacity to mobilise a literate and politically aware and active population. In chapter 7, we shall survey protests against and attempts to prevent environmentally damaging policies of the government or the private sector. These include the campaign to protect Kerala's Silent Valley, attempts to limit overfishing, the Muthanga struggle of tribal people for greater control over their environment, and the now famous struggle to stop the environmentally destructive policies of Coca-Cola and Pepsi in Kerala's central foothills in order to protect the local water supply.

A key turning point in the recent history of Kerala's political activism came with the Left Democratic Front (LDF) victory in the state assembly elections of 1987. Faced with a sense that further land reforms and other wealth redistribution activities would yield only marginal benefits; and faced with growing concerns over economic stagnation, LDF leaders and activists decided to try involving Kerala's literate, politically aware public in direct action for development. Chapter 8 summarises the Panchayat Resource Mapping Programme—the main environmental element of the New Democratic Initiatives—a major breakthrough in development thinking. The New Democratic Initiatives included the creation of elected district councils to decentralise part of the state planning apparatus, the organisation of a Total Literacy Campaign, intended to awaken idealism and commitment as well as to teach the illiterate 10 per cent of the adult population to read; a stepped up campaign to install high-efficiency wood stoves in as many households as possible; and the Panchayat Resource Mapping Programme (Franke and Chasin 1994:xiv–xxi; Thomas Isaac, Parameswaran, and Franke 1997). As these campaigns were underway, activists in Kalliasseri village in northern Kerala were developing a model for local village planning.

These various processes and experiments led by the late 1990s to a set of concepts and techniques that enhanced the People's Campaign and have left a mark on Kerala's society that will likely endure for some time. The People's Campaign is the subject of chapter 9.

Chapter 10 surveys a number of creative local initiatives to use human resources to promote greater sustainability as an integral part of development planning: biodiversity registers, local ecological management initiatives such as pond rehabilitation, watershed master plans, micro-hydroelectric projects, the Kanjikuzhy sustainable vegetable production project, a biological disease control programme in Koyilandy, self-sufficient local water supply schemes, an integrated coastal zone management project, the Pathiramanal Island Biosphere and Conservation Site, and environmentally sustainable microenterprise anti-poverty projects. We conclude the book with chapter 11 in which we attempt to place Kerala's environmental problems and the experience of its people to date within the main international debates about the scope and seriousness of the environmental threats described in chapter 1. We believe the Kerala experience indicates that activism, participation, and egalitarianism offer substantial lessons for environmental activists and concerned citizens throughout the world.

7 Movements to Defend Kerala's Environment

The awareness and activism of Kerala's people have brought about several inspiring campaigns to protect threatened natural resources.

The Save Silent Valley Campaign

The 'Save Silent Valley Campaign' during the late 1970s and early 1980s, culminated in the abandoning of a hydroelectric project that included building of a dam, approved and on the brink of construction, in favour of a biosphere reserve. This campaign to conserve a unique piece of tropical rain forest by defying the decisions and actions of the State Government was the first of its kind in the third world. The successful campaign to save Silent Valley constitutes a determined and collective action of people of Kerala from all walks of life in environmental conservation and the positive response of a political leadership for a cause effectively espoused by common people throughout the society.

Silent Valley is located in the northern segment of the western Ghats mountain range that runs the length of the Kerala-Tamil Nadu border forming part of the Nilgiri hill ranges. Figure 7.1 shows the location of Silent Valley. It is the catchment area of the Kunthipuzha river, a tributary of the Bharatapuzha, one of Kerala's longest and most important rivers that we described in chapters 2 and 5. A well defined watershed characterised by plateau and hill slopes, Silent Valley is surrounded by hill ranges all around with its outlet in the southern part where the Kunthipuzha plunges into the Mannarkkad plain. The eastern boundary of the watershed is higher and is dotted with peaks including Kovittola Betta (2336 m), Anginda (2383 m), Mukkali Mudi (2193 m), and several other peaks over one thousand metres high.

This valley encompasses one of the highest rainfall zones in Kerala. The Kerala State Electricity Board (KSEB) recorded 5045 mm

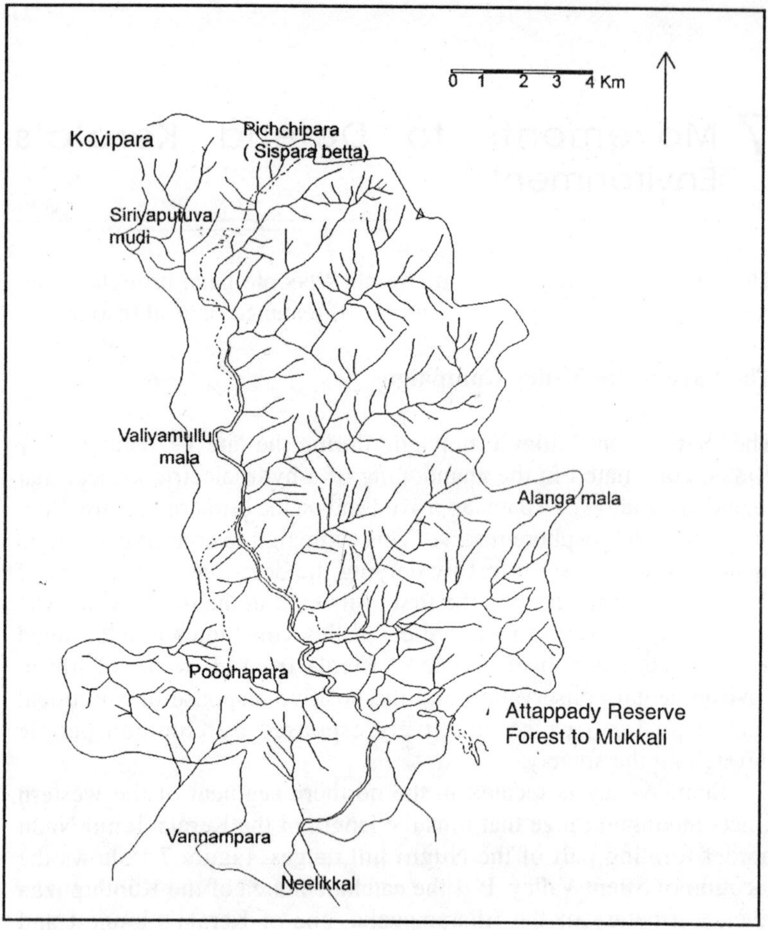

Fig. 7.1 : Silent Valley National Park

of rainfall at Arthala at 1200 m elevation. There are reports of still higher rainfall at higher elevations. Neelikkat at an elevation of 1005 m in the southern boundary of the watershed receives 4,364 mm of rainfall on average, of which 79.5 per cent precipitates during the southwest monsoon from June to September. Due to deep dense vegetation cover the Kunthipuzha river is perennial, although the size of the catchment spreading over the entire Silent Valley park is only around 90 sq km.

Silent Valley has no record of human habitation, not even of tribal peoples that otherwise live in the highlands of Kerala. Due to its sheltered nature and its inaccessibility from all sides Silent Valley represents an 'ecological island' preserving the characteristics of a typical humid tropical forest of peninsular India, not disturbed by anthropogenic activities at any time. Because of this pristine condition it can serve as a reference for assessing the status of other humid tropical forest areas in peninsular India. The isolation of Silent Valley is an important factor for studies of evolution (Sinha et al. 1999). Apart from being a storehouse of a large number of valuable endemic, endangered and rare plants and animals, this valley offers immense possibilities for future research on unknown flora and fauna. Singh et al. (1984) indicated that the basal value (area covered by trees at ground level in a unit of forest area) of the tropical forests in this valley is 102.7 m^2/ ha, well above the range of 35–73.6 m^2/ha obtained in similar forests elsewhere. It was further brought out in this study that the evergreen forests of Silent Valley exhibit a high degree of species richness and diversity. According to a study by the Zoological Survey of India (1986) many species which were available in the Western Ghats 50 to 100 years ago and were not collected before disappearing in those areas still exist in Silent Valley. References to Silent Valley appeared as early as 1901, when the area was identified for selective felling of timber by the forest department. The timber was used for railway car construction. A hydroelectric project was initially conceived in 1921 and the first reconnaissance survey was conducted in 1951 followed by a technical investigation in 1958. The project was approved by the National Planning Commission in 1973 and the Kerala State Electricity Board (KSEB) started preliminary work.

Origins of the Save Silent Valley Movement

Opposition to the dam and hydroelectric project developed over several stages. Steven Green, a scientist from the New York Zoological Society working on primates, and Romulus Whitaker, an American born snake expert and founder of the Chennai snake park, brought to the attention of the Indian scientific community the importance of conserving Silent Valley as a natural habitat. Among those initially drawn to the issue were the environmentalists Zafer Fatehally of the Indian section of the World Wildlife Federation, Salim Ali, world renowned ornithologist associated with the Bombay Natural History Museum, H. M. Patil, the then Union Minister of Finance of the Government of India and Chairman of the Wildlife Board of India and scientists along with environmentalists from Kerala (Nair et al. 1999). Considering its geological antiquity, physiographic isolation, unique habitat, species diversity, pristine condition and exquisite natural beauty they proposed to declare Silent Valley a World Natural Heritage site (Asari 1999).

The Silent Valley Campaign Takes Off

The National Committee on Environmental Planning and Coordination (NCEPC) set up a task force under the chairmanship of Zafer Fatehally to study ecological problems in the Western Ghats region including the Silent Valley area in 1976. This task force recommended abandoning the hydel project and declaring the Silent Valley area a Biosphere Reserve, an area without any local administrative structure but otherwise similar to a national park (Prasad 1999). The task force also suggested some safeguard measures in case the hydroelectric project could not be abandoned. The Government of Kerala chose to implement the hydroelectric project with these safeguards. In 1978 the Government of India approved implementation of the project subject to enactment of legislation to ensure the safeguards. By this time, popular opinion was starting to grow against the dam. People from all walks of life came forward. Prominent individuals expressed their opposition and urged the Government to abandon the project. Several Non-Governmental Organisations (NGOs) participated in the protest movement. The Kerala People's Science Movement (KSSP) took the lead in creating mass awareness of the importance of Silent Valley and the necessity of preserving it. Silent Valley Samrakshana Samithis (Silent Valley Defense Committees: SVSS) were formed in many cities. Later the SVSS in Kerala was merged with the Prakriti Samrakshana Samiti (Society for Conservation of nature), formed by prominent scholars, writers and concerned citizens of Kerala. A genuine people's movement developed. Sugatha Kumari, the noted Malayalam poetess opined, "Let us remember that it was not by the effort of one group or a handful of persons that this battle was won. Not only major organisations but also dozens of small groups came up to fight for Silent Valley and hundreds of people took up this challenge" (Sugatha Kumari 1999:15). From a successful protest demonstration against installation of a hydel power project the Save Silent Valley movement came to convey a strong message for environmental conservation and the larger role society can play in development. Over a period of 6 years people attended thousands of meetings and hundreds of lectures, seminars, exhibitions, and theatrical performances. They lobbied political leaders, wrote articles in newspapers, and instigated court actions.

Activists and scientists raised several objections to the project, despite its promise to supply 522 million units of electricity to Kerala and the South Indian grid :

- The dam would submerge 830 hectares including some of the most important areas of Silent Valley;
- (As noted earlier) The Valley's flora and fauna are complex and not yet well studied;
- The gene pool is of unknown but possibly immense diversity;
- It contained a major remnant of the gene pool of the Lion-tailed Macaque, one of the most threatened primates in the world;
- Because ecological systems are highly integrated, the dam could harm other nearby environments in ways not easy to predict in advance.

(Thomas Isaac, Franke, and Parameswaran 1997:37)

The massive public campaign, the strength of their arguments, and the participation of hundreds of thousands of Kerala's people combined with the consensus of scientific opinion reflected in several studies. In 1983 Silent Valley was declared a national park under provisions of the Wild Life Protection Act of 1972. In 1985 the then Prime Minister, the late Rajiv Gandhi dedicated the park. In 1986 Silent Valley National Park was incorporated into the Nilgiri Biosphere Reserve.

New Threats to Silent Valley

Proponents of the original hydroelectric project never gave up their hopes of turning the Kunthipuzha river's power into electricity. In June 2004 the Kerala State Electricity Board (KSEB) proposed construction of a dam 3.5 kilometres downstream from the western boundary of the national park. Immediately, public opposition began to surface against this apparent attempt to undo the environmental victory of earlier years.

The 2004 proposal to reopen the issue and to plan for constructing a hydroelectric project at Pathrakadavu is an attempt to ignore the issues raised during the Save Silent Valley Campaign in the 1970s and 1980s. The larger implications are also not given proper attention. The Pathrakadavu dam is being touted as "eco-friendly" since it is small and does not totally block the river flow. But Silent Valley National Park does not coincide exactly with the ecological unit it was designated to protect, and the absence of a legally protected buffer zone makes the valley susceptible to degradation from measures taken near its borders (Surendranath 2004).

The Monsoon Trawling Ban : Complexities of the Movement to Save Marine Resources

Kerala's 570 km long Arabian Sea coastline is more productive than India's eastern coastline along the Bay of Bengal. The Kerala coastline contributes around 50 per cent of the sea-based fish production coming from the entire west coast. Marine fisheries of Kerala witnessed a phenomenal growth in output over the last 3 decades as a result of technological innovations. Considerable emphasis was given to modernising the fishing sector by the Government of Kerala. Until the early 1970s the traditional fishermen constituted 80 per cent of total fishermen, and contributed 70 per cent of the catch and 60 per cent of the value (Nayak 1993:6). The mechanised sector, which was mainly oriented to shrimp catches experienced a decline in per unit catch by the end of the 1970s and the competition began between the traditional and modern sectors for the same fishing grounds. This has led to conflict over the years. The struggle of the traditional fishermen reached a peak during the mid 1980s. Trawling by the mechanised boat operators particularly during the monsoon was considered harmful for fish resources. Traditional fishermen demanded a ban on trawler fishing during the monsoon months – June, July and August – the spawning months for fish when trawling has a particularly destructive effect on fish populations (Kurien and Sebastin 1980).

Origins of the Trawling Ban

The Government of Kerala banned mechanised fishing for the first time during June, July and August in 1981. Subsequently, due to the intervention of the Kerala High Court two committees were formed in 1981 and 1984 to look into the ban issue (Thomson 2002). Discontent prevailed with frequent clashes among different sections of fishermen. Responding to growing pressure from the traditional fishermen, the Government of Kerala took a decision to ban trawling during the 1988 monsoon and constituted the N. B. K. Nair committee to examine the ban issue scientifically. The committee recommended a total ban on trawling by all types of vessels during the major monsoon months of June-August and suggested a study to assess the impact of banning. The Silas committee was appointed in 1994 to make an independent study of the ban on trawling. It recommended setting apart an exclusive fishing zone for non-motorised craft and small motorised craft of less than 15 HP. This type of fishing could continue throughout the year. The area lying to the west of this zone and within the territorial waters should be declared a restricted fishing zone where trawling would not be permitted during June, July and August. There would be no ban on trawling outside the territorial waters. These recommendations could not be implemented due to objections from the mechanised fishing lobby and lack of governmental enforcement resources.

The ban on mechanised trawling continued after 1988 with varying durations. In 2004, the Government of Kerala decided to ban trawling in the territorial waters for 45 days beginning on midnight of June 14 (*The Hindu*, 25 May, 2004). At about the same time, the Government of India imposed a 64 day ban. This Government of India ban is intended to coordinate all the coastal states in the country to conserve and regenerate stock during the monsoon months. All states are empowered to declare their own bans within the framework proposed by the Central Government. While the Government of Kerala imposed the ban for 45 days, the Government of the nearby state of Karnataka did so for 60 days.

Debating the Trawling Ban

The Kerala trawling ban is hotly debated and the conflicting interests are complex and shifting. During the 1970s and 1980s the argument was that the traditional fishermen, who formed the bulk of the operators could not operate in the deep sea: therefore, the mechanized fishers should be subject to the ban but not those in boats propelled by oars or sails. During the last two decades however, mechanisation has become widespread, and a majority of the fishing folk now use mechanised gear and more efficient nets. A study of Thiruvananthapuram district indicates that by the 1990s half of the traditional boats were motorised, covering around 3/4 of the fisher folk. In some areas motorisation is total (Nayak 1993). As more traditional fishers come to use mechanised gear they begin to question the trawling ban. The Kerala Fishing Boat Operators' Association (KFBOA) raised this issue from time to time and in 2004 they announced their intention to defy the ban. (*The Hindu*, June 4, 2004). There were sporadic responses in support of the KFBOA's call. However no major actions took place and the Government maintained a low profile approach to the issue.

To what extent the marine resources are actually protected by the trawling ban is yet to be conclusively documented. It has been documented however that trawling has a destructive effect during the spawning season coinciding with the first couple of weeks during the monsoon. The Government of India's 64–day ban is based on certain findings by the Department of Ocean Development and the Central Marine Fisheries Research Institute. Modern methods employed by the traditional fishermen may be as harmful or more harmful than trawling. Notwithstanding the present debate it is important to note that the agitations by traditional fishermen in the 1970s and 1980s were forceful and determined interventions by the primary resource users to protect the resource base. The Kerala agitations were mirrored by significant but equally inconclusive struggles along the more southerly coast in the Kanyakumari area of Tamil Nadu (Sundar 1999).

Muthanga : Assertion of Tribal Rights on Forest Land

With 84,326,240 tribals, India possesses the largest such population in the world which accounts for 8.6 per cent of the national population according to the 2001 Census. The tribals mostly live in the forests and draw their livelihoods from forest related activities including the sale of minor forest products. The Constitution of India provides for reservation in education and jobs to facilitate their cultural and economic betterment. In spite of various attempts the fruits of development have not reached most tribal people. The problem lies perhaps in the official prescription of suggesting non-forest livelihoods to forest dwellers and in pursuing a uniform development package for all sections of the society irrespective of their social and cultural backgrounds and the environmental milieus in which they have been living for generations. The paternalistic approach of the Government and the political parties, bureaucratisation of tribal affairs, the marginalisation of academics and specialists in the matters of tribal development, and limited application of tribal self rule are some of the underlying causes hindering tribal development (Roy Burman 2003). Several Conservation Acts promulgated by the Government of India such as the Wildlife Protection Act of 1972, the Forest Conservation Act of 1980 and the National Forest Policy Act of 1988 restricted tribal activities within the forest areas and treated tribals as outside communities in the matters of relocation and encroachment of forest land (Damodaran 2003). Finally tribal people were effectively made illegal encroachers on their own land (Joshi 2003). There is widespread discontent among the tribal people that often leads to the outbreak of political disturbances. However, what happened in Kerala at a place called Muthanga in 2003 is unique and carries implications powerful enough to suggest a reconsideration of our whole approach to tribal development.

The 364,189 tribals in Kerala constitute around 1 per cent of the total state population. The maximum concentration of 36 per cent live in Wayanad district situated within the Western Ghats with a sizable forest cover. This area was a destination for internal migrants from other parts of the state – similar to the case of Idukki district that we analysed in chapter 6. Encroachment in the tribal areas dates back to the 19th century with the introduction of plantation agriculture in the virgin forest lands. In January 2003 Wayanad district witnessed a unique case of a tribal rebellion against the state. A group of tribal people belonging to the *Adiyan* and *Paniyan* sects, and members of the Adivasi Gothra Sabha (Assembly of Indigenous People's Clans) marched to the Muthanga Wildlife Sanctuary and occupied around 300 hectares of land in that part of the sanctuary which they consider as important to their religion and culture. This part of the sanctuary was mostly eucalyptus plantations with little or no vegetation cover. The Government viewed the action as a clear violation of the 1972 Wildlife Protection Act. The State machinery was pressed to forcefully evict the tribal activists. On February 19, 2003 a fight between police and agitating tribals resulted in the deaths of a police constable and a tribal (unofficial reports put the toll at five). More than 50 people were injured in this incident. More than 300 tribal people were arrested. It has now become a long drawn out battle between the tribal people and the State. Meanwhile the State Government's efforts to restore land rights to the tribal people in this area failed to obtain the required approval from the Government of India. There are now serious efforts to allocate land to the tribals.

The Muthanga incident is perhaps an expression of assertion of tribal rights to the forest and the first major struggle challenging forest laws and establishments. It symbolises the second homecoming for the Adiyan and Paniyan longing for their traditional habitat, which they lost to the British in 1805 (Damodaran 2003). Wayanad district formed part of Madras Presidency before Independence. With the expansion of teak and tea plantations the colonial rulers started encroaching upon the tribal lands. The planters, both English and Indian, brought Tamil labourers from the plains, cleared forest lands and started plantations. The movements of the original forest dwellers were restricted. The dawn of the 19th century witnessed the first tribal outburst when the Pazhassi Raja, the then king of the Wayanad region, waged a war against British rule. Pazhassi Raja lost the battle and died in 1805. With his death, the tribal people in Wayanad lost their independence and were forced to work as labourers. A good number of non tribal people moved in as a result of internal migration and settlement laws during the British period. After Independence, the scenario did not change much. With various development projects, expansion of plantations and declaration of forest areas as reserve forests and protected areas, tribal people felt constrained in their home area. Policies of the erstwhile Madras state to allocate land jointly to the tribals and to migrants, considered as outsiders by the tribals, in the 1950s created discontent among the tribes. In the 1970s and 1980s migrants took advantage of the KLRA (see chapter 6) and succeeded in evicting tribals from their own lands. At the end of the 1960s and early 1970s the Adiyan and Paniyan tribes, who were effectively enslaved by the migrant landlords, rebelled and tried unsuccessfully to free themselves. This struggle could be considered as the first home coming attempt by the tribals after they lost their habitat in 1805.

The present struggle at Muthanga has a far reaching impact in the matter of forest policy versus tribal development programmes. It is a matter of great concern that since November 2001 there were 35 violent conflicts involving tribal settlers and forest departments throughout India. Serious doubts were raised about actual availability of land to distribute to the tribes in Kerala as had been promised by the Kerala Government. It seems that official statistics do not reveal the on-the-ground reality. In this context a review of all forest notifications may be called for. Joint Forest Management (JFM) is considered an important step to tackle the problems arising in the sectors of forest dwelling tribals and forest management. However in Kerala, non tribals are also members of Joint Forest Management Committees (JFMC) as stakeholders. The issue of safeguarding tribal interest through forest management is not being properly addressed. Tribal people use forest land in perceptibly different ways than do the migrants who are non tribals. Tribals rarely alter the forest cover in sharp contrast to the non tribals, who replace forest vegetation by plantations and other agricultural crops. Wayanad district experienced widespread land use change. Natural vegetation cover came down from 50 per cent of the geographical area of the district in 1954 to 29 per cent in 1989 (CESS 1997). During the last two to three decades Wayanad district has come to face serious water scarcity and recently people had to move to adjoining areas in Karnataka state in search of water. Water scarcity is partly related to land use change. The struggle of the tribal population to restore their land and habitat directly contributes to environmental conservation and safeguarding the natural resource base.

Plachimada: A People's Struggle to Protect Their Water Supply from Multinational Corporate Misuse

Background to the Struggle

Plachimada hamlet is a section of Perumatty panchayat in Chittur taluk, lying close to the Palakkad Gap. Despite the area having only a marginal water supply, Hindusthan Coca Cola Beverages Private Limited (HCBPL), an Indian unit of the US based soft drink Coca Cola Company started a bottling plant there in 2000. The venture was initially welcomed as part of the industralisation process in the fast emerging Palakkad-Coimbatore industrial belt in south India. Palakkad is one of the draught prone districts in the state with normal annual rainfall of 2329mm (state average: 3063mm; See table 2.3). Farmers depend on canal irrigation to raise a second crop of rice or vegetables. Most of the slightly more than 1,000 families in Plachimada get only 100 to 120 days agricultural labour per year. Nearly all are *Adivasis*, or indigenous tribal people. As we noted in the discussion of Muthanga above, adivasis are among the groups that have benefited least over Kerala's many decades of improving education and health. The company was bringing in up to 250 jobs.

With acquiescence from the local panchayat council, Coke acquired a little over 15 hectares of paddy land (about 38 acres) for locating its plant and started operation, although it failed to obtain the necessary clearance for converting paddy fields into an industrial area. This clearance is mandatory under the Kerala Land Utilisation Act of 1967, which prohibits diversion or conversion of paddy land for other uses. The company also failed to obtain permission from the panchayat for digging borewells and to use electric pumps to draw water. According to the Kerala Panchayat Act of 1994, among other responsibilities, the panchayats are empowered to control and maintain traditional drinking water sources, issue permits for building construction and installation of machinery, and issue licenses for industries within their jurisdiction.

The Factory Opens: The Water Turns Bad

Perumatty panchayat granted Coke a license to operate on 27 January 2000, and soon thereafter the plant began operation. It eventually hired 70 permanent employees and 150 part time labourers. Critics observed that the plant was turning out 85 truckloads per day of cases of bottled Coke (Jananeethi 2002:4). When the company was denied access to local irrigation water that the panchayat felt would interfere with the rice harvest, Coke dug 6 borewells on its newly acquired property to add to two ponds that had already been in use.

Within six months, people living close to the plant began noticing a deteriorating well water quality and a drop in the groundwater level. Local farmers noticed a decline in crop yields. One farmer with a pump claimed he could previously run it for 14 hours per day, but now it would only bring up water for 30 minutes; he further claimed the water level had declined from twenty-two molas deep to just one mola — a mola being a local measure approximately equal to the distance from the finger tips to the elbow (Christian Aid nd:44). Others claimed the water was brackish and sometimes turned milky white and was not fit for drinking, cooking, or bathing. Nearly 100 people complained of stomach aches that they related to the water (Wramner 2004:33). It was reported that three wells in the area had become depleted and water levels in 11 wells out of 19 wells in Plachimada declined considerably between April 2002 and May 2003 (Vijayalakshmi 2003:18).

Scientific analysis supported the villagers' claims. One test found the total dissolved solids (TDS) as high as 1700 ppm in the wells located in the plant's vicinity whereas it was less than 500 parts per million (ppm) in the observation wells located away from the plant (Vijayalakshmi 2003:18). Other studies showed that the water was hard and had high amounts of chlorine – two features often associated with intestinal irritations. One study found the chloride content to be 540 mg per litre while the desirable standard is 250 mg for drinking water. The District Medical Officer found chloride values of 770, 910, and 860 mg/litre in three wells belonging to residents living near the factory (Surendranath 2003a:1).

After two years of operation it was estimated that the plant had utilised 141,015 cubic metres (m^3) of water to produce 37,604 m^3 of beverages and 103,411 m^3 of effluence which was discharged in the vicinity of the plant (Mathew, *The New Sunday Express* 4 April, 2004). Total withdrawal of water was 510 m^3 per day.

Why should water quality decline? Observers have suggested two reasons. First, as the water table drops, water is pumped up through more layers of subsoil. The moving water breaks off and picks up more particles that cause changes in taste and consistency of the water. Secondly, drawing the water table down causes the water to be concentrated from different rock layers than previously, leading to the accretion of some chemicals over others.

The Slurry Sludge Mystery

In July and August of 2003, Coke was suddenly faced with a further environmental concern. As a goodwill gesture (or perhaps to get rid of production by-products cheaply), the company was offering slurry from the plant free of charge to local farmers for fertiliser. Unexpectedly, BBC Radio 4 announced on its "Face the Facts" programme, that one of Britain's leading poisons experts, Professor John Henry, and a senior scientist, David Santillo, had analysed samples of the sludge and found them to contain unacceptably high levels of cadmium and lead (Venugopal 2003). Cadmium is a known carcinogen and lead is basically poisonous to humans. Greenpeace India conducted a parallel study showing high amounts of cadmium, chromium, phosphorous, zinc, and aluminum (Ipye 2003). Already aware of the British study, the Thrissur based NGO Jananeethi had sent a team to Plachimada on 22 July. Their report states that the sludge was "a cream coloured and tar black residue...with 28 to 30 per cent calcium in it and 4 to 5 per cent phosphorous" (Jananeethi 2002). The lack of nitrogen confirmed the British laboratory finding that the sludge was of little to no use as fertiliser: in fact, adding too much calcium and phosphorous could have harmful effects on farm output.

Coca-Cola (2003) replied that the test results conflicted with their own data that showed the plant effluents to be completely safe. Their complete reply to all the various charges has been posted on a website listed in our bibliography (Coca-Cola 2003). The Kerala State Pollution Control Board supported Coke's stand, based on samples taken in January of 2003, but in response to the new data, the KSPBC took new samples on July 29 and found them to contain dangerously high levels of cadmium, although acceptable levels of lead (Venugopal 2003). The KSPCB Chairman announced that it was surprising to find high levels of toxic heavy metals in the sludge of a soft drink factory: "It is a mystery how cadmium is present in the sludge from the Coke factory. If our tests confirm our initial findings, we may have to insist that Coca-Cola follow the rules for secure handling of the hazardous waste produced at the plant" (Krishnakumar 2003). KSPCB scientists speculated that the metals could be a result of the heavy utilisation of the groundwater and the consequent lowering of the water table. Studies by the Swedish Environmental Protection Agency (not in Kerala) had indicated that changes in groundwater quality due to lowering of the table could enhance the solubility of cadmium and lead in water. Other international studies show a relationship between phosphate fertilisers and cadmium poisoning. The Coke sludge might be interacting with local soil characteristics to generate the dangerous byproducts (Krishnakumar 2003).

April 2002 : The Struggle Begins

On 22 April 2002, before the sludge debate but based on the water shortages diminishing the lives of the villagers, local activists had begun protesting Coke's water use. Marches, picketing and appeals to the panchayat council were used. Eventually the protesters launched a *dharna*, or silent vigil, at the factory gates. Interspersed with occasional larger actions, this dharna continued 24 hours daily for 640 days.

Intervening Variables

The Coca-Cola bottling plant interacted with a variety of factors to generate the battle over water rights. While rainfall in the area had been 2,137 mm in 2000, the year the plant opened, in 2001 the rains dropped to 1,147 and to just 670 mm in 2002 according to a report of the Kerala State Ground Water Department. Scientific critics pointed out that the Ground Water Department report used only 8 months of rainfall data to estimate the annual amount and that it included no study of runoff or storage aspects of rainfall as we have described in earlier chapters of this book (Surendranath 2003b:1).

On the political front, the Plachimada campaign converged by coincidence with a campaign against the U.S. attack on Iraq. In early 2003, as the Bush Administration cranked up its war talk, leftists in Kerala chose a Coca-Cola boycott as one of their forms of protest. Telling Coke to Quit India, in a reference to the famous campaigns led by Mahatma Gandhi in earlier decades, the left organisations were joined by usually anticommunist Muslim organisations who felt solidarity with Islamic Iraq. As a result, the Coke boycott across Kerala became exceptionally powerful with some spokespersons claiming a 40 per cent to 70 per cent drop in sales. When evidence from New Delhi suggested that Coke might have pesticide contamination, additional groups joined the boycott including even the Kerala Hoteliers Association. The State Assembly banned Coke from its cafeteria and students got Coke machines removed from campuses and dormitories around the state.

On insistence of the local people the panchayat took up the issue with the plant and subsequently the Government of Kerala and the Kerala High Court came into the picture. The Kerala High Court endorsed the panchayat's autonomy and authority in controlling utilisation of natural resources including water under its jurisdiction. The expert committee appointed by the High Court had worked out the amount of water that the plant can draw in normal monsoon years. However, about the question of affluent discharge, deteriorated water quality, land and soil pollution there were hardly any suggestions.

The Struggle Goes International : The World Social Forum and the Plachimada Declaration

Initially media coverage of the Plachimada dharna was scant. As word crept out from Palakkad of the fight by poverty-stricken indigenous farm labourers against one of the world's biggest multinational companies, however, environmentalists and water activists around the world began to take notice. As of April 2005, a Google search of Plachimada turned up 5,570 hits (some duplications) with reports on the struggle in at least ten languages, including English, French, German, Spanish, Dutch, Portuguese, Italian, Turkish, Indonesian, and Japanese. On November 15, 2003, the US Public television show NOW, with Bill Moyers, featured an interview with the famous Indian environmentalist Vandana Shiva, and a publicly televised reading of the letter from the Perumatty Panchayat Secretary to the local plant manager, laying out the various charges, and the (brief) reply from the manager. All three documents are posted on the PBS website (NOW 2003).

In April of 2002, hundreds of demonstrators joined the small dharna in Plachimada in a one-day solidarity protest. By this time several NGOs had become active supporters of the campaign. Kerala's left political movements had also joined, thus giving the Plachimada campaigners a voice in the state assembly in Thiruvananthapuram. The Congress-led ministry tended to support Coke on the grounds that the state needed the jobs, but local party activists sometimes sided with the water campaigners. The workers at the plant and some unions were hesitant to criticise a job-producing plant, and some workers actively opposed the campaign.

In January 2004, Plachimada came to the attention of the international movement to protect water as a human right and to oppose water privatisation. This campaign has been active in several countries and has built upon the now famous struggle in Cochabamba, Bolivia, where thousands of protesters eventually forced the government to withdraw a contract with the US-based firm Bechtel that claimed ownership of the city's water and had raised prices exorbitantly. With the World Social Forum taking place in Mumbai, activists responded to an invitation from the Perumatty panchayat and the Plachimada Water Defense Committee to visit the area for a three day World Water Conference. The noted Indian environmental activist Vandana Shiva helped arrange the contacts. Among those present were Maude Barlow, Canadian water activist and author of *Blue Gold*, Heide Kautala, a member of the Finnish parliament and former EU parliament member, Diana Green of the Canadian Service Employees International Union, US environmentalist Ward Morehouse, Jose Bove, leader of the Confédération Paysanne (a small-farmers association in France), MP

N. N. Krishnadas, and M. P. Veerendrakumar, President of the Indian Newspaper Society.

The Plachimada Declaration

After three days of presentations of papers and discussions, the participants at the World Water Conference assembled in front of the Coke plant entrance to read the Plachimada Declaration, which follows :

> **Box 7.1 :** The Plachimada Declaration of 23 January 2004
>
> Water is the basis of life; it is the gift of nature; it belongs to all living beings on earth.
> Water is not a private property. It is a common resource for the sustenance of all.
> Water is the fundamental right of man. It has to be conserved. Protected and managed. It is our fundamental obligation to prevent water scarcity and pollution and to preserve it for generations.
> Water is not a commodity. We should resist all criminal attempts to marketise, privatise and corporatise water. Only through these means we can ensure the fundamental and inalienable right to water for the people all over the world.
> The Water Policy should be formulated on the basis of this outlook.
> The right to conserve, use and manage water is fully vested with the local community. This is the very basis of water democracy. Any attempt to reduce or deny this right is a crime.
> The production and marketing of the poisonous products of the Coca-Cola, Pepsi Cola corporates lead to total destruction and pollution and it also endangers the very existence of local communities.
> The resistance that has come up in Plachimada, Pududdery and in various parts of the world is the symbol of our valiant struggle against the devilish corporate gangs who pirate our water.
> We, who are in the battlefield in full solidarity with the Adivasis who have put up resistance against the tortures of the horrid commercial forces in Plachimada, exhort the people all over the world to boycott the products of Coca-Cola and Pepsi Cola.
> Coca-Cola - Pepsi Cola "Quit India."
> (*The Hindu*, 24 January 2004 for most of this text; complete text in Vandana Shiva 2004)

Court Orders and Political Change : A Temporary Victory

The sustained actions of the Plachimada protesters had drawn political support from the opposition LDF legislators in Thiruvananthapuram. They had garnered international recognition. A judicial battle ran parallel to the popular actions. On 31 March 2003, Coke's license with Perumatty panchayat expired and on 7 April, the panchayat decided to reject the renewal application on the grounds of overexploitation of the local water resources. Hindustan Coca-Cola Beverages Pvt. Ltd. (HCBPL) responded by filing a petition with the Kerala High Court. The court ordered Coke to proceed through the Local Self Government Department (LSGD), which led to a hearing on 11 June. After a second hearing on 13 October, the LSGD issued a decision in favour of Coke. The LSGD argued that the panchayat had not conducted a scientific investigation sufficient to prove its case against Coke (summarised from Wramner 2004:41).

Now the panchayat went to the Kerala High Court, which ruled on 16 December that Coke could continue to function at Plachimada but that it had to find alternative water supply sources. The court also held that groundwater is public property and that Coke could draw only as much water as any typical resident owning a similar amount of land (Wramner 2004:42; Krishnakumar 2004a).

As the court battles proceeded, the popular struggle went on with marches, speeches, and the presistent dharna. In March 2004 then Congress Party Chief Minister Anthony ordered the Coke plant temporarily closed, ostensibly because of water shortages during the long Kerala dry season that runs from December to early June. In June 2004 the Kerala High Court received a report from an expert panel suggesting that Coke be permitted to draw 510,000 litres of water per day, about one-third of what critics claimed the company was using. Activists continued the dharna demanding the plant remain closed. Scientific supporters of the activists suggested that the plant could process and bottle coconut water at much less cost, better health for the customers, little water uptake, and with no slurry. As of June 2004, the Coke plant remained closed due to an interim order by the High Court. The struggle entered its 1000th day by January 15, 2005. Inaugurating this 1000th day of dharna and picketing Ms. Medha Patkar, leader of the Narmada Bachao Andolon (Save Narmada Campaign) described that this struggle by the tribals and local people of Plachimada symbolised the struggle worldwide against commercialisation of water by multinationals. Vandana Shiva, who participated in this programme said that the Anti Coca-Cola Action Committee would hold an agitation demanding 'Coke–Pepsi Quit

India' in front of all 87 plants of Coca-Cola in the country. Tony Clarke's book 'Inside the Bottle' was released on this occasion. (*The Hindu*, 16 January, 2005).

At the end of January, 2005 the High Court issued an order allowing the company to draw 500,000 litres of groundwater every day and asked the Panchayat to grant a licence to the company. Perumatty panchayat filed a special leave petition in the Supreme Court challenging the High Court order with a plea that there were substantial questions of law of public importance regarding the rights of a grama panchayat to drinking water and for agricultural purposes in contrast to the rights of a multinational company to extract ground water for manufacturing soft drinks. On May 6, 2005 the Supreme Court issued notices to the Coca-Cola Company, the Kerala Government and the Kerala State Pollution Control Board. The matter is now in the Supreme Court.

The battle is thus not over, but the sophisticated and activist culture of Kerala had generated a popular campaign similar to that of Silent Valley. The power of the people to protect their natural resources was being felt in Kerala again and this time – with internet links and international conferences – it was reverberating internationally at an even more rapid rate.

Kerala's activist traditions have spawned more than protest. The energy and awareness inherent in the Silent Valley, Muthanga, and Plachimada struggles have also been applied directly to actions for sustainable development. In chapter 8 we shall examine statewide actions involving thousands of ordinary citizens, with an emphasis on the Panchayat Resource Mapping Programme, a Kerala innovation.

8 State Level Initiatives for Environmental Resource Based Planning

Why Planning Changes Are Needed

The change from centralised to decentralised planning is widely felt across the globe. In the 1970s, it was realized that in spite of some success stories over the first 30 years of international development, the hard core of mass poverty in third world countries remained mostly untouched (Higgins 1981). Since then there has been an attempt to emphasise disaggregated planning at the regional and local levels from below. This change creates openings for participation of the largest possible numbers of people both in planning and in the implementation and evaluation of development projects. Since the 1970s experiments in integrated area development programmes, growth centres and spatial planning have come into vogue, although the success of these experiments remains limited.

Simultaneously the negative impact of environmental degradation started drawing international attention. In 1987, the World Commission on Environment and Development (WCED) brought out its report on *Our Common Future*, in which the concept of sustainable development was first defined. That report was a milestone in the growing awareness of the need to create environmentally appropriate uses of resources as we illustrated with several examples in chapter 1.

Growing economic disparities, poverty, overcentralisation and gradual but continuous degradation of the earth's environmental resource base together pose major challenges to the future of humanity. As we argued in chapter 1, the strategy for sustaining the environment brings together decentralisation, greater reliance on the local resource base, democracy and diversity, all of which create paths to open system thinking and diverse options and actions instead of reductionism, linear thinking and standard solutions (Chambers 1994).

In chapters 4, 5 and 6 we described the changes in the environmental resource base of Kerala, their possible impact, and the drivers behind the changes. We stressed that environmental changes are manifested at the micro level and therefore must be addressed in a decentralised mode through local actions, bringing the actual resource users to the forefront of the process.

Kerala's experiment in local self government is perhaps one of the very few current experiments worldwide with the potential to present an alternative system of environmental resource management and planning that could lead to sustainable development through empowerment of ordinary people.

The Macro Policy Framework of Local Self Government

Community rule at the village or hamlet level has existed in India since early days. In spite of strong caste divisions villages functioned as close-knit units in the pre-British period with little attachment to rulers. The British introduced revenue villages and a system of governance, the main purpose of which was to collect revenue for the British Government. Resources were siphoned from the large rural hinterland to the British coffers through the colonial port cities.

Gandhiji's stress on 'Gram swaraj' or village self-rule is legendary. Panchayati raj—village self-government—was established immediately after Independence. The Constituent Assembly, while drafting the National Constitution, included village self-government in the directive principles of the Constitution but did not make it part of the government structure. This was primarily to avoid institutional oppression of the mostly illiterate lower castes in the hand of the upper castes, comprising the elite class in those days. Many progressive thinkers were influenced by the famous *dalit* (former untouchable) leader Dr. Ambedkar who feared that sudden imposition of local self-rule would inadvertently turn over power to the local high castes and effectively undermine democracy (Thomas Isaac and Franke 2002:11).

Community Development Blocks, Panchayats and Growth Centres

Community development blocks (CD Blocks) started functioning in 1951 to provide economic stimulation to the rural areas. Article 40 of the Indian constitution reads :

> The state shall take steps to organise village panchayats and

> endorse them with such powers and authority as may be necessary to enable to function as a local self-government.

In keeping with Article 40, panchayati raj started in 1959. Then the Ashok Mehta committee was appointed by the Janata Government in 1978. The committee identified three phases of panchayati raj in the country. The 1st phase had started in 1959 with considerable promises to uplift the rural masses. By that time the 2nd five-year plan was already underway with emphasis on agriculture and industrial development. The country started feeling the impending food crisis in the early 1960s compounding the gigantic task of uplifting the rural areas.

The 2nd phase, marked by the committee as a phase of stagnation, took place from 1965 to1969. The third phase from 1969 to 1977 was a period of decline (Mathew 1994).

By the end of the third five year plan (1962–1967) it was evident that the community development programme, introduced through CD Blocks in a bureaucratic mode, was not succeeding (Fisher 1979).

Part of the failure was attributed to the continuance of planning along sectoral lines – that is, through departments such as forest, major irrigation, Scheduled Caste and Tribe that often did not coordinate at the village level (Sen et al.1977). Other factors included inadequate databases, and inadequate attention to spatial planning at the higher levels. By the end of the 1960s it seemed clear that the community development programme would benefit from improved understanding of spatial planning at the Block level. A pilot project of growth centres with support from the Ford Foundation was initiated in 20 blocks distributed over 18 states with technical support from 12 institutes – two agricultural universities, two Indian Institutes of Management (IIMs), one Indian Institute of Technology (IIT), the National Institute of Community Development, National Institute of Rural Development, Indian Institute of Public Administration, Council for Social Development, the Gandhian Institute of Rural Studies, the Warda Institute, specialising in Gandhian studies and rural development and the Indian Statistical Institute. This was preceded by exploratory studies of two Blocks undertaken by the Council for Social Development and the Ford Foundation. The Planning Commission approved the growth centre project as part of the fourth five-year plan under the direction of the Ministry of Agriculture (Roy 1973; Fisher 1979).

In the mid 1970s development boards such as the Sunderban Development Board and Dandakarahya Development Board were set up to promote the growth centres. Comprehensive Area Development Programmes (CADPs) were also introduced in a number of states in this period. None of these experiments produced the hoped for results.

Problems with the application of spatial planning and problems with growth centre theories had cropped up both at the conceptual level and on the implementation front. When the concept of spatial planning got applied to planning, especially in providing a scientific base to planning, the discussion mostly centred around high level decision making regarding land allocation. Social needs were considered local. This idea resulted in postulating planned space – landscape – as a pure isolated objective which is far from reality. Since development must address social welfare, the landscape cannot be isolated from people and their needs. Space is shaped and molded from natural and historical elements through political processes (Lefebvre 1976). The spatial planning for a given area cannot succeed without considering the political economy. Thinking about landscape without the social context leads to problems with institutions and organisations, problems of coordination and integration, and problems with planning techniques and with planning machinery (Chattopadhyay and Aniyan 2000). By the late 1980s it was clear that a new approach was needed. This new approach was fostered in part by two important constitutional amendments.

Constitutional Amendments and New Panchayats

With the 73rd and 74th Constitutional Amendments in 1992, development moved to 'fresh woods and pastures new' by arriving at the grassroots. Establishment of panchayati raj became a constitutional obligation and at the same time a step towards local area planning, use of local resources, and decision making at the level of actual resource users. This opened a significant opportunity both for the people and planners to act together at the village level. As Sundaram (1997) pointed out, local level planning had come a long way from its beginning as a partial and fragmentary effort about three decades back to its present role as a pragmatic tool for multilevel integrated planning for the rural people and their resource base.

West Bengal became the first state to avail itself of the opportunity brought about by the constitutional amendments. It strengthened the grama sabhas (village assemblies), subsequently making it obligatory for them to propose plan activities. But soon it was realised that holding elections for elected bodies at the lowest level did not automatically bring about meaningful local self-government. Actual local self-governance requires not only elections but also a series of actions that empower the local people to take planning decisions and to implement and monitor them. West Bengal's experience offers valuable lessons.

Lessons From West Bengal

The government of West Bengal appointed a committee in 1993 to evaluate the activities of the grama panchayats – the elected village councils. This committee (Mukherjee and Bandopadhyay 1993) conducted a thorough analysis of the functioning of panchayats in West Bengal and made specific recommendations for strengthening panchayati raj institutions in the state. Some of the committee's observations are worth noting for our understanding of what happened later in Kerala:

(1) In 1979, the panchayats in West Bengal elected Left majorities and became a new generation of panchayats. These panchayats achieved significant land reforms that helped to generate social and political awareness among the people and facilitated the development of new leadership. This intangible achievement strengthened local democracy.

(2) It was of great concern that the initial enthusiasm had faded and panchayats were engrossed in routine work mostly under Jawahar Rojgar Yojana (JRY).

(3) Somehow the panchayats became locked in a development bind and the wider purpose of local self-government remained far removed. The fundamental policy issue of local self-government must be stressed in the panchayats from the very beginning and the pace of change should be set to blend development functions with self-governance.

(4) If the panchayats are truly self-governing, their functional and financial autonomy should be more prominent than merely to assist in administering higher-level decisions and spend devolved funds.

(5) The panchayats are at present not concerned about financial self-reliance. With near zero financial self-reliance, there is

near zero autonomy and correspondingly near zero self-government. By failing to mobilise resources, the panchayats had increased the cost of governance, because now their own cost has to be added to the cost of the pre-existing systems.

(6) The panchayats have little control over development staff from the line departments. Self-governing panchayats with wider functions need strong administrative and technical support.

(7) Effective multi level self-government requires that the different layers do not operate in isolation from each other. Panchayats should be organically linked with each other and with the state government.

(8) Panchayats without programmes are as meaningless and potentially as dangerous as governments without policies. The panchayat should remain armed with programmes to meet the growing hopes and wants of the people.

(9) A statewide programme is required to consolidate the fragmented plots of the tillers. This is a logical next phase of land reform.

(10) A major weakness in local development planning is the absence of seasonal fieldwise crop surveys. Without a record it is not possible to compute a particular season's output for any particular crop. A plotwise record in a prescribed form of the crop grown on each field in every crop season should be maintained to develop an authentic database. The Panchayat Resource Mapping (PRM) model (to be discussed shortly) adopted in Kerala from 1990 onwards illustrates this point.

(11) The panchayats must prepare and vigorously implement area specific plans for agricultural development.

(12) West Bengal should consider launching a land literacy programme involving science for and by the people for which the panchayat should mobilise the people.

(13) Growth centres should be developed to maximise the benefits of decentralised planning. Marketing facilities should be developed as a part of spatial planning.

Resource Management and Local Self-Government

The brief discussion on West Bengal above brought out certain important points concerning ecological transformation and the role of local self-government. Ecological transitions, particularly related to land, water and biomass take place locally on operational holdings. Only with the maintenance of a database, monitoring it over the years and recording the changes that are taking place in agriculture, fisheries, fresh water etc. at the local level can local governments acquire the information needed for sustainable development. The brief review of West Bengal, the state with the most vibrant panchayat system in the country, shows that this database and relevant programmes are missing at the panchayat level. Establishing a democratic set up at the lowest level does not necessarily translate into local self-management of resources. Devolution of political power should be followed by financial devolution, and capacity building for resource management, planning and taking action.

Panchayats need to take active roles in managing common property resources. Land is privately owned but water and forests are common property administered by the government. Ecological functions operate through land, water and forests but as the management systems are vested with different departments, integrating the actions is a major challenge. Panchayats have marginal roles in water and forest management. Due to the fact that water channels and forests cut across panchayat, block and district boundaries, it may not be prudent to leave the management of these resources solely at the discretion of local panchayats.

The water policy now adopted by the government of India is not very clear in this respect although the watershed development programmes now being introduced are more participatory in nature (Kolavalli and Kerr 2002). However, this participation, which is at best an attempt to facilitate community management of forests and to consider a larger role for users of irrigation water, does not provide any scope for the panchayats. The macro policy decision of the Government of India to introduce participatory watershed development based on the recommendation of the Hanumantha Rao Committee without anchoring it in the panchayat is perhaps a modern version of introducing CD Blocks or growth centres without developing local self-government. A tangential introduction of community based watershed development programmes may even undermine the panchayats as the watershed committees will have separate funds, separate committees and separate technical guidance. West Bengal's achievements are noteworthy, but the brief discussion above shows they are subject to many shortcomings. In Kerala a similar approach developed but planners and activists were able to incorporate geographical and landscape concerns more directly into their programmes. This is shown most clearly in the Panchayat Resource Mapping (PRM) Programme in Kerala. A considerable portion of the following section has been adopted from Chattopadhyay et al. (2000).

Participatory Panchayat Resource Mapping (PRM) in Kerala

PRM came about as a result of the intersection of (1) concern about the environment and sustainable development, (2) the urge of the science and technology community to be useful at the ground level of resource management and development, (3) interest of the people's science movement (KSSP) to be effective in development work and (4) last and most significantly the strong determination of intellectuals in the left parties to apply modern science and technology inputs to mobilise the masses at the grassroots level to generate a more egalitarian, participatory and environmentally sustainable form of development.

From Models of Development to Maps for Development

As we indicated in the introduction to Part II of this book, the 1987 LDF election victory brought to the forefront of Kerala politics a new generation of activists and idealists whose thinking expanded earlier Marxist notions of class struggle to add both gender equality and environmental concerns. Their main efforts to empower women awaited the 1996–2001 LDF ministry (Thomas Isaac and Franke 2002: esp. 165–66 and 190–94), but their environmental work started immediately. A key component was maps.

The value of maps was becoming apparent to people at all levels of Kerala society. In 1989 A. P. Chandran, a local primary school teacher in Vazhayoor panchayat, Malappuram district, along with his associates, prepared a sketch map showing water resources in the village.

The Centre for Earth Science Studies (CESS) had brought out *Resource Atlas of Kerala* in 1984. The maps were in 1:1,000,000 and 1:2,000,000 scales with districts and taluks as the lowest units. In 1989, with support from the Kerala State Planning Board a series of maps for district planning of Kasargod district was brought out on a scale of 1:250,000.

By the first half of 1990 on the initiative of Sri Subrata Sinha, the then Director of CESS, a microlevel resource survey on a scale of 1:12,500 for Ulloor panchayat near the CESS offices was completed with collateral plotting on a 1:3960 scale cadastral (revenue survey) map (CESS 1991). The cadastral maps were necessary to involve local volunteers for mapping. Since the cadastral maps show revenue plot boundaries of individual land holdings it is much easier for the local people to plot on these maps than on topographical maps or other thematic maps. The Ulloor maps aroused considerable interest among all sections of society seeing that for the first time a set of basic resource maps was available at the panchayat level on such a convenient scale. This exercise was popularly known as the "Ulloor Model." The Director of CESS held discussions with the Vice-Chairman of the State Planning Board and with functionaries of KSSP. It was agreed that KSSP would mobilise volunteers at the panchayat level for resource mapping and CESS would provide technical support and train the volunteers. By this time the 1989–1991 total literacy programme involving thousands of volunteers was underway (Franke and Chasin 1994 : xv–xvi; Thomas Isaac, Franke and Parameswaran 1997:41–42; Thomas Isaac and Franke 2002:31). It was felt that volunteers mobilised for the literacy campaign could possibly be drawn into a massive campaign for local level development by initiating PRM programmes covering the entire state. The 28th Annual Conference of KSSP at the Cochin University of Science and Technology in February 1991 gave a call for activists to participate in the development process of the state. The situation was ripe for taking a major step. A micro-level participatory resource mapping module was available at that time and the volunteers were willing to take the next step. Kerala was attempting to use the energy of the literacy movement to generate total word and number literacy and to carry that energy forward to create land literacy.

To begin the process of statewide implementation of PRM, CESS and the KSSP jointly submitted to the Government of India's Department of Science and Technology (DST) in New Delhi a project entitled "Panchayat level resource mapping for decentralised planning with people's participation." A DST team visited CESS and witnessed a field demonstration of participatory resource mapping by the volunteers in Attipara, a panchayat with high landform diversity, located very near to the CESS campus. Attipara had a vibrant KSSP chapter with talented and dedicated local volunteers. The project was formally sanctioned on the 27th of March 1991 and an amount of Rs. 1.6 million was allotted.

Meanwhile, the Government of Kerala issued G.O. period No. M/S 78/90/plg dt. 15.12.1990 in December 1990 for conducting PRM in all the state's panchayats within 2 years. CESS was identified as the nodal agency and KSSP was entrusted with the job of mobilising volunteers. By this time the Kerala State Landuse Board had also become a partner in the programme. Subsequently through another government order, the time frame for completion of the work was extended to 5 years and the responsibility for mobilising volunteers was given to the panchayats along with voluntary organizations such as NGOs as collaborating agencies.

In March 1991, a 7 day state level camp was held at the KSSP IRTC campus in Mundur, Palakkad district for training master trainers. About 100 volunteers from all over the state were trained by the CESS scientists and by specialists from the Kerala State Landuse Board. A Malayalam handbook was prepared that detailed the background of the programme, mapping procedures and mapping symbols.

With completion of this training the stage was set for launching the statewide programme. This stage began with resource mapping in a couple of panchayats among which Kalliasseri in northern Kerala emerged in the forefront. Kalliasseri's success was due to the strong local panchayat leadership ably spearheaded by T. Gangadharan, a panchayat member, local primary school teacher, and KSSP activist. In Kalliasseri the resource mapping by the volunteers and technical experts under the guidance of CESS scientists was completed on schedule. The problems were identified and maps were presented to the panchayat. A group of experts from CESS, KSLUB and the Centre for Development Studies (CDS) including the then CESS Director and Planning Board Vice Chairman visited the panchayat and held detailed discussions on the outcome of the resource mapping. An action plan map was prepared. The local volunteers moved beyond the mapping and undertook various surveys to assess vegetable consumption, egg consumption, electricity distribution (mapping the house connections), and drainage mapping. All these generated considerable data which were finally put together as a development plan for Kalliasseri Panchayat (Thomas Isaac et al. 1995). Perhaps this was the first panchayat in India to come out with a complete plan document based on local level data analysis. It generated enormous interest in all sections of the society. The "Kalliasseri model" is now well known throughout Kerala and all over India (Thomas Isaac, Franke and Parameswaran 1997:43–44; Thomas Isaac and Franke 2002:40–41 and 63–66).

Conceptual Frame

The theoretical and methodological principles on which PRM was initiated can be summarised as follows :

(1) To be sustainable, development has to rely upon environmental resources or renewable natural resources, such as land, water and biomass.
(2) Resource use and intervention for development require knowledge of local level details about the site where the intervention is attempted.
(3) Spatial integration of environmental opportunities and knowledge of the socio-economic realities should begin from the local level.
(4) A micro level database is necessary to assess the ground realities. These data can be synthesised at various levels and in the process a hierarchical data structure for different planning requirements can be built up, which is essential, but has been missing in past development efforts.
(5) People can effectively participate in the planning process, only when they participate in gathering data, analysing the data, developing a plan and executing that plan.
(6) Traditional knowledge – sometimes referred to as Indigenous Technical Knowledge or ITK (Franke 1995) – is not being adequately mobilised for planning. It needs to be reinvented, upgraded and integrated into planning through blending modern science with technological inputs. Active participation of local people will facilitate this process.
(7) Empowerment of people is possible by providing them with factual information and drawing them into the process of information generation, management and utilisation.
(8) Participation of local people in planning will strengthen democratisation of the society.
(9) Action at the ground level has to be taken based on ground level data. Information at the higher levels is of limited use for ground level decisions.

It should be noted that participation has become a common buzzword among government agencies and NGOs. Buzzword participation is often passive participation which is no participation at all. The participation envisaged here is rather different. Table 8.1 provides three perspectives and brings out the unique qualities of PRM.

Table 8.1 : Three Approaches to Planning

Item	*Conventional Approach*	*Traditional Participatory Approach*	*Panchayat Resource Mapping (PRM) Approach in Kerala*
Movement of the Ideas	From government department to the people	From non governmental organization or government department to the people	From Science and Technology Organisations and NGOs to the people and back to the organisations.
Awareness Campaign	Routine	Routine or through a field officer or local leader	An integral part of the programme
Surveys	Department or organisation to the people	Organisation and people	Trained people and experts
Project Formulation	Department	Organisation	Local people and experts
Project Design	Department	Department	Trained resource persons and experts
Project Execution	Contractor	Contractor but local body might contribute	People's organisation and panchayat
Maintenance	No follow up	Left to local people without skills	Local people with acquired skills
Benefit Goes to	Contractor (money)	Local leader or organisation (name and fame)	People
Goal	Fulfilling target	Fulfilling target	Empowerment
Results	Dependence on government	Dependence on organisation	Self-reliance

Source : Chattopadhyay 2003

Implementation

The initial approach paper to the mapping programme (CESS 1991) dealt elaborately with the modalities to implement PRM. During the early stages CESS and KSSP jointly identified the panchayats with the best potential to succeed at mapping. In later stages the panchayats took the initiative in requesting PRM by informing CESS. In addition to KSSP, the Centre of Science and Technology for Rural Development (COSTFORD), *Desiya* (National) *Sasthra* (Science) *Vedi* (Platform) (DSV)—a local voluntary organisation—and the People's Community Organisation (PCO) also participated.

Once the panchayat was selected, the next phase was environment building and selection of volunteers. The various activities required to implement PRM are given in table 8.2. When the panchayat was ready with a minimum of 5 volunteers for each ward technical personnel from CESS visited the panchayat to initiate the training programme. The training included mapping techniques, how to conduct a socio economic survey, and how to replot the details from the field map and diary to a fresh map. The replotted maps were sent to CESS for final drafting by trained experts. During the volunteer mapping and replotting activities, professional earth scientists (geographers and geologists) conducted a land and water resource survey at a scale of 1:12,500 and took note of all the data generated through voluntary mapping mostly on a scale of 1:3960. In some parts of Malabar voluntary mapping was done on a 1:7920 scale. Copies of all the maps were handed over to the panchayat and the scientists explained any technical issues to members of the panchayat and to members of the general public who attended the open meetings to evaluate the significance of the maps. This initiated the process of preparing a panchayat development plan.

Table 8.2 : Activities in Panchayat Resource Mapping

Activities	*Participants*	*Expected Output*
1. Environment building	People, panchayat and NGOs	Congenial situation to introduce the programme
2. Organising volunteers	Panchayat, NGO and technical personnel	Selection of volunteers
3. Training for mapping	Technical personnel and volunteers	Trained volunteers
Resource Mapping		
4. Land use map	Volunteers	Draft land use map
5. Assets map	Volunteers	Draft asset map
6. Land resources map	Technical personnel	Draft landform and surface material map
7. Water resources map	Technical personnel	Draft water resource availability map

(Contd..)

(Contd..)

Activities	*Participants*	*Expected output*
8. Environmental appraisal map	Technical personnel and volunteers	Environmental appraisal map
9. Action plan map	Technical personnel, volunteers, panchayat functionaries, departmental experts and planners	Action plan map
10. Project formulation and implementation	Technical personnel, panchayat functionaries, departmental experts and planners	Projectwise action details
11. Monitoring *	Volunteers and technical personnel	Regular monitored data
12. Panchayat information system (proposed) *	Technical personnel and panchayat functionaries	Information at the panchayat level

* These activities were not taken up because of political developments in Kerala. The Information Kerala Mission (Thomas Isaac and Franke 2002:175–76), a project to install computers in all the villages, proposed to take over these functions, but they have not yet been able to realise their goals.

Implementation of the 12 PRM activities required four basic groups of tasks :

Task 1: Landuse and asset mapping by the volunteers after due training on the 1:3960/1:7920 scale cadastral map (depending upon the available map scale) and collection of details about all water sources including wells, ponds, tubewells, and borewells.

Task 2: Preparation of relief, landform, surface material and water availability maps by the earth scientists on a 1:12,500 scale through field surveys and collection of data at 250m intervals (16 samples per sq km.)

Task 3: Preparation of the environmental appraisal map through collection of the relevant data by the earth scientists.

Task 4: Collection by the volunteers of household level socio economic data.

During the implementation phase, preparation of landform, surface material, water availability and environmental appraisal maps were altered to become based on field investigations. Relief and drainage maps were prepared from topographical maps (prepared by the Survey of India depicting contour lines, drainages and other essential information on a scale of 1:50,000).

Collection of socio-economic data had been a part of the programme in the first phase. However, the Department of Economics and Statistics raised certain objections and socio-economic data collection was subsequently discontinued. Attempts were made to use trained volunteers for collection of land related data covering slope aspect and surface materials in Thrissur district. This could not be achieved at the desired level. In all, 222 panchayats mapped their resources in this project up to 1998, including the entire district of Kollam. By 2004 another five districts had completed the mapping.

At the closing of the PRM Project at CESS in March 1998 an attempt was made to conduct post mapping exercises for selected panchayats and also to make use of PRM data for watershed management, coastal zone management and similar applied problems. The PRM had been originally envisaged to produce a set of eight maps: landuse, assets, landform, surface material, basement configuration, water availability, environmental appraisal and action plan. An additional map on relief was prepared from the Survey of India topographical maps for those panchayats having appreciable relief variations. Table 8.3 provides the types of maps prepared under this programme, information available in these maps and their utility.

Table 8.3 : Resource Maps and Their Utility

Map Number and Name	*Information collected*	*Utility of the map*
1. Landform	Geomorphic units, slope categories and direction of slope, erosion-prone areas	Land capability assessment, action plan for soil conservation, landuse planning, and carrying capacity assessment
2. Basement configuration	Lithology, depth to bedrock and geologic structure	Assessment of the thickness of weathered or unconsolidated materials, assessment of water holding potential, erosion and landslide management
3. Water availability	Perennial and seasonal areas, depth to water table, location of springs, approximate water potability	Identification of potential water sources, development of ponds and springs, guidelines for optimal well depth and location, planning for multiple use of water resources
4. Surface material	Rock outcrop, cover material, soil type, soil texture, soil thickness and depth, soil drainage	Assessment of productivity, soil stability, geohydrologic investigation of availability of minor minerals of (sand, clay, rock)

(Contd....)

(Contd....)

Map Number and Name	*Information collected*	*Utility of the map*
5. Land use	Existing landuse, plotwise details of crops, sites of paddy field reclamation, distribution of waste lands, marshes and wetlands, water bodies	Assessment of existing landuse pattern, computation of area under different landuse categories/crops, formulation of sustainable landuse pattern, planning for landuse consolidation
6. Assets	Ward boundary and administrative divisions, roads, railways, canals, schools, colleges, hospitals, dispensaries, power lines, transformers, markets, ration shops, recreational facilities, post and telegraph/telephone facilities, ponds, springs, wells and other water structures	Assessment of spatial distribution of infrastructural facilities, help preparing estimates for construction and engineering purposes, delineation of gaps and under-developed areas
7. Environmental appraisal	Erosion and landslide proneness, water logging, water deficit areas, waste lands, flood prone areas, extent of salinity infestation, degraded land, reclamation/ land conversion, deforested areas and any other problems	Environmental hazard management, Landuse planning in conformity to natural setting base to initiate discussion at local level for action planning
8. Action plan	All problem areas warranting intervention like ponds to be desilted, drainage requirement, low productivity zone, erosion affected areas	Preparation of action programmes, project preparation, priority assessment and planning

Resource Mapping in Mezhuveli Panchayat : A Case Study

Mezhuveli Panchayat in Pathanamthitta district provides an ideal case study for illustrating the strengths and weaknesses of PRM. Resource mapping in Mezhuveli was taken up in the first phase of the programme during 1990–91, when 25 panchayats were identified for a trial run. Our account is based on a report typical of those that CESS sent to the respective panchayats after completion of the first phase of PRM.

Mezhuveli lies in the midland region at the confluence of the Pamba and Achankovil rivers. The relief of this panchayat is generally moderate varying from 10m to 132m above mean sea level. Malanthevarkkunnu at 132 metres altitude (kunnu means "hill" in Malayalam) located in the western part of the panchayat is the highest point. The southern and southwestern portions of the panchayat and the valley floors are characterised by low relief. The general slope is towards the north. Most of the streams flow in that direction and join the Pamba river. Mezhuveli panchayat covers an area of 14.44 sq km with 14,121 persons according to the 1991 census. In 2001 the total population of the panchayat grew to 15,223 persons.

The first activity was to train 50 volunteers organised by the panchayat and by KSSP. Project scientists supervised the work. Eight maps were prepared. We shall describe maps 1, 4, 5, 6, and 7 and describe in narrative or with tables the information from the other maps.

Map 1. *Landform.* Landform can be broadly classified into 3 groups: laterite crests or mounds, laterite slopes, and valleys. Eight landform units were identified in Mezhuveli. There are two units with lateritic crests (elongated ridge crest and hill crest), five units with lateritic slopes – steep side slopes (greater than18°), moderately steep side slopes (10–18°), moderate side slopes (5–10°), gentle slopes (less than 5°), foot slopes (toe part of the slope above the valley), and one unit of alluvial lowland (valley). A brief description of these units follows.

Lateritic crest. The crest is the topmost part of a topographic sequence, and is composed of a ridge crest and a hill crest. The area covered by this unit in Mezhuveli is 1.17 sq km or 8 per cent of the total area. The slopes are less than 5°. Ridge crests are elongated and broad while hill crests are isolated and limited in extent.

Laterite slopes. The crests are encircled by slopes which gradually incline to the valleys. This intermediate topographic unit can be further divided into 5 mapping units on the basis of degree of inclination as described above. The slope angles range from 5° to 18°. The steep slopes are prone to erosion. The lateritic slopes are retreating, therefore the flat bottom valleys are enriched with loose material. Due to hard crust formation, pediplanation (parallel slope retreat) dominates the process of landscape evolution in the lateritic terrain. The total area covered by the sloping surfaces is about 57 per cent of the panchayat area.

Alluvial lowlands. This unit covers about 4.97 sq km or 34 per cent of the total area. The surface material is alluvium. The majority of the valleys in the panchayat are formed by the tributaries of the Pamba river draining the northern part and the others by the Achankovil river that drains the southern part. The largest valley occupying the central part of the panchayat covering about 2.49 sq km is formed by the Kozhi thodu (thodu means "canal" or "channel" in Malayalam), a tributary of the Pamba river.

Map 4. Surface material. Surface materials of the panchayat can be classified into three groups. Riverine alluvium mostly found in valleys and stream banks covers a little over one third of total area of the panchayat. Based on texture and other characteristics the riverine alluvium as observed in the panchayat can be classified into two types: sandy clay loam and loamy sand. The water holding capacity of sandy clay loam soil is higher than the loamy sand and therefore crop suitability is different in these two soil types. Rice grows well in the land with sandy clay loam soil, while loamy sand is ideal for cultivating vegetables like cucumber, bitter gourd and cow pea. Laterite soils are found along the side slopes. Hard crust laterite surfaces and rock exposures characterise the ridge crests and hill crests.

Availability of surface and subsurface water. At Mezhuveli the volunteers provided the bulk of the information on surface water availability while the earth scientists collected information on subsurface water conditions as the latter activity requires greater technical knowledge.

Surface water. The ponds, streams, canals and springs constitute the surface water source (table 8.4). There are 47 ponds of which 27 are perennial. They are utilised mainly for irrigation. Many ponds were found to be in a state of eutrophication. Excess weed growth or other biological elements caused an excess of dissolved nutrients leading to low levels of oxygen. The ponds have been abandoned.

Table 8.4 : Existing Surface Water Sources in Mezhuveli Panchayat – 1991

Water source	*Number of sources or length in kilometres*	*Notes*
Ponds	47	About 795 persons depend on these ponds, mainly for irrigation.
Streams	15.10	Drain into Pamba and Achankovil rivers
Springs	23	About 1,000 persons depend on these springs
Canal	3.20	Leakage from the canal causes water logging especially in ward 3. This leakage helps recharging of the water table in some areas

Source : Resource mapping by the volunteers.

Of the 47 ponds listed on table 8.4, one is owned by a local Hindu temple, three by the panchayat and 43 ponds are private. Most of the ponds are small, and many are just shallow water pockets in the paddy fields. Due to eutrophication many ponds provide no water for drinking or domestic purposes.

Part of Mezhuveli panchayat falls within the irrigation command area of the Pamba irrigation project. The project recharges the wells, but it creates water logging which has a negative impact on land resource development. Prolonged water logging results in soil salinisation. This has been reported from many places in India and abroad where continuous irrigation has rendered fertile agricultural land totally unusable. Soil salinisation and lack of aeration badly affect agricultural operations.

Subsurface water. In addition to the surface sources, the volunteers identified visible subsurface water. Wells are the major source of drinking water in Mezhuveli. The panchayat has 172 wells per sq km. Around 73 per cent of the households have their own wells and the rest depend on public wells. About 11,000 persons depend on the 2,469 wells for drinking and other domestic purposes. In this panchayat, 356 wells were found to dry up in the peak summer months of April and May. There are 16 borewells, used by some 20 people. Sinking a borewell is costly for individuals and the government sponsored borewells are subject to failure, ill maintenance and are not properly located to cater to the neediest sections.

Potential areas of water availability. Water availability of an area is directly linked to its geomorphic set up. The panchayat was divided into four zones indicating areas of high, moderate and low water potential and dry areas (i.e. no water potential). The information was available in map form. Water availability in high areas is relatively low. The four hills, Kanjirakkunnu, Chandanakkunnu, Malamehevarakkunnu and Ullanarodikkunnu form part of the dry areas. The areas having moderate water potential lie along the slopes and foot slopes. The valley floors usually yield more water. On the crests and slopes the surface runoff is high and infiltration is low.

Map 5. Landuse. The landuse survey was conducted by the volunteers. The general land utilisation pattern of the panchayat is similar to that of any midland region. Agriculture is dominated by cash crops: coconut, arecanut, rubber, mixed tree crops and cashew. Landuse in Mezhuveli can be categorised in four groups: (1) the area planted in seasonal crops, (2) the area planted in tree crops and orchards, (3) the area in plantation crops and (4) wastelands.

Area in seasonal crops. This includes the area under paddy, tapioca, betel leaf, plantains and vegetables. Seasonal crops cover 2.14 sq km or 15 per cent of the total area of the panchayat. Paddy is the dominant crop covering 1.72 sq km followed by tapioca. Most of the area that was previously under paddy is now under plantain, betel leaf and tapioca. Of the 4.97 sq km of valley floor mostly suitable for lowland agriculture, 3.25 sq km has been diverted for settlement, tree crops, road construction and other non-agricultural activities. To assess the performance of agriculture, the productivity of selected crops was compared to the state average (table 8.5). The productivity of paddy was higher while that of plantains and tapioca was lower. Still, the area under paddy cultivation has been declining over the years. The various factors were discussed in chapter 4. Agricultural planning in this panchayat should take note of this high production potential for rice. The analysis show that the mapping can be important for microlevel planning.

Table 8.5 : Area and Productivity of Selected Crops in Mezhuveli Panchayat in 1991 Compared to Kerala State Average Productivity

Sl No.	*Crop*	*Area in hectares*	*Percentage of Total Panchayat Area*	*Productivity in Kg per Hectare in Mezhuveli*	*State Average Productivity*
1.	Paddy	172.40	12.20	2,425	1,973
2.	Rubber	709.00	49.42	1,250	794
3.	Tapioca	36.00	2.65	10,000	19,112
4.	Cashew	19.00	1.32	1,000	899
5.	Plantain	2.00	0.19	2,000	7,592
6.	Coconut	17.00	1.18	*1,750	*5,362

*Coconut productivity is given in numbers of coconuts per hectare.

Area in tree crops and orchards. This category includes coconut, arecanut, and fruit bearing trees such as mango, jackfruit, and non fruit bearing trees. An area of 4.88 sq km or 33.8 per cent of the total geographical area of the panchayat comes under this category. Coconuts cover an area of 17 ha. But the productivity of coconut is very low compared to the state average. The low productivity of coconut is possibly due to lack of irrigation, inferior traditional varieties of trees, pest attacks and crop diseases.

Area in plantation crops. This category includes rubber and cashew plantations. Rubber, the dominant crop in the panchayat, covers an area of 7.09 sq km or 49 per cent of the total area. The slopes and valleys are covered by rubber plantations and the popularity of the crop is due to the high and quick economic returns from it. The higher productivity compared to the state average is one of the incentives for expanding rubber plantations. Cashew is also raised on 19ha.

Open and fallow lands. This class of landuse includes cultivable and uncultivable wastelands. It covers an area of 14 ha. The uncultivable wastelands comprise sacred groves, parks and open spaces in crest areas.

Since settlements are dispersed in the agricultural areas an exact figure could not be calculated. However it is estimated that around 64 ha. of built up area is merged with agricultural landuse.

Map 6. Assets. This map was also prepared by the volunteers. All the infrastructural facilities found in the panchayat are depicted on this map. The spatial distribution of wells and water sources is also noted. Well data have been used for assessing water availability.

Map 7. Environmental appraisal for landuse planning. This is a derivative map prepared by synthesising the information obtained through both the scientific investigation and the voluntary mapping. Integrated mapping units (IMUs) have been identified by combining landform and surface material.

For each IMU, erosion, water availability, existing landuse and desirable landuse and environmental evaluations and suggestions have been worked out. For example, the elongated ridge crest has a problem of erosion along the fringe of the unit; intervention is required for water ´supply; tapioca cultivation should be phased out and groundwater potential needs to be enhanced. These suggestions evolved through interaction of experts and local villagers.

Preparing the Action Plan

Once the resource mapping is completed all the maps are handed over to the panchayats and explained. The panchayats begin the exercise of consulting concerned persons to prepare an action plan for implementation. The Mezhuveli action plan presented below is based on the resource survey and subsequent interaction with the panchayat. The following intervention proposals emerged :

- The areas of Chandanakunnu, Ullanthamodi colony (a

Scheduled Tribe area), Kaniramkunnu and Maladevankunnu experience drinking water shortages. All these places are located on the ridges. Water harvesting could help to solve this problem. Roof water harvesting may be one of the options. Rainwater harvesting is an ancient technique known in many parts of the world. It is now being investigated in a major way to improve the efficiency of water use in a potentially water short world. All rainwater systems, ancient or modern, involve a catchment surface, storage reservoirs, and a delivery system to bring water from the catchment to the reservoirs and from the reservoirs to the points of usage (Gould and Nissen-Peterson 1999:21). Three common types of rainwater catchment systems are rock catchments, surface catchments, and roof catchments. If massive unjointed rock surfaces are found on the site, rock catchments can be created by constructing small holding dams from which a gutter system can deliver the water using gravity to lower points of storage or usage. Rock catchments are the cheapest and most easily designed and built of rainwater catchments. The effects on the local hydrological cycle and on local ecological relations need to be investigated before construction. South Indian tanks are often rock catchments. PRM is an ideal device for conducting the research on the hydrological cycle. A second type of catchment uses surface runoff guided to a canal linked to an irrigation system. Many ponds in Kerala already effectively constitute systems of surface runoff capture. Roof catchments make up the third typical rainwater catchments. Roof catchments require roofs with impermeable surfaces. Kerala is well endowed for this type of harvesting because a high percentage of roofs are tile or cement, but the SC/ST communities that often lie on the ridge tops are also the poorest and are the last in the community to shift from thatch roofs that are not good water harvesting roofs. This situation includes Mezhuveli. With an impermeable roof a substantial amount of rainwater should be "catchable" for storage in above and/or below ground tanks from which it can be pumped to outlets for uses other than drinking. Some modern systems include filtration and purification setups that even allow for drinking water to be derived from the rain catchments. Captured rainwater can be used at least for washing and irrigation. Rooftop rainwater catchments have minimal effects on other ecological relations since the water that is diverted from rooftops would otherwise go into a drainage system anyway. During the 1996–2001 People's Campaign (described in chapter 9) Chellanam panchayat developed an innovative system of rooftop rainwater harvesting that could probably be adapted to conditions in Mezhuveli.

- Due to water leaking from the aqueduct constructed for the Pamba Irrigation Project, valleys traditionally planted in paddy become waterlogged. Proper drainage facilities need to be provided to improve conditions for raising a third crop of paddy and for halting excessive soil wetness.
- An area of nearly 1 sq km distributed in the steep slopes is susceptible to erosion. Soil conservation measures, such as contour bunding, vegetative screening, and developing a basal cover (such as grass or trees) can be tried.
- Because rubber plantations and mixed tree crops cover more than 90 per cent of the total area of the panchayat, a cover crop should be introduced around the bases of the trees so that heavy rains that pour down the tree trunks cannot cause soil erosion around the bottom of the trunks.
- At present a number of ponds in the panchayat are silted up resulting in reduction of their storage capacity. These ponds need desilting and their catchments should be properly managed. This will not only improve the sources of groundwater, but also facilitate aquaculture.
- There are 23 springs found along valleys and valley heads which need to be better managed as well as tapped for drinking water.

This information is now available at the panchayat for purposes. Once a broad plan is prepared, the next step involves preparation of projects and setting priorities for implementation. Mezhuveli panchayat was one of the panchayats that successfully executed several programmes under the People's Plan Campaign that we shall describe in chapter 9.

Impact of the PRM Programme

As it can be seen from the example of Mezhuveli, Panchayat Resource

Mapping can be a useful planning tool (CESS 2000; Chattopadhyay et al. 2000). Panchayats with resource maps were well placed to work out their plans especially for land and water management. The programme initiated a new culture and new thought processes in the sphere of resource based planning. Common people became aware of the fact that resource mapping is vital for local level planning. In many places the actual users of land participated in the mapping. They could bring in some of their traditional practices while discussing the action plan. This gave them a higher level of confidence, which has long term implications for capacity building. A sense of leadership and of working together for the common good came to characterise the volunteers who participated in the programme.

Across Kerala more than 2000 people were trained in resource mapping and another 10,000 people were indirectly involved in the process (CESS 2000). Political leaders at various levels, officials and voluntary organisations became aware of the spatial elements in planning and the need for micro-level resource mapping and map based resource data. Socio-economic data alone are not sufficient.

The Kerala government responded positively by issuing a government order for resource mapping to be taken up by all the panchayats. Many panchayats made budgetary provisions to conduct resource mapping. During the 1996–2001 People's Plan Campaign – described in chapter 9 – the panchayats used the maps wherever they were available. The water conservation societies in north Kerala particularly in Kannur district, adopted PRM base maps for their water conservation work. The asset maps are used by many departments and panchayats for planning tertiary sector development like establishment of primary or secondary schools, health centres, drinking water facilities, and construction of roads. In many cases the community gained confidence and started mapping additional items on their own. In Kalliasseri panchayat, for example, as we noted earlier, local people mapped the power lines to find the most efficient hookups for each house and thus reduce electrical line loss.

PRM created a database which is essential for immediate plans and also for long term planning. Preparing plans for 3 years, 5 years, and 10 years is required along with each annual plan in India. This is necessary so that each year a certain portion of the total plan is covered; otherwise, annual plans will stand apart as isolated activities. Panchayats have to depend more and more on local environmental resources and concentrate on productivity enhancement. Towards this end, total resource assessment is perhaps the first step.

The PRM methodology evolved in Kerala was used by the Bharat Gyan Vigyan Samithi, a New Delhi based voluntary organisation to implement drinking water and sanitation programmes in West Bengal, Bihar, Orissa and Tamil Nadu under the Rajiv Gandhi National Drinking Water Mission. Madhya Pradesh, Karnataka, Pondicherry and Tripura are utilising this technique for various land and water resource management projects. Two Government of India departments, namely, the Department of Rural Areas and Employment and the Department of Rural Development recommended PRM for conducting resource inventories as part of undertaking watershed management and wasteland development programmes. Mapping has been found to be an important tool for mobilising people (CESS 1991).There are instances in Bihar where interest in literacy among the tribal groups could be generated following resource mapping. Therefore it can pave the way for larger interventions.

Resource mapping does not succeed automatically. As we have tried to show in this chapter, it requires technical support which in turn requires effective coordination of local government, NGOs, and science and technology institutes. Converting PRM to a government programme threatens to dampen initiatives by the ordinary participants. The key elements are a strong political will and mobilised and active organisations at the panchayat level. In Kerala the People's Campaign for Democratic Decentralisation that we describe in the next chapter has strengthened these elements.

9 The Kerala People's Campaign for the Ninth Plan of 1996–2001

A New Surge of Environmental Action

A Remarkable Experiment

Kerala's many movements and experiments in democracy, equality, and participation culminated in a remarkable experiment in local democratic planning. Living up to the constitutional spirit of the 73rd and 74th amendments, and performing the true role of guiding the development process in the state, the State Planning Board (SPB) in Kerala spelt out an innovative course of activities. The brochure issued by the SPB in 1996 (GOK 1996), immediately after its reconstitution following the establishment of the 1996–2001 LDF ministry, stated that:

> 35–40 per cent of the Ninth Plan program will consist of schemes formulated and implemented by the local bodies within the respective areas of their responsibilities. In order to empower the panchayats and municipalities to undertake this task in a scientific and participatory manner the board resolved to organise a people's campaign for Ninth Plan.

To implement this programme, that Bandyopadhyay (1997) termed as innovative, audacious and original, the details were meticulously worked out and acted upon. It was initiated through various phases having clear objectives, involving various actors and bringing about a concrete output (table 9.1). Details pertaining to all these steps have been elaborately discussed in Thomas Isaac and Franke (2001 and 2002). Certain important features of the campaign have contributed to strengthening local democracy and generating the potential to initiate sustainable development.

The Kerala People's Campaign for Democratic Decentralisation constitutes a remarkable radical experiment in democracy. Initially called the People's Campaign for the Ninth Plan—referring to the ninth 5 year plan since Indian Independence—it substantially decentralised the functions of the government bureaucracy; and it decentralised planning while mobilising the energy of hundreds of thousands of activists and volunteers to go beyond what government-funded projects could accomplish.

Table 9.1 : Major Phases of the Kerala People's Campaign for Democratic Decentralization

Phase	*Dates*	*Objective*	*Activities*
1. Grama Sabha/ ward convention	August-September	Identify felt needs of the people; generate participation and interest in the Campaign	Meet at local school or other location; plenary sessions and small group discussions, listing of local problems
2. Development Seminars	October-November	Assessment of resources, development problems and tentative solutions	Collect data at local offices; carry out transect walks; write local development reports; conduct development seminar
3. Task Forces	December	Turn proposals from Development Seminars into formal projects	Meet, discuss, and draft project proposals
4. Village and Municipal Councils	January	Finalise local plan	Prioritise projects; draft and approve local plan
5. Block and District Councils	February	Finalise block and district plans	Analyse local plans; integrate them and develop appropriate complementary projects
6. Volunteer Technical Corps-Expert Committees	February-March	Appraisal and technical corrections to projects	Read, evaluate, suggest improvements to projects. Assist panchayats to make viable projects
7. Release of Funds, Project Implementation, Monitoring, and Evaluation	March-June	Implement and monitor the projects in the local plan	Open accounts, draw checks from treasury, set up beneficiary committees, hire contractors, activate monitoring committees

Source : Thomas Isaac and Franke 2002:4

Note : Dates during the first year of the Campaign, 1996–97, were significantly delayed.

The campaign generated procedures to insure delivery of project funds to the former untouchable Scheduled Caste and Scheduled Tribe populations who previously often did not receive them. It brought about special projects for women and simultaneously encouraged women elected officials and activists to participate more fully and effectively in public life. It created conditions for eliminating or drastically reducing corruption in several areas of government spending. It rationalised procedures for all kinds of projects and programmes. It promoted and supported the spontaneous release of local initiatives in areas such as neighbourhood groups that hold promise for even deeper decentralisation and for strengthening of community and of civil society. The campaign gave training to thousands of local officials and activists in what must be one of the most extensive adult education and empowerment programs conducted anywhere in the world. Most importantly for our study the campaign raised consciousness about environmental issues in development planning in Kerala. It built upon the PRM we discussed in the previous chapter and inspired several local communities to create innovative projects in environmental protection and sustainable resource management. In this chapter we shall give an overview of the campaign. In chapter 10 we shall consider some of its further consequences in the form of sustainable development projects and programmes. Here we shall focus on capacity building, resource management, project formulation, identifying ecozones, and micro-watershed development.

Capacity Building and the Emergence of Grassroots Planners

One of the typical problems encountered in planning is the lack of trained personnel who can work side by side with the people in the village and not sit in the office at the district headquarters or in the state capital. It was realised that there should be a drastic change in what planners do and how planners should be trained. The massive training programme undertaken for implementing the people's campaign initiated the process of building up Grass Root Planners (GRPs) or Local Level Planners (LLPs). During the campaign, training was imparted to one hundred thousand local resource persons (LRPs) and an additional one hundred thousand task force members at the grama panchayat level. As a result, on average there are around 100 LRPs and 120 task force members in each panchayat. This adds up to one trained person for roughly every 150 people in the state. These persons from the panchayat were involved in leading discussions, explaining possibilities and programmes, identifying problems, discovering the aspirations of the panchayat people, reshaping their attitude to development, making the aspirations conform to the potentials and helping them learn how to use their social, political and administrative organisations for development purposes.

The French "Social Animateur" and the Chinese "Barefoot Planners" bear certain resemblances to Kerala's Local Resource Persons. The major difference is that in the French and Chinese cases an outsider is formally trained to work in a village he or she might not know intimately. The Kerala local planners emerge from the local population. This greatly increases the chance of success because they know the local conditions. The GRP/LLPs should have a certain amount of technical knowledge so they can talk to the experts and be able to interpret the experts' ideas to ordinary people. They should have an understanding of the local social structure. There is every possibility for this group to emerge as a mainstay in panchayat level planning. Table 9.2 indicates the topics taken up at all levels of the training programme.

Details of the Training Programme : Kerala 9th Plan Campaign 1996

Training Module Topics

Challenge of development, Philosophy of decentralisation, Local rules and statutes, Role of Grama Sabhas, Participatory Rural Appraisal (PRA) techniques, secondary data collection and analysis, preparation of development reports, project preparation, plan document preparation, plan appraisal, sectoral development perspectives, Implementation of projects, micro-level development models, SC, ST and gender issues, Resource mobilisation.

Training personnel

(a) 600 Key Resource Persons (KRPs) or state level experts got 20 days training,
(b) 15,000 District Resource Persons (DRPs) got 14 days training,
(c) 1,00,000 Local Resource Persons (LRPs) got 3 days training,

(d) 10,000 Volunteer Technical Experts enrolled.

Composition of the trained personnel

Women	24.82 per cent
SC/ST	2.75 per cent
Government Officials	5.47 per cent
Others	66.96 per cent

Note : Composition per cents are overlapping.
Source : Kerala State Planning Board, Thiruvananthapuram 1998; cf. Thomas Isaac and Franke 2002:68–69, 76–77, 164, and 177–79.

Resource Management

At the very beginning of the campaign activities centered on a dialogue at the grama sabha level to bring out the felt needs of the people. More than two million persons attended the first grama sabhas of August-December 1996. Six months later in the second round, 1.8 million participated and 2.1 million attended the third round in September 1997. These figures represent about 11 per cent of the voting age population and perhaps 20–30 per cent of all households. Female participation ranged from 26 per cent to 38 per cent while Dalits ranged from 9 per cent to 18 per cent. (About 51 per cent of Kerala's population is female while 11 per cent are Dalits.) The assemblies were designed to maximise participation: speeches by officials were supposed to be limited and the large plenary meetings of 150 to over 800 persons were broken into smaller subject groups that became the basis for the data collection and planning of later stages.

Trained facilitators attempted to push the small group discussions beyond immediate felt needs by encouraging participants to think socially, historically, and environmentally and not just in terms of prices and incomes. In the discussions about rice farming, for example, facilitators who got statements about low market prices and high wages as barriers to investment, tried to get ideas about irrigation, soil fertility, waterlogging, and other elements of the local scene that might also play roles in the low returns to farmers. Objective assessments of resources by the people with guidance from trained resource personnel were intended to help in linking the problems to local resource availability. Following the grama sabhas, teams of volunteers collected secondary data, organised transect walks or Participatory Rural

Appraisal (PRA) campaigns to promote the perception of ecozones as operational planning units. Further activities included understanding local history and geography, reviewing ongoing schemes, and consolidation of grama sabha reports at the panchayat level. All these activities served to create a local level database, and helped build confidence for participation in a new development culture. This had a tremendous psychological impact on the local populace. Building up the data sets from the grama sabha level with required consolidation at the panchayat level set in motion the process of information driven planning at the lowest level. The seeds of spatial planning were sown in these actions. Local democracy was given new dimensions.

Box 9.1 : What Is a Transect Walk?

Although the People's Campaign activists desired to carry out PRM in all the villages of Kerala (see chapter 8), the rush of campaign events and requirements rendered it impossible. Instead, they chose to adapt an element of the technique called "Rapid Rural Appraisal." This technique is called a transect walk. Using any large-scale outline map of the panchayat, the volunteers, sometimes with the help of experts, would choose a single more or less straight path that would take them through the largest possible number of ecological zones. If necessary, a second transect could be added. The teams would walk this line, making observations, notes on the maps, and stopping to interview residents or workers along the line. In addition to landform features, activists were instructed to note soil and water resources and their utilisation, cropping patterns, productivity, and environmental problems. Campaign organisers estimate that about 75 per cent of panchayats carried out the transect walk in the first weeks following the grama sabhas of September to December 1996 (Thomas Isaac and Franke 2001:82–84).

The development reports prepared by all the panchayats and municipalities cover a wide range of issues. The reports were organised into two major sections. The first part is a general appraisal of the panchayat and the second part presents various development sectors in 12 chapters. Both problems and possible solutions are discussed. Task forces for various development sectors were identified. For every panchayat and municipality in Kerala a development report is available today.

Project Formulation

The Task Forces formed in the second phase of the programme were supposed to prepare projects for consideration of the elected panchayat councils or municipal councils. There were 12 task forces in each local body to look after all the development sectors. The sectors for those projects were prepared are :

(1) Agriculture and Irrigation
(2) Animal Husbandry, Social Forestry and Fisheries
(3) Industry
(4) Energy and Housing
(5) Transport and Tourism
(6) Education
(7) Health
(8) Drinking Water and Sanitation
(9) Women and Development
(10) Welfare of SC/ST
(11) Culture
(12) Resource Mobilisation

Because project preparation is a technical job, the participation of officials, locally available and technically qualified people and subject experts were mobilised through various measures. The project reports were prepared following a common format which includes introduction, objectives, activities, organisation, financial analysis, identification of beneficiaries, achievements and monitoring mechanisms. Having a task force for each sector and a common format for each project gives equal importance to all aspects at the initial phase. Which sector needs priority and what are the strengths of the panchayats can be objectively assessed only after assessing each sector. Intersectoral coordination, although not separately promoted in the campaign, emerged during the discussions at the panchayat level.

Plan Finalisation and Formulation of a Development Strategy

With the completion of the third phase of the campaign each grama panchayat and municipality had a shelf of projects corresponding to the development problems identified by the people at the grama sabhas. Altogether more than one hundred thousand projects emerged in the process. These projects were then prioritised and those deemed most important were incorporated in the final plans of the panchayats. At this point each panchayat had to clearly spell out its local development strategy on the basis of problems identified, resource potential, priorities, mechanisms for implementation and capacity to mobilise financial resources for project implementation. A monitoring mechanism for each selected project was also decided on. The plan document prepared by the panchayat comprised eight chapters : introduction, development strategy, resource mobilisation, sectoral programmmes, integrated development, welfare of scheduled castes and scheduled tribes, women's development and monitoring.

Plan Integration, Evaluation and the Voluntary Technical Corps

The plans prepared by the panchayats were supposed to be integrated at the block and district levels. This integration was supposed to be organised in such a way that the plans of the higher tiers complemented those of the lower tiers. Integration was also planned for the ongoing centrally sponsored schemes implemented through the community development blocks.

At this point, it was realised that the District Planning Committees did not have sufficient technical staff to undertake proper scrutiny of the vast number of projects that emerged from the grama panchayats. People in the panchayats had difficulties in understanding the technical issues in project formulation, and especially in budgeting, comprehending engineering drafts and formulas, and in presenting statistical tables required by the state government. To deal with this problem, campaign organisers came up with the idea of appealing to Kerala's many retired but active experts. The Volunteer Technical Corps (VTC) was born in 1997 through newspaper advertisements that led to almost 10,000 applications. Eventually 4,000 volunteers were mobilised who gave at least one day per week of free time to help evaluate and improve the project proposals. The VTC later merged with other expert groups to become the Expert Committees that helped bypass contractors and reduce corruption. The Voluntary Technical Corps were expected to provide the expertise required at the Block or Corporation or District level. The VTC had no right to change the priorities set by the local councils. Their function was confined to technical and financial appraisals of the projects and providing suggestions for modifications so that the projects could be made viable. Table 9.3 shows the contrasting levels of education between the Voluntary Technical Corps members and the Key Resource Persons who were the highest level of trainers.

Table 9.3 : Per cent of Voluntary Technical Corps (VTC) and Key Resource Persons (KRPs) at Various Educational Levels

Educational Level	*Per cent of VTC*	*Per cent of KRP*
Below degree	7.88	24.16
Degree, P.G., B.Ed., M.Ed.	43.95	64.17
M.Phil and Ph.D.	5.95	0.16
Professional Degree Holders	41.30	11.03
Other	0.92	0.48

Source : Kerala State Planning Board 1998

Identifying Ecozones

As we argued in earlier chapters, identifying ecozones at the panchayat level is a first step to initiating local natural resource based plans. Each ecozone has its own resource potential. Crop suitability and agricultural productivity depend on the characteristics of ecozones. Complete ecozone identification could not be fully realised due to the lack of resource maps covering land, water and biomass for many grama panchayats. At the time of the campaign such maps were available for some 100 of Kerala's 991 panchayats. In place of the maps, transect walks were used as described earlier to gather some initial data and to encourage local people to see that they could identify local problems through participatory methods. Terrain and watershed maps at the block level had not been integrated. This is one of the tasks to be attempted in the future. However, panchayat people did become aware of the importance of ecozonation and the interlinked character of land, water and landuse through the transect walks. Realising the necessity of resource maps for panchayat level development activists and officials identified Panchayat Resource Mapping as one of the major activities required to be given a panchayat budget allocation. By 1998 222 panchayats had been mapped including the entire district of Kollam. The programme was subsequently transferred to the Kerala State Landuse Board from CESS. Another five districts have been covered as of 2004.

Project Implementation

The first result of the campaign was that project implementation at the panchayat level took place and monitoring mechanisms were developed. Plan guidelines for 17 sectors were prepared by the experts for the use of local bodies.

In four instalments per year, the State Planning Board released funds for the impllementation and monitoring of the projects. In the first year of the campaign more than 1,00,000 projects were proposed and 67,766 projects or about 56 projects per local governmental unit, were selected and implemented.

The People's Campaign and Sustainable Development

The people's campaign generated substantial action on behalf of environmental rehabilitation and sustainable development. Agriculture and watershed management received considerable attention. Ecology restoration programmes were initiated as a part of watershed management and agricultural development.

R. Rukmani (2004) has described an ecorestoration programme in Mulamkunnathukavu panchayat in Thrissur district. The study covers three micro watersheds – Malawi, Theyyathankundu and Wadakkechira with areas of 456 ha, 288 ha, and 200 ha respectively. The programme was carried out in participatory mode under the people's plan campaign. A coaliton of stakeholders including community based organisations, the government bureaucracy and elected representatives of the local council helped shape and implement the programme.

Major parts of the catchments of the Malawai and Theyyathankundu micro watersheds had been under thick forest until the 1960s. Streams had been perennial. Pepper cultivation was introduced in the 1940s. By the 1980s the Rubber Board started promoting rubber cultivation in this area. Paddy, which used to be raised in the low lying valleys, was completely abandoned by the 1990s. Moderate to severe soil erosion was reported from all three micro watersheds. The Soil Conservation Department constructed Puerto Rican type stone pitched contour bunds during the 1970s. Local people had taken an interest in contour bunding since 1977 and extended the construction on their own. With the introduction of the people's campaign in 1996 micro watershed based environmental restoration was given a new emphasis in this area. Agronomic, structural and biological measures were envisaged. The aim was to make optimal use of rainfall, which requires proper development of all types of land, agriculture, grasslands, forest lands and uncultivable land as an integral part of the system.

The interventions had to cover both individual property and common property and had an area as well as a drainage line dimension. Area treatments include continuous or staggered trenches, contour bunds, and terracing. Drainage line treatments include check dams and gully plugging. The techniques used are bunds, terracing, trenching, better tillage practices, mulching, and reforestation.

Interventions on private property were undertaken by the owners. Neighbourhood group discussions and grama sabhas were used as platforms to encourage participants to select, design and carry out the interventions on the concerned holding. Common property treatments were reforestation, plant nurseries and environmental awareness. Altogether 4,500 bare root (without soil cover) seedlings and 5,000 packeted seedlings were produced. Plantings were done in catchments of reservoirs, on school premises, near forest fringes, on canal bunds, on roadsides and on common lands.

Participation varied from area to area depending on land scarcity, degree of dependence on agriculture, amount of property, amount of irrigation, and political loyalties. Project effectiveness was also correlated with agricultural practices such as amount of tillage, stoppage of clean cultivation, and introducing zero burning. Earlier all unwanted biomass had been removed to clean the field before introducing a crop. This was known as clean cultivation. In some areas, unwanted biomass was burnt to clear the field. Ending the practice of clean cultivation and introducing zero burning (no burning of unwanted biomass) are part of the new practices leading to organic farming in which all the biomass is ploughed under to form mulch as a part of field preparation.

Rukmani's study shows that conscientisation is the single most important prerequisite for participatory environmental restoration activities; only through constant dialogue and communication could the concept of sustainable land husbandry diffuse into the minds of the inhabitants. The grama sabhas, and neighborhood group meetings were crucial for promoting participation.

The interventions resulted in better water and soil management, more appropriate crop intensity, less invasive tillage practices, and increased green manure use. There are perceptible differences between the treated and untreated holdings. Rukmani found that on the fields of participants mean crop intensity increased in all three micro watersheds by 29 per cent, 43 per cent and 44 per cent respectively. In the case of dropouts and non-participants the increase of crop intensity was nil or very low (table 9.4).

Table 9.4 : Changes in Crop Intensity after Watershed Interventions

Watershed	*Participants*	*Dropouts*	*Non Participants*
Wadakkechira	+29%	0	0
Malawai	+43%	-16%	+23%
Theyyathankundu	+44%	-8%	+1%

Source : Rukmani 2004:55

Participants were able to reduce fertiliser consumption substantially while non-participants recorded increases in fertiliser consumption (table 9.5).

Table 9.5 : Changes in Fertiliser Consumption in Kilograms per Hectare after Watershed Interventions

Watershed	*Participants*	*Dropouts*	*Non Participants*
Wadakkechira	-69	-6	0
Malawai	-75	-12	+1
Theyyathankundu	-74	-13	+29

Source : Rukmani, 2004: 56

Participants also increased consumption of green leaf manure in their fields, while there was little or no change for non participants and for dropouts. Eighty-three percent of participants reported an increase in soil moisture; 48 per cent felt there was a rise in the water table and 94 per cent thought that soil erosion had decreased. Detrimental runoff declined in the opinion of 92 per cent of respondents while 65 per cent felt they now had achieved resistance to drought. Water availability increased in local wells and water consumption increased. Rubber yields went up and tapping times increased. Treated plots recorded yields of 18 kg per tapping with 120 days of tapping against 8–12 kg per tapping with 90-100 days of tapping in the untreated plots.

Rukmani's study seems to show that by carrying out an environmental restoration project as part of the people's plan campaign, Mulamkunnathukavu panchayat experienced long term improvements in the capacity of its watersheds to provide drinking water, agricultural renewal, and better cost benefit ratios for farmers.

Significance of the People's Campaign

The panchayat-level planning experiment in Kerala is the latest in a history of people's actions that have occurred over many decades. The people's campaign is primarily policy-driven compared to the usual individual programme-driven development initiatives. Sustainability calls for policy level interventions rather than the usual practice of developing individual projects.

Plans were drawn up using databases at the grama panchayat level. This was perhaps the first time in India that such an attempt succeeded across an entire state. The Kerala people's campaign initiated a new development culture to tap the local resource base for solutions to local problems. People's participation in enhancing productivity in agriculture, fishing and forestry will help in lowering the energy intensity of crop production and animal husbandry. As McKibben (1995) has pointed out, humans need to live more "lightly" on the earth – using fewer resources and creating less waste.

At the community level the campaign brings back camaraderie and neighbourhood solidarity. Due to the polarisation of people along political lines, age-old village unity had previously been undermined, resulting in conflict and tensions. The concept of a development society at the panchayat level with neighborhood groups to supplement it sets a new agenda and brings the people back towards a common agenda. It offers a new challenge for providing more detailed science and technology inputs for land, water and biomass resources.

Despite the many advantages of the people's campaign, a number of the development projects envisioned by individuals or NGOs did not produce the desired results or could not be upscaled. Successful development projects nurtured by well intentioned Non Governmental Organisations (NGOs) could not be replicated. When a panchayat takes up a development programme it becomes part of their activities. Some panchayats may function well and some may not live up to the mark, however it is a democratic institution over which people have eventual control through elections. NGOs are often not democratically controlled. A well meaning NGO could help in development but it cannot be viewed as a replacement for the panchayat. Considering that the overwhelming majority of the people live in the rural areas of Kerala, people's planning and local empowerment could create a critical mass that may neutralise the forces of consumerism and individualism that have been unleashed in recent years (Sinha and Ghosh 1997).

In the process of democratisation, the initiatives of the people's campaign may have fostered empowerment of the people and a reconnection to their environmental resource base. The campaign has broadened the understanding of local environmental problems. The Kerala experiment in democratic decentralisation may have larger socio-political dimensions in that it can provide important lessons in pursuing the path of sustainable development. In the next chapter we shall analyse some of the local level initiatives that occurred in the buildup to the campaign and some that came as consequences of it.

10 Local Initiatives
The Potential of Local Democracy and Participation for Sustainable Development

From People's Campaign to Local Sustainability

Kerala's long history of caste and class activism has combined with the stimulus of the People's Campaign for Democratic Decentralisation to foster a new generation of local participatory initiatives. Many of these initiatives are aimed at promoting environmental reconstruction and sustainable development practices. This chapter will describe selected programmes undertaken by the district and grama panchayats for environmental resource management.

People's Biodiversity Registers: The Ernakulam District Campaign

Among the most important local actions for sustainable development are the campaigns to list and protect local resources whose deterioration might not be obvious in daily life. As we have noted in chapter 1, biodiversity refers to the variety and abundance of life forms. It represents the total number of species found in an area. Biodiversity studies have assumed considerable importance over the last two decades due to strides in molecular biology enabling the transfer of genes from one organism to another. The question of intellectual property rights allowing the patenting of living organisms and their products has generated further interest in documenting the biodiversity of a locale. Given the importance of biodiversity and its potential for furthering development, it is of paramount importance to identify the species in a biological community and to classify them into the three categories of rare, endangered and endemic. This monumental task can best be accomplished with the participation of the local community as in many places local residents are the custodians of significant amounts of traditional knowledge of their plant and animal communities. One site for biodiversity research in Kerala is the *kavu*, or sacred grove, where small Hindu shrines are kept amidst ancient plots of plant and animal life. The national Indian Centre for Science and Environment estimates there are 2,000 such groves in Kerala. Data for 761 of them indicate 722 distinct species of flowers. The long established root networks of the ancient trees help in water absorption while the snake population helps to control rodents (*Down To Earth* 2003). The Convention on Biological Diversity developed at the Earth Summit in 1992 accepted the sovereign rights of countries over their biodiversity resources, acknowledged the need to respect the knowledge of sustainable use of living resources by local communities and supported the need to share with them the benefits of the commercial application of such knowledge. Such local community knowledge needs therefore to be documented, not anonymously as in the traditional Western biological literature, but with specific acknowledgement of the local persons or communities involved.

The People's Biodiversity Register (PBR) is a programme for collecting information available in the domain of folk science on the status, uses and management of local biological resources, which are normally guarded as secrets and almost always orally transmitted across the generations. Such information may disappear when people lose control over the local resource base or when their livelihood activities shift away from the local environmental resources. The PBR of Ernakulam district represents a novel attempt to create a biodiversity database at the panchayat level with local people's participation (Ernakulam 1999).

The programme consists of creating a local database on :

- Resources of the locality including medicinal plants, cultivated plants and their wild relatives, vegetables, fruit trees, ornamental plants, cash crops, timber trees, fodder crops, domesticated animals, animals of medicinal importance, fish varieties and pests;
- Human management of plants such as harvests and changes in other agricultural practices;
- Widely shared knowledge of properties and uses of biodiversity resources, e.g., use of certain medicinal plants in the treatment of human or animal ailments;

- Potential local knowledge such as a claim that a particular medicine-man/woman knows a cure, for example, for snakebite;
- Volume of trade and prices of economically important resources.

The programme envisions making use of the local database to:

- activate the perceptions of local people on participatory management options;
- create a network of decentralised databases giving full credit to the origin of information at the individual or community levels; and
- protect local biodiversity and local knowledge from being privatised by commercial interests who might file patents on modified products, processes, and bio-materials that are built upon local resources and knowledge.

The Ernakulam district panchayat – the tier of the local government above the village and block – carried out this programme as part of the people's plan campaign described in chapter 9 during 1997–99 using a standard methodology from the People's Biodiversity Programme of the Centre for Ecological Sciences of the Indian Institute of Science in Bangalore.

The field investigations at the panchayat level unfolded as follows :

(1) Building rapport between the local volunteers and the subject experts;
(2) Identifying biodiversity user groups along with the key resources to be investigated;
(3) Identifying and training local resource persons and volunteers;
(4) Identifying village elders as sources of information;
(5) Interviewing members of user groups;
(6) Mapping the landscape of the study area;
(7) Discussions in the grama sabhas;
(8) Field visits for truth testing of medicinal claims where appropriate;
(9) Discussions with traders, marketing people, Ayurvedic doctors and others likely to be involved with local biodiversity, and concerned about resource use at the study site;

(10) Documenting the knowledge and perceptions of the local community in a list or compendium kept at a local library but also made available to a wider scientific audience where feasible.

The programme envisaged training students and unemployed youths in the techniques of questionnaire research. The questionnaire was administered to 10,000 people altogether in 86 panchayats covering 885 wards. Items covered were food crops, oil seeds, cash crops, tuber crops, vegetables, pulses, fruit crops, medicinal plants, animals of medicinal use, spices, fumigants, chewable plants, fodder crops, plants collected from the forest, timber trees, fruit trees, wild relatives of cultivars, ornamental plants, domesticated animals, birds, fish, pests, animals attacking crops and plants, weeds, ethnic groups, knowledgeable persons as sources of information, and traditional knowledge. Location details, use, variety, present state and similar information were collected for each of these items. The information was then tabulated under 23 heads and recorded on data sheets that were bound into books. These books contain information on the history, geography, demography, and ecology of the villages involved.

Thus there are 86 Bio-Diversity Registers (BDRs), one for each of the 86 grama panchayats that are kept in the respective panchayats. The district panchayat office in Ernakulam has a complete set of copies of these 86 BDRs. These documents are available to anyone who is interested both at the grama panchayats and the district panchayat. Although some action plans were drawn up based on these reports they were not mostly executed. Recently an effort has been taken under the leadership of M. K. Prasad of the Kerala People's Science Movement (KSSP) to consolidate the reports and bring out a comprehensive district level BDR.

The knowledge base of the local people – Indigenous Technical Knowledge or ITK – was found to cover the names and uses of 656 important living organisms (Ernakulam 1999). The inventory contains 177 species of medicinal plants, including seven plants being used in the treatment of snake bites. Traditional experts use a single herb, *ottamooly*, to treat diabetes, jaundice, and migraine headaches. Medicinal plants are also used for pediatric ailments. Many local experts refused to show the plants or discuss their expertise. The medical knowledge is usually kept secret and is transmitted from generation to generation orally. The biodiversity inventory also registered 2 food grains, 3 oil seeds, 4 cash crops, 7 tubers, 25 vegetables, 11 pulses, 33 grasses, 50 ornamental plants, 22 fruit crops, 23 spices and essential oil plants, 82 timber or fruit trees, 16 wild relatives of cultivars, 24 fumigants, 2 chewing plants, 23 birds, 41 types of animals (snakes except python are not reported), 82 species of fish, 10 types of pests, and 19 weeds. Paddy and Ragi are two food grains documented in this district. 66 varieties of paddy had been under cultivation. Fifty-two are traditional varieties; however, cultivation of many of these traditional varieties has been discontinued. Among the 41 animals, 17 animals are useful for medicine. Out of 64 animals and birds 28 are domesticated. There are 23 ayurvedic physicians and another 45 persons who claim to be able to cure snake bites. The use of plants for treating snake bite (*visha chikilsa*: visha=poison, chikilsa=treatment) is one of the areas having a number of local experts. As many as 521 farmers are conversant with traditional farming operations. Out of 177 medicinal plants there are 19 plants in the threatened and rare group.

The study indicated that the households that earn their livelihood by traditional occupations like the potters, weavers, basket makers, mat weavers, and blacksmiths are fast disappearing. Mat weavers used to depend on locally available biomass including bamboo grown around wetlands and river banks. Tall grasses of some varieties were used for manufacturing of certain types of brooms and ropes. People engaged in these vocations find it difficult to survive. The lands where these vegetations grew have been diverted for other uses. The biodiversity registers helped people realise that biodiversity is a function of landscape diversity. Biogeography is an emerging field to study these relationships. Mapping and descriptions of the landscapes in terms of vegetation cover using the terms of landscape ecology appropriate to the local people helped understand the patterns of use and management of the different landscape elements. The strategies for conservation and sustainable use of the biological resources of the local communities are spelt out on the basis of this. The history of the user groups, the landscapes and the biological resources of the study area were also recorded for understanding the current status, use patterns and management options for the future. Rejuvenation of sacred groves by planting indigenous varieties and enhancement of homestead biodiversity are two immediate outcomes envisaged from the PBR project.

The PBR contains information on the uses of bio-resources for cash incomes; traditional protection measures for sacred groves, self regulated fishing strategies and the development aspirations of the local people. Sustainability options include prioritisation by the local people for conservation of different elements of various landscape elements and of the various species of plants and animals along with documentation of the strategies they adopted in the past.

Other PBRs in Kerala

In northern Kerala, a biodiversity study of Kalliasseri panchayat identified 106 types of trees, 69 types of shrubs, 162 medicinal plants (four used to cure snake bites), 103 ornamental plants, 397 species of flora, 27 types of climbers and creepers, 10 varieties of mangroves, and nine varieties of mangrove associates (Sreedharan 2004). Kalliasseri panchayat has 20 rare plants. M. Jayarajan (2004) identified 578 sacred groves in Northern Malabar. An intensive study of five groves distributed across different ecological zones found dozens of rare indigenous plant and animal species. An inventory of sacred groves in the Mararikulam area of central Kerala was also done. We shall discuss elements of the Mararikulam BDR towards the end of this chapter.

Pond Management and Irrigation in Thathamangalam

Local sharing of natural resources is a traditional practice in many tropical countries. Although defunct in some areas with the advent of modernisation, there are still local initiatives to revive or reinforce the traditional practices. Kerala is noteworthy for such initiatives. Here we recount two resource sharing experiments: pond management and sharing of the fish catch.

As we have noted in chapter 2, ponds occupy unique positions in the rural ecosystems of Kerala. Traditionally ponds have been used to meet the local demand for both domestic and irrigation water. Due to prolonged negligence ponds have fallen into various stages of dereliction. Kerala has up to 995 large ponds each with a water retention capacity of more than 1,500 cubic metres. On average a panchayat in Kerala has 15 to 16 ponds. Palakkad is one of the districts in Kerala where there are a large number of ponds with sizable capacity. It is also a drought prone district – as was shown in the discussion of the Coca-Cola plant in chapter 7. There are five irrigation projects in this district to cater to the needs of agriculture. However, they have not yielded the desired results and quite often crops whither due to lack of a timely supply of water. The introduction of canal irrigation systems has had an adverse impact on pond management. Most of the ponds have become silted in and thus rendered unusable. Some ponds may have been filled to accommodate other activities.

During the past two decades, Kerala's farmers have developed initiatives to revive the traditional systems and to strive for decentralised practices. Farmers of the Parithikkavu Nellulpadhaka Padasekhara Samithy near Thathamangalam in Chittur taluk composed of 102 members succeeded in maintaining four ponds of one to four acres (0.4 to 1.6 ha) that supply irrigation water. Their actions saved their paddy cultivation on 75 hectares during a severe drought in the early part of 2004 when the second paddy crop was lost in most parts of the district for want of irrigation water.

Founded around 20 years ago, the Padasekhara Samithy took care to conserve traditional ponds and wetlands. Utilising pond water they could convert single cropped areas to double cropped rice lands making rice cultivation profitable. Non-profitability is one of the underlying reasons farmers divert rice fields to other uses in Kerala. Membership of the Samithy grew from the 25 founders to 102 farmers and cultivation spread over 75 hectares of paddy fields. Observing the success more farmers joined the Samithy and now an additional 280 hectares of the nearby paddy fields have come under its fold, raising the total operational area to 355 ha. Both the state and central governments recognised this pioneering activity with awards.

Recently, the Samithy was selected as one of the eight farmer's organisations in the country by the Union Ministry of Agriculture as part of its 'Integrating Front Line Extension System' in the country for meaningful transfer of technology to farmers. It was given a grant of Rs. 4 lakhs by the Central Government for the purpose (*The Hindu*, March 7, 2004).

The Samithy followed the group farming approach. Group farming was encouraged in Kerala since the 1980s to enhance productivity of individual crops. Due to fragmented landholdings individual farmers found it difficult to enhance agricultural productivity to the desired level. Apart from cost escalations individual operations face various hurdles related to land and water management. For example a farmer in the downslope will face problems if the upstream farmers neglect soil conservation. Similarly the flow of irrigation water from the valley head downstream needs cooperation among the concerned farmers. Group farming is a voluntary collective operation for crop husbandry where individuals share common agricultural practices while maintaining private ownership. GALASA (Group Approach for Locally Adapted and Sustainable Agriculture), a KSSP sponsored programme is one such successful attempt (Pathiyoor et al. 1990; Thomas Isaac and Franke 2002:32–33).

Setting up a community plant nursery (sometimes called a "seed bed") in the Padasekharam (a group of land parcels) reduced the farmers' losses. Earlier, each farmer would have a private rice plant nursery, which remained fallow once the seedlings were transplanted onto the rice fields. For small landholders to leave this area fallow was a significant loss. The Samithy also updated soil maps of the area through regular soil testing. By introducing high-yielding rice varieties such as *Kairali* and *Kanchanan* as the second crop in place of local varieties such as *Thavalakkannan, Chenkazhama,* and *Anakkomban*, the average yield increased from 1,750 kg per ha to 5,000 kg per hectare. Consequently, the profits have increased contributing to improved standards of living according to the secretary of the Samithy. Apart from acting as a coordinator for 16 farm organisations in the area for developing and implementing farming activities the Samithy started an agricultural club in the local school, giving emphasis to vegetable cultivation. It also established a library on agriculture. A "single window system" is maintained for the members of the Samithy for receiving subsidies, gifts, loans and credits made available to the villages by development agencies. The single window system refers to a single place from where farmers can draw all agriculture related services. This has made the entire operation hassle free for the farmers. They need not run to different places for various services.

The experience of the Samithy shows that local initiatives and decentralised resource planning can have positive effects on production, profits, the standard of living, and sustainability. Given the dependence of society on agriculture, the fact of micro-level variations of agro-climate and the importance of crop management at the farm level, the lessons learned from this experiment include maintenance of local water resources (ponds) and provisioning agro-services through the single window system. These lessons can help in moving towards sustainability, as it is increasingly realised that sustainable development requires micro-level interventions in enhancing the local resource base and providing easy access to resources for the users of the environment.

The Karanila Traditional Fish Sharing System

Karanila (*kara* = shore; *nila*= share) is a local institution created by traditional fishermen to share their catch with the fishermen who stay back onshore for whatever reasons. It is basically a group of fishermen using common fishing gear consisting of a fishing net, boat and accessories. All the households attached to the karanila, even if some of the fishermen could not go to the sea on a particular day, share the catch. Traditionally all fishermen are attached to a karanila. The central maritime districts of Kerala—Alappuzha and Ernakulam—are well known for Karanila practices (Kurien and Vijayan 1995).

The encircling net is called a *thanguvala* (thangu = a local dialect term to connote encircling; vala = net) and its craft the *thanguvallam* (vallam = canoe or open boat). Motorisation permitted increasing the size of the net and fitting rings to the top and bottom, making it function like a purse-seine and thus bringing in larger catches. This new modified net is known as a ring-seine. Fishermen involved in this operation are known locally as *thanguvallam* units. Prior to motorisation these units had crews of about 15 persons. Now the large motorised units can accommodate crews of 35 to 40. Working and non-working crews are attached to a unit. The non-working crews are also entitled to a share. There are also temporary crews who work for up to a year but do not have ownership rights. The fish catch is still divided through the traditional sharing system. This system ensures an adequate supply of labour for the fluctuating needs of the fishery and it maintains a system of income sharing. The socioeconomic and cultural logic of this karanila system is the age-old belief that the resources of the coastal waters are the community's heritage (Kurien and Vijayan 1995). The karanila can teach some lessons with long term implications in natural resource management and sustainability :

- Local household consumption of fish gets first priority as all people involved in a karanila system, whether for fishing at sea or helping in shore receive a portion for household consumption. This automatically takes care of the nutritional needs of the larger community.
- The mechanism of onshore resource sharing can possibly be extended to the sea, minimising the number of trips and reducing competition. This will have a significant and positive impact on the economy and on energy and resource conservation.

- At present the catch is shared by all people involved in a karanila group. But competition at sea has become fierce and there is no control on the number of trips. As fishing goes on more or less within the same area and number of fishing boats increases the catch per trip has significantly come down. Many trips turn out to be futile. This escalates the economic burden as fishing trips are quite costly now. If the sharing system under karanila is extended farther out to sea, the trips could be rotated and in a given area all the units would get an equal opportunity to fish. The overall cost per trip would come down and an element of equitability would be brought into the operation.
- This karanila system has the potential to contribute to general fisheries management Kerala. It has already facilitated the creation of powerful fishworkers organisations in the state. Ernakulam and Alappuzha districts have the largest trade union of artisan fishworkers in the state. The karanila can possibly be extended to other locations. This is necessary because state-sponsored increases in fishing assets and high capital flows are threatening to break down the traditional institutions leading to overfishing and consequent depletion of fish stocks.

Preparing a Watershed Master Plan – The Perambra Block Experiment

Panchayats in Kerala have taken up several initiatives at their own behest that hold the potential to contribute to long term sustainable development. Some of those are outcomes of the 9th Five Year Plan initiatives introduced in the state during the People's Campaign that we described in the previous chapter (Thomas Isaac and Franke 2002:177).

Perambra block of Kozhikode district in the foothills of the Western Ghats took upon itself the task of operationalising government guidelines regarding the preparation of integrated watershed management plans. A key problem in creating sustainable development is that political units do not usually correspond with the appropriate ecological unit for planning the most sustainable resource use and maintenance. Small-scale experiments across India have shown that planning based on watersheds – natural units for managing water and other resources – can bring dramatic improvements in drinking water availability and crop productivity (Rao 2000). Agricultural scientists have divided Kerala's 44 river basins into 151 sub-watersheds (10,000 to 50,000 ha) and 960 "milli" watersheds of 1,000 to 10,000 hectares. Below this are several thousand "micro" watersheds with areas of less than 1,000 hectares (Kerala State Landuse Board 1998).

The Perambra block panchayat got in touch with KSSP, the people's science organisation in Kerala that we have mentioned earlier. To develop the watershed plan KSSP and block panchayat members proceeded according to the following sequence :

(1) Training 101 local master trainers from different walks of life such as farmers, activists, politicians, students, professionals, officials and elected representatives for 4 days at KSSP's Integrated Rural Technology Centre near Palakkad, a campus the organisation frequently uses for training local implementation teams.
(2) Preparing the maps of 46 micro watersheds by the master trainers.
(3) Training selected local persons from each watershed.
(4) Forming two survey teams of seven persons each for each micro watershed.
(5) Forming Watershed Committees at the level of the village panchayat.
(6) Creating a favourable political environment for watershed planning. This included special grama sabhas, discussions in school assemblies, writing banners and posters, and conducting a *jala jatha* (water procession or parade) to spread the message of the necessity of water conservation. People marched in these jathas with banners and posters describing various aspects of water conservation and the value of water in our life.
(7) Surveying gullies and streams
(8) Conducting transect walks (see chapter 9)
(9) Intensive meetings of Neighbourhood Groups (NHGs) of farmers.
(10) Holding a Watershed Mahasabha (major assembly) in each micro watershed. These gatherings were exclusively to discuss watershed development and water related issues. There was one assembly in each micro watershed. The

problems were listed, possible interventions discussed and information was gathered to fill in a questionnaire prescribed by the Department of Rural Development, Government of India.

(11) Preparing an action plan for watershed based interventions including wasteland improvement, groundwater recharge, sustainable cropping patterns, soil erosion amelioration, and reforestation.

(12) Expert interventions in crop selection, choice of fish types and methods of soil conservation. Expert opinion was also sought to assess the technical viability of projects proposed by the local people.

(13) Finalisation of the watershed master plan detailing the specific intervention requirements such as the appropriate types of contour bunding, excavation of pits, and choice of crop types.

The whole exercise took 16½ months. In order to plan and implement the programmes, farmers' groups called *Karshaka Kootaima* were formed. (Karshaka means farmer; Kootaima means common interest group). To give a fillip to the activities on June 5th (World Environment Day) a call was given to dig one lakh (100,000) contour pits.

In response people came forth to dig 82,000 pits 90 cm x 60 cm x 50 cm (deep) as voluntary contributions to the project. Digging pits along contour lines is one of the most commonly used indigenous practices for groundwater recharge. These pits collect and store water during rainy days, when there is sufficient rainfall and surface runoff. Stored water in the pits slowly infiltrates and enriches the groundwater. This is one of the objectives of watershed management where the aim is to increase the residence time of water on land surfaces. Given the rapid depletion of groundwater and the restricted recharge due to the hard rock terrain, constructing these trenches and pits is a significant step forward.

Subsequently activists participated in an innovative action to trace the Kuttiadi river in partnership with the Kerala Forest Research Institute (KFRI) as part of the Know Your River programme. This programme was intended to equip local people with knowledge about the nature of the river bed, flow patterns and bank erosion, biodiversity and the role of river bank vegetation in protecting river banks, the relation between river water levels and water tables in nearby wells, the need to preserve the river, and the complex relationship between the river and livelihood opportunities. This exercise concluded with planting bamboo on both sides of the river to hold the banks in place and prevent erosion.

Recognising the innovative venture of the Perambra Block Panchayat project, in 2000–01 the Kerala Institute of Local Administration (KILA) as a part of their training programme arranged field visits by 120 Block Panchayat Presidents (of a total of 152 in Kerala) to Perambra, for studying the experience with an aim towards replicating it in their respective areas. The Perambra experiment has to date not yet been followed in practice in other areas.

A Participatory Water Supply Project in Adat Grama Panchayat

Adat Grama panchayat in Puzhakkal block is located around 15 km northwest of Thrissur town. The topography is lateritic midland bordering on low lying *kole* land that we referred to in chapter 4. The kole land, a local term, connotes low lying seasonally water logged clay dominated areas in Ernakulam and Thrissur districts. The western borders of these lands are sandy stretches merging with the coastal plain. Parts of the kole lands are penetrated with saline water. Scarcity of drinking water is a perennial problem here. About 20 years ago the Kerala Water Authority (KWA) set up a drinking water supply scheme with an underground reservoir with a capacity of 36,000 liters and a surface reservoir of 1.5 lakhs litres capacity with distribution through 75 public taps along the main roads in three wards. It covered around 25 per cent of the people in need of safe water. The panchayat then decided to try a self-help approach. A two-phase scheme for Rs. 221 lakhs was formulated. The first phase consisted of putting up a water tank of 3.5 lakhs litres capacity and a new underground reservoir of 42,000 litres capacity. The outlay was Rs. 139 lakhs of which the beneficiary contribution was expected to be Rs. 90.25 lakhs or 65 per cent. The distribution system was envisaged to provide 2,500 house connections and 260 street taps at a cost of Rs. 82.5 lakhs to be fully borne by the beneficiaries. Thus out of a total project cost of Rs. 221.50 lakhs the beneficiary contribution would come to Rs. 172.75 lakhs or 78 per cent.

The project began with a household survey to identify the water needs of the prospective beneficiaries. This was followed by grama sabhas to create awareness of the need for beneficiary contributions, to identify locations for the street taps and to establish a monitoring committee.

The panchayat level monitoring committee was headed by the President, and included the elected panchayat council, the Kerala Water Authority (KWA) Executive Engineer, the Public Works Department (PWD) Executive Engineer, the Head of the Department of Civil Engineering at the Government Engineering College in Thrissur and representatives of beneficiary groups. A ward level monitoring committee was headed by the Ward Member with representatives of Residents' Associations and Beneficiary Groups, with KWA officers as members. The ward level committee was entrusted with monitoring the implementation of the distribution system.

The panchayat got technical support for implementation both from the KWA and the Government Engineering College in Thrissur. The first phase of the project was completed in 2002 at Rs. 132 lakhs—a saving of Rs. 7 lakhs over the originally planned outlay. A Rs. 30 lakhs contribution had already been paid by the beneficiaries and the remaining Rs. 60.25 lakh of the remaining Rs. 90.25 is currently being collected. For the second phase, beneficiary contributions have been set at Rs. 5,025 per connection. Once the beneficiaries make this payment, implementation under the second phase will start. All people in the panchayat needing drinking water will be covered by this two stage programme.

Olavanna Panchayat's Water Supply Initiative

Many villages in Kerala's lush green and wet environment face acute drinking water shortages despite the state's receiving more than 3,500 mm of rainfall annually. Pantheerankavu and Kambiliparamba areas within Olavanna panchayat in Kozhikode district would have remained two such water shortage villages but the local communities decided to take matters into their own hands. In 1987, when villagers of Olavanna grama panchayat started collecting funds for a mini piped water scheme, just three small water supply schemes constructed and operated by the Kerala Water Authority existed in the village. About 1,000 households (around 25 per cent of the total households of the panchayat) were served through 25 public taps and 42 private connections. Operation and maintenance of these systems were always problematical.

Faced with an acute water supply problem, the people of Olavanna decided to experiment with a rural piped water supply scheme with private resources and funds. The scheme, initiated in 1987, works on fairly simple principles. A cooperative society was formed, which charged a one time membership fee of Rs. 4,500 to Rs. 12,500 per family, depending on the cost of the scheme. A typical scheme has an intake well, an overhead tank and a distribution system with house pipe connections. Expertise, material and labour are found locally. Poorer families pay their contributions in instalments.

Various setups have distinct, but simple procedures for consumption, maintenance and billing. The technology has been geared to meet local needs. In Kambiliparamba, the society eliminated the need for a pump operator by installing a sensor-based switching system that starts the pump when water in the tank falls below a certain level and stops it when full.

The by-laws of the typical cooperative society running the water supply schemes have evolved with time and experience. The number of households in a coop ranges from ten to eighty-two. Every year, the general body (GB) elects an executive committee (EC) to run the scheme on a daily basis. The EC's accounts and annual report are scrutinised by the GB. The societies have worked out their own procedures for holding GB meetings, managing finances, installing new connections, even dissolution of the society. Normally the assets of a dissolved society are handed over to a newly registered society and not divided amongst the members.

It has been the experience of the societies that once an initiative is launched, funds are not a major problem. They come in from private sources. Their success is now a shining example of what participation and activism can do where government agencies failed to deliver. The experiment has caught the popular imagination in Olavanna and water cooperatives in this panchayat are not only catering to the needs of local people but also proving to be a practical water resource management alternative. The first twenty-two cooperatives brought safe drinking water to 1,500 of the approximately 7,000 households in need (Thomas Isaac and Franke 2002:37). By 2004 there were 70 cooperatives functioning in this panchayat supplying the water needs to just over 4,000 households.

Panchayats like Olavanna have proved that 'willingness to pay' and 'willingness to charge' for better services go hand in hand and that local initiatives can play a significant role in natural resource management. The Olavanna experiment illustrates the importance of managing water resources in decentralised mode. Water is a decentralised resource and sustainable management calls for a decentralised approach. The best sites for this type of water management in Kerala are the undulating terrain areas with a predominance of hard rocks such as the midland valleys. As these valleys are narrow and small each valley has a limited capacity and therefore small cooperatives are appropriate to the layout of the resources. Having a direct role in managing their water supply makes it easier for local coops to see their common interests. Drawing groundwater by all the cooperatives simultaneously can pull down the water table. This can be compensated for through recharging. Trenches, recharging pits, contour bunding and infiltration galleries along the slopes will help recharge the groundwater. Each family can take part in this recharging programme.

The community does not have to run after politicians and bureaucrats since it can put up its own commercially viable services. These experiments show that small, resident-owned cooperatives can be better constructed, better maintained, and more cost effective than those heavily subsidised or run by the state or by large outside investment concerns such as multinational corporations.

The Microhydel Electric Project at Upputhara

Idukki district panchayat established a microhydel project on a tributary of the Periyar river in the remote Upputhara area using appropriate technology obtained from Centre for Appropriate Rural Technology in Mysore. Though the project was conceptualised in 1998, the panchayat had to wait for two years to go through the formalities to standardise various processes and procedures.

Using the technology of a reverse functioning water pump (instead of sucking water it allows flow) and utilising a 7.5 HP kirlosker motor as generator the project was commissioned and within 10 months implemented by the district unit of *Nirmithi Kendra* (Nirmithi means construction; Kendra means centre), Idukki, an organisation promoted by the Government of Kerala to introduce appropriate technology for rural development.

Right from the beginning the stakeholders were assured full ownership of the project. A nine member management committee was set up. This committee appointed an operator at Rs. 1,000 per month. Each of the 45 member households contributed Rs. 30 per month for which each house is allowed to use 4-6 lamps. But there is a collective social decision to save energy by using only compact fluorescent lamps. The total amount received by the committee from individual family contributions is Rs. 1,350 (45 x Rs. 30), of which Rs. 1,000 is paid to the operator and Rs. 350 is pooled as a maintenance fund. The people conduct a social (public) audit of the accounts every year. Currently they get power only for 10 months in a year. If an additional check dam is constructed 12 month service is possible.

The Upputhara example shows that a local government can meet a local need through the use of appropriate technology and a community management system can ensure the sustainability of the project.

Idukki district now has 212 household level micro hydroelectric units spread over 25 grama panchayats supplying minimum electrical needs for 870 households (*The Hindu*, September 28, 2004). A typical microhydel project consists of a source of water, often a stream, a check dam, or a storage tank, a plastic or PVC flexible hosepipe and a rewound motorbike or motorcycle dynamo which acts as the generator. A majority of these units have installed capacities of less than 150 watts per hour, which can provide lighting and also power a radio, tape recorder or television set. A little over 50 per cent of these units supply round the year power. It is reported that 150 of the projects cost less than Rs. 10,000 each. There are 20 installations incurring expenditures of less than Rs. 2000 each. These small microhydel units were established mostly by individual households to supply their own needs. There are also group installations and setups by local governments. A recent report (Chathukulam and Gireesan 2004 and 2005) indicated that 95 per cent of these units were established after 1996 as a follow up of the 9th Plan Campaign. It is inspiring to see the use of local resources to meet the electrical needs, but it is important to look into the safety measures and other technical and logistic issues that may require some outside expertise.

Small-scale generation of electricity like the Upputhara experiment has several advantages over large dams, coal generation, or nuclear power – all of which pose serious long term environmental dangers that are well known internationally. By tapping a portion of the water from a fast-running river, or naturally existing waterfall, the microhydel generator system has only a slight effect on the natural processes of the watershed. The water that is tapped off the river to turn the generator is sent immediately back to the river just a few hundred meters downstream. Microhydel infrastructure is also highly efficient in the sense that the power generation takes place near the homes of the end users. Thus, little transmission loss occurs. The rugged mountainous spine of eastern Kerala is difficult to reach with traditional power line technology where households are widely scattered among the slopes and valleys, but the microhydel plant can be built right in such isolated areas. With more than 50 major waterfalls and hundreds of associated rapids, Kerala's Western Ghats range can be an electricity supply source without endangering the natural environment. Observers believe that up to 10 per cent of Kerala's electrical needs could be met with such small, dispersed projects. Much of the technology has already been developed in China, which has built nearly 90,000 such microhydel plants since 1950, now accounting for perhaps 6 per cent of that country's electrical supply (Franke and Chasin 1997:3064). Along with Upputhara, Chapparappadavu panchayat in northern Kerala constructed a micro-hydel plant in 1998 that provided current to twenty-three households at a startup cost of Rs. 200,000. At Rs. 8,696 per household, this is a prohibitive outlay, but the panchayat gave a 50 per cent subsidy from the people's campaign plan funds (Thomas Isaac and Franke 2002:148). In Palakkad district, Karimba panchayat members – with assistance from KSSP – put up the Meenvallam microhydel plant that connects to the statewide grid, another variation on this promising technology (Franke and Chasin 1997:3063).

The Koyilandy Integrated Biological Mosquito Control Project

(This account is adapted from Elamon, Franke, and Ekbal 2004.) Koyilandy Municipality in Kozhikode (Calicut) district is a lowland town of more than 57,000 inhabitants. The town has a dense population, poorly drained ditches filled with stagnant water, open septic tanks, and puddles that fill up near houses in the monsoon seasons. For ten years the municipality had been combating mosquitoes primarily by spraying Bitex, an organo-phosphate compound. People in heavily sprayed areas complained of shortness-of-breath and other health problems and the mosquito population had mutated resistant forms and seemed hardly to be declining. In 1996 a local husband-wife team of professional biologists, Drs. K. K. Anilkumar and Reena Anilkumar, proposed a multi-pronged biological control approach to the mosquito problem that the local people's campaign activists decided to promote as an experiment.

The project revolved around six elements. First, herbal and bacterial larvacides were sprinkled over drainage canals and other places holding stagnant water. The bacterial component, *Bacillus thuringiensis*, is considered safe for humans, as was the herbal concoction of several elements including neem (*Azadirachta indica*), a tree from which several bioactive elements are produced that repel, suppress appetite, sterilise and/or kill insects but are harmless to humans. Both mixes degrade rapidly after use thus limiting the likelihood of the mosquitoes developing genetic resistance. In a second pass over the breeding areas, volunteers in the project sprinkled an herbal repellent intended to prevent female mosquitoes from laying eggs. The third component of the project was to place traps in the houses in three wards and in the government hospital. The traps contain an organic solution that attracts female mosquitoes to lay the eggs in places where the *pupae* can be monitored and destroyed after hatching. Fourth, a programme was initiated to cover the vent pipes of septic tanks with nets to prevent mosquitoes from entering and laying eggs. A campaign to encourage homeowners to fix cracks in all tanks and walls made up the fifth component that was followed by a general awareness campaign including local disease prevention classes that made up the sixth phase.

While conducting the campaign, activists also organised for monitoring and evaluation. Two control areas were chosen: one where no actions were taken and one where traditional insecticides were employed. Over an eight-month period of the most detailed monitoring, it was found that the six-pronged integrated biological approach was reducing the mosquito breeding by more than 100,000 per month. In one month, more than a million mosquitoes were destroyed at one or more phases of their life cycle. This was estimated to be an 85 per cent improvement over areas where no interventions were made or where toxic chemicals were in use. The Koyilandy experiment generated substantial interest among other coastal and lowland communities where mosquitoes are a problem. Kumarakom panchayat, lying along a coastal lagoon, organised 1,600 volunteers to clean algae along the lakeside. They also went house to house encouraging residents to install nets on latrine exhaust pipes and to clean their house compound areas of puddles. In 2002 the city of Ernakulam, Kerala's largest, contracted with the Anilkumar team to try to reduce the mosquito population. The Koyilandy people's campaign project has generated mosquito control data that may have international implications in view of recent discoveries in Mexico and other countries indicating that combined community participation and a mix of carefully chosen safe environmental practices can offer promising alternatives to toxic chemical spraying in combating malaria (McGinn 2003). The development of environmentally neutral mosquito control has become more urgent with the growing evidence that overuse of antibiotics is producing increasing numbers of strains of malaria resistant to medicines (Pennyback 2004:67). A detailed account of the Koyilandy mosquito control project can be accessed at : *http://chss.montclair.edu/anthro/Koyilandi.htm.*

Water Self-Reliance: The Erimayur Jalanidhi Programme

Kerala's rural water supply and sanitation programme has taken novel steps promoting self reliance in water management. With training, village housewives work as pump operators, office managers, and entrepreneurs. The Jalanidhi (Jala = water; nidhi = property) programme of the Government of Kerala is being implemented in Thrissur, Palakkad, Malappuram and Kozhikode districts with the support of the World Bank. With a total outlay of Rs. 4,510 million the aim is to cover 100 grama panchayats between 2001 and 2006. By the end of 2004 86 projects had already been taken up, covering 78 per cent of the target population for 2006 (SPB 2005:144). The project is designed to bring sustainable safe drinking water to the rural poor. Erimayur panchayat was part of the first batch in 2001. The project covered 4,901 out of 6,300 households. Considering geographical proximity, availability of common water sources and other factors 31 Beneficiary Groups (BG) were formed covering all 4,901 households. Out of the 31 BGs, 26 with 1,180 households opted for the new technology of rainwater harvesting (see chapter 8) combined with pumped water. This is perhaps first case in India where rainwater harvesting and pumping systems are clubbed together for a domestic water supply (Ajai Kumar 2003). All the households have 1,000 litre ferro cement tanks into which rainwater from their roof is directly funneled during the monsoons. On dry days the tank is filled by pumping water from the pump house installed for each BG. The ferro cement tanks serve a dual purpose: they store harvested roof water during rainy days and hold pumped water on non-rainy days.

During the rainy season the excess water from the tank is let out to recharge pits to augment ground water resources. As the households have rainwater harvesting with a storage tank, the storage requirement of the pump house is reduced. This arrangement helps decentralised water storing and links rainwater harvesting to the regular water supply. Based on the estimated household usage of water and the predicted amount of rain, planners have calculated that a 20 m^2 roof is sufficient for harvesting rainwater for domestic purposes in Erimayur.

All 26 BGs now can avoid 120 to 125 days of pumping in a year during the monsoon months, thereby saving considerable electricity. This also cuts down on operation and management expenses. During dry days pumping starts at a fixed time. The storage tanks in every house make big overhead storage tanks unnecessary. Having water available in their own house reduces the strain on the people who are mostly agricultural labourers, comprising 90 per cent of the population. They are now self reliant in water and need not bother about the sometimes erratic nature of piped water coming from long distances. Construction of large centralised water tank systems that depend on a single water source–usually reservoirs or pumped groundwater – leads to high stress on stocks and on water management capabilities. By contrast, the Erimayur setup has the potential to contribute to sustainable water resource management particularly in drought affected areas like Palakkad.

The local community carried out all the implementation actions. This kept much of the money in the community as wages. Trained women were involved in the construction of the rainwater harvesting tanks, the pipe lines, and in pump installation. As noted earlier, housewives now operate the pumps, maintain the accounts, carry out plumbing activities, and do system maintenance. In Erimayur, a team of 216 trained women constructed 800 tanks and 80 km of pipe. The value of their labour was Rs. 9,58,400. The trained women now even go out of the panchayat for plumbing, latrine construction and similar jobs. The workers earned an annual average Rs. 12,000 each from the project. Now they take plumbing work in and around the panchayat for a daily wage of Rs. 200. This wage compares favourably with the Rs. 188 per day earned in 2004 by skilled workers in the agricultural sector who are mostly males and the Rs. 102 earned by female agricultural labourers (GOK 2005:360).

The Erimayur experience illustrates the potential of community self-reliance to run the water supply and the benefits of women's involvement and empowerment. The government is now convinced to upscale such community driven programmes to the entire state.

Sustainable Vegetable Cultivation in Kanjikuzhy

Kanjikuzhy panchayat in Cherthala taluk lies in the sandy coastal tract of Alappuzha district. The panchayat covers an area of about 17 sq km. The Central Kerala coastal plain lies within a couple of metres above mean sea level. It is characterised by ridge and runnel topography. There are silica sand deposits in Kanjikuzhy. Within the sandy belt runnels are depressions where the soil contains slightly higher proportions of silts. As a result the soils here are moist and are suitable for paddy. The ridges dominated by sands are relatively elevated and accommodate mostly coconut or other tree crops. A landuse survey in 2003 (CESS 2003) discovered that around 73.5 per cent of the total area of the panchayat is under coconut and mixed tree crops. The paddy area is limited to 12.8 per cent while another 4.8 per cent which was earlier planted in paddy has been diverted to other crops and to buildings. In general the soils are poor in fertility and in water retaining capacity. This area suffers from environmental problems such as crop diseases, groundwater pollution, wetland siltation, low productivity and loss of agricultural biodiversity. However, sandy soils with low moisture holding capacity are suitable for some types of vegetables commonly eaten in Kerala. Many such areas were traditionally well known for growing vegetables. Despite these apparently severe environmental limitations, traditionally Kanjikuzhy farmers produced substantial vegetable output. Over the years, vegetable production declined, but in 1997, the panchayat revived the traditional vegetable cultivation utilising Rs. 60,000 from untied funds—a small grant from the state government for local communities to spend on locally decided projects – for seeds, fertilisers and pesticides. Nearly 6,000 households participated in the cultivation of bitter gourd, cowpea, brinjal, chilly, ladies finger and snakegourd. In order to create the motivation, planting was done on a single day called *Pachakkari Krishi Dinam* (Vegetable Planting Day). The campaign encouraged traditional varieties of vegetables, use of organic fertilisers, biopesticides, and physical and biological methods of pest control rather than widespread chemical spraying (Thomas Isaac and Franke 2002:33). In the next year 5000 people took part in the programme and only seed packets were distributed. Overall beneficiaries contributed Rs. 2.50 lakhs out of Rs. 3.5 lakhs spent and 50 hectares were covered resulting in the production of 100 tons of vegetables. The scheme tapered off in the next two years and in 2000–01 it was not implemented at all.

But riding on the success of the Kudumbashree (Family Prosperity) network, tuber planting was initiated in 2001–02. In 2002–03, 101 self-help groups and 123 Kudumbashree units took up cultivation on 50 hectares and produced 120 tons of vegetables. Simultaneously the departmental scheme of organising farmers into *Haritha Sanghams* (Green Groups) was launched. Five Haritha Sanghams are each cultivating five hectares of land. In 2002–03, these Haritha Sanghams produced around 175 tons of vegetables. In spite of bottlenecks like marketing problems and frequent change of key officials, Kanjikuzhy panchayat has succeeded in sustaining a novel project over a period of seven years. Unlike many panchayats it has integrated the scheme with the departmental scheme with the panchayat spending Rs. 5.9 lakhs, and the Agriculture Department Rs. 5.2 lakhs in six years from 1997. About 120 hectares were brought under vegetable cultivation resulting in the production of 295 tons of vegetables in 2002–03. The Kanjikuzhy experiment has now become a component of Kerala's most ambitious local sustainability project: the Integrated Coastal Zone Management Project in Alappuzha district.

Vegetables, Coconuts, Fish and Development – The Alappuzha Integrated Coastal Zone Management (ICZM) Project

In several places throughout this book we have emphasised the watershed as one of the most important geographical units inside which sustainable development planning should occur. Another significant area involves the contact zones between different types of environmental units. Coastal areas fall into this category and Kerala is a region with substantial coastal resources. Managing the transition zone between land and sea has become a critical need and Kerala is developing a major initiative along the Chertala-Alappuzha coast in central Kerala. Fig.10.1 indicates the panchayats covered by ICZM activities in the Alappuzha-Chertala area. The strength of this project derives partly from the careful research and high levels of scientific input; it also makes use of Kerala's traditions of political mobilisation and citizen involvement.

The Conceptual Basis of Integrated Coastal Zone Management

Integrated coastal zone management is a strategy to develop all relevant areas simultaneously in order to achieve sustainability of coastal resources. This means socio-economic improvement and livelihood security, reduction of poverty, reduction of resource conflicts, the spreading of environmental consciousness and empowerment of local residents as genuine stakeholders in the future of their environment.

Ecologically ICZM aims to identify and classify areas in need of conservation or regeneration, protect against coastal erosion, conserve or regenerate biodiversity, restore ground and surface water resources, and undertake related environmental maintenance. In Alappuzha-Cherthala more specifically, this requires :

- Protection and enhancement of coastal resources;
- Restoration of the Alappuzha – Cherthala canal;
- Maintenance and upgrading of sacred groves;
- Promotion of micro enterprises to reduce the development footprint on the fragile coastal ecology;
- Spatial planning to further reduce the developmental impact;
- Rainwater harvesting (see chapter 8) and other innovative techniques where possible to create the most efficient use of existing resources; and
- Creating economic activities that reinforce rather than pit the fishing and land-based communities against each other.

In particular, the ICZM project in this area will attempt to merge the coconut, fishing and vegetable traditions into a set of environment friendly adaptations and linkages. In the next section we shall provide an overview of the ICZM area, followed by a list of the most important ecological degeneration indices, and then give an outline of how the project is aimed towards rejuvenating and then sustaining the environment while bringing the poorest households above the poverty line.

The ICZM Experiment Area

As we saw in chapter 2 (table 2.2), 77 per cent of Alappuzha district is in the lowland zone with the remaining area in the midlands. The Cherthala-Alappuzha area is a sandy plain lying between the

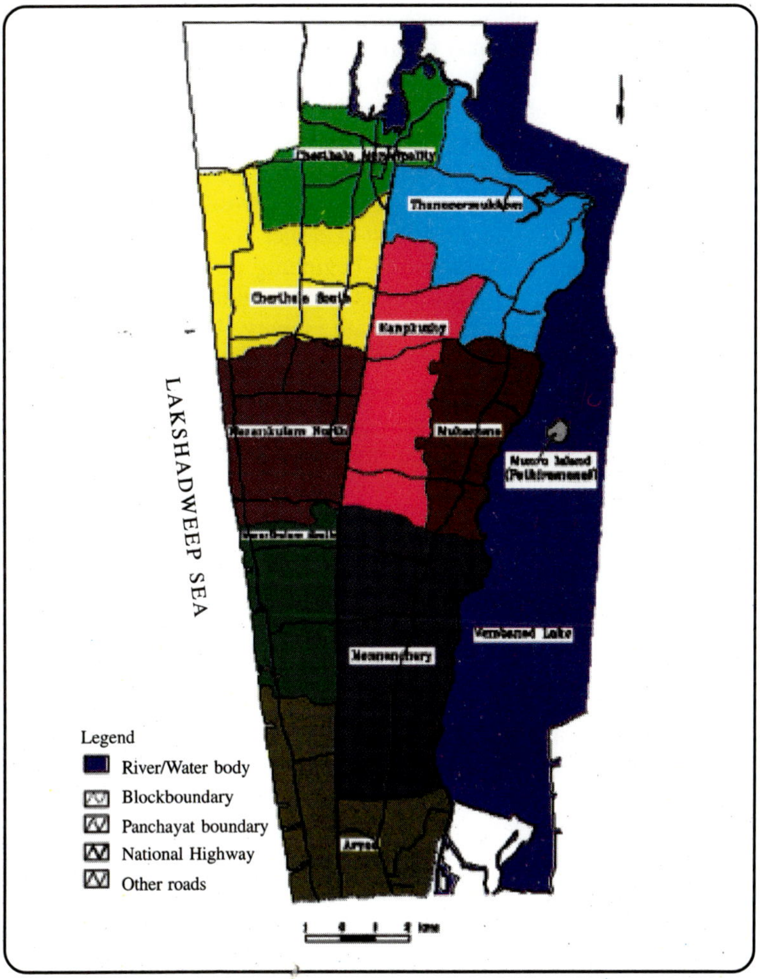

Fig. 10.1 : Alappuzha - Cherthala area

Lakshadweep Sea and the Vembanad lake. The 161 sq km area is spread over eight panchayats: Cherthala south, Thaneermukkom, Kanjikuzhy, Muhamma, Mararikulam north, Mararikulam south, Mannancherry and Aryad. The Thaneermukkom barrage across Vembanad lake connects the area with Kottayam district in the east. Average annual rainfall is 3140 mm. There is some rainfall in every month except January. This area is known for its coir industry, fishing, and smallholder vegetable cultivation.

Coastal features include beaches, sandy coastal plains and ridges and runnels within the sandy plain. As noted in chapter 3, runnels are shallow areas between the ridges and are used for cultivation. Ridge and runnel topography occupies nearly 94 sq km (59%) of the total area. The coastline runs for 20 km. This coastline is affected both by erosion and accretion. A stretch of about 14 km in Aryad and Mararikulam south panchayat is affected by erosion. A sea wall had been constructed all along this coast in earlier years. However, during a major sea surge in the monsoon beginning in May, 2002 the sea wall got damaged in certain sections of Mararikulam south resulting in damage to houses in the immediate vicinity. The northern stretch of coast is currently stable with broad beaches. A remarkable feature of this area is the *chagara,* or seawater-suspended mud bank. The mud bank is basically suspended colloidal mud 7 to 15 metres thick that forms during the southwest monsoon. Apart from Kerala mud banks are reported only in a very few locations: off the Surinam coast in northeast South America, off the South Korea coast and off the Louisiana coast of the USA. The causes of mud banks are yet to be conclusively established. The Kerala mud bank lies in the water about 8 km from the shore and is typically about 4 km in length (Balachandran 2004). The mud bank has a powerful dampening effect on the ocean waves. The calm sea surrounding the mud bank allows the fishermen to work their boats even while the high monsoon waves make fishing outside the mud bank almost impossible. The organic matter in the mud is rich in nutrients which attracts schools of fish from all around. As the wave energy is dissipated away from the coastline and the sea remains calm, the shoreline parallel to the mud bank is naturally protected from erosion. The effects of the mud bank on the shrimp catch are so profound that the term *chagara* has taken on the additional meaning of "the great ocean harvest – the time of plenty." The chagara became part of the dramatic setting for one of Kerala's most famous novels, *Chemmeen (Shrimp)* by Thakazhi Siva Sankara Pillai. A national award winning motion picture of the same name was made based on this novel and there was a song starting with the word 'Chagara'.

From the shore to a distance of 500 m inland (the Coastal Regulation Zone) coconut palms cover 81 per cent of the area. Beaches cover around 12 per cent. Mixed settlement with vegetation is the other important category. There are wetlands all around. The average population density was 1,607 persons per sq km in 1991 and in 2001 it grew to 1,788 persons per sq km.

This area has both modern and traditional industries. There are 41 modern units of coir related industries and 438 small-scale coir factories. Figure 10.2 shows a weaver making a coir mat, one of Kerala's major products for many decades. There are around 36,000 fishermen in 12,200 households. Some migratory fisherfolk also operate here.

Health, drinking water and sanitation facilities are not adequate. People are prone to water borne diseases. Households with septic tanks vary from 19 per cent in Muhamma panchayat to 32 per cent in Cherthala south. In some panchayats 45 per cent of the households have piped water, but most people still depend on well water, which is not very safe because solid waste seeps through the porous soil contaminating the water.

Recent Activism and Participation

Alappuzha has one of the most activist histories in Kerala. Farmers and coir workers in this area are one of the backbones of the left movements and have fought for decades for land reform and other improvements in their lives. In recent years Alappuzha became a success story in transferring its political energies from protests to participation in development activities. With the assistance of UNICEF an Urban Basic Services Programme was launched in Alappuzha town. A set of nine indicators was developed for monitoring poverty reduction at the local level. One hundred and fifty neighbourhood groups of 20–40 poor households were formed, with each household being represented by a woman. Institutional credit was provided under a National Bank for Agriculture and Rural Development scheme to help the groups start small businesses currently referred to in the development literature as micro-enterprises. In spite of the partial success of this local level anti-poverty initiative, poverty in most of the panchayats in Alappuzha persists.

Low Biodiversity and High Environmental Stress

The area is under significant environmental stress due to water pollution and coastal erosion. The level of degradation is indicated by the low level of biodiversity. The ecosystem has lost its pristine biota. The productivity of inland fish ponds and agriculture is low.

Fig. 10.2 : Man Weaving a Coir Mat in the Mararikulam Area. Photo by Richard W. Franke

There are 18 native tree species, 13 cultivated tree species, 9 sand binders (creepers that grow on undisturbed sand dunes and inhibit erosion of sand by strong winds and tidal waters), 10 hedges (beach grasses) on the sandy coasts, 15 wetland binders, 9 hydrophytic (water loving) plants, 44 wetland graminicids (grasses), 12 mangrove plants and associates, 21 homestead garden plants and 33 medicinal plants (CESS 2003). The potential for at least some ecological regeneration is indicated by an inventory of sacred groves showing that each panchayat has a large number of groves of varying sizes.

Agricultural biodiversity is declining. A study of landuse changes indicated that the wetlands once used for rice cultivation have been diverted for coconut and for settlement. Monocrop coconut cultivation covers as much as 80 per cent of the gross cropped area (Aniyan et al. 2002). Coconut trees are affected by root wilt disease. Coconut productivity here is 4,052 nuts per ha compared to the state average of 5,843 nuts per ha in 1992–93. Rice is the second crop covering 10 per cent of the arable land. Cashew, pepper, bananas, and vegetables account for the remaining cultivable land. About 14 per cent of the population works in agriculture.

Unsustainable agricultural practices such as extending coconut cultivation up to the sea front in water logged areas and planting coconut trees near the high tide line have destroyed many mangroves resulting in eutrophied water bodies (having such an excess of dissolved nutrients that oxygen levels decline). Stake nets (nets spread within the stakes marked by wooden or bamboo poles within the backwaters and estuaries by the fishermen) along the estuaries destroy the hatchery grounds of the fish that breed there. Large-scale availability of coconut byproducts has led to coir retting, spinning, and weaving becoming the main industrial employment which currently covers nearly 45 per cent of the work force. The root wilt disease mentioned has so lowered coconut production that many spinning and weaving establishments have to import raw materials from the nearby state of Tamil Nadu. This results in extra transportation costs and all the associated environmental costs.

Mining of backwater lime shells and sand including silica sand are two other assaults on the natural resource base. The manual lime shell mining faces competition from mechanical dredgers. As the shallow water deposits get exhausted, economically marginal households find it increasingly difficult to depend on this resource.

Marine and backwater fishing employ 17 per cent of households. The marine fish catch is stagnating. There has been a decline in the karanila system that we described earlier in this chapter. Backwater fishing is also declining as Vembanad lake water is polluted and over exploited. We shall provide some details shortly.

Coastal Erosion

Part of the coastline is subject to erosion in the monsoon months. Wave heights of 3.2 metres along the west coast of India are higher than in the east coast. During the monsoon high waves can bring on substantial erosion. Erosion is a natural phenomenon but the vulnerability of the coastline increases due to the absence of coastal vegetation. The sand binding cover vegetation – originally grasses that held the dunes in place – has been replaced by coconut palms, whose shallow root systems are not suitable for preventing erosion (See fig. 5.7). A stone and concrete sea wall is a common modern protection measure but it obstructs the free movement of fishermen and restricts beach related activities. The sea walls frequently get damaged and need repairs. Ultimately the sea walls only transfer erosion problems from one place to another. The cost to construct one km of sea wall is between Rs. 200 and 300 lakhs.

Decline of the Alappuzha-Cherthala Canal

A significant feature of the ICZM area is a 22 km long canal that connects the Vadai canal at Alappuzha in the southern end to Vembanad lake at Cherthala in the north. Excavation of this canal started in 1940 as a famine relief operation but remained unfinished. Because of the sandy terrain the water level fluctuates with the tides in the Vembanad lake. In the dry season, low tides actually bring down nearby wells and lower the water table enough to kill many plants and create drinking water shortages.

As an experiment, in the recent past one local community constructed a bund at Thiruvizha to preserve the water table and promote agricultural activities. This had a positive impact. A series of bunds was constructed in quick succession. At present, there are 20 bunds across the canal dividing the entire canal into 22 segments. The canal is now totally unsuitable for transport. Dense aquatic weeds – especially *salvinia* (African payal), cover most of the canal and thus severely limit survival of the fish population. Up to 169 toilets open directly into the canal. Certain sections in the southern part act as an open air toilet. The canal has become a major breeding ground for mosquitoes and rats. The canal water in most stretches is too polluted for use.

Degradation of the Lake Vembanad Water System

Vembanad lake lies east of the Alappuzha-Cherthala plain. It is the largest backwater system in the state and has provided opportunities for fishing and lime shell collection to the people living in the adjoining panchayats. In 1976 the Thaneermukkom barrage was constructed across the northern border of the lake. This barrage has divided the lake into two parts.

The southern part has now turned into a completely fresh water body with limited outlet to the sea. The stagnant water in this southern part has become polluted due to waste disposal from the surrounding settlements and to an accumulation of fertiliser and pesticides from the vast paddy fields on the southern and eastern sides of the lake. Another effect of this change from saline or brackish water to fresh water is a heavy growth of weeds and water hyacinth. This has spread all over and many of the narrow watercourses feeding into the lake are completely congested hindering the free movement of boats.

Water quality surveys of the southern part of the lake indicate that the salinity increases from the south to the north. Similar trends can be found in the distribution of pH – a measure of salinity or alkalinity. Seawater, which is alkaline, has a higher pH value. A lower pH value indicates an acidic nature. The eastern part of the lake adjoining Kottayam town has a pH value less than 5, well below the pH value of freshwater which is around 6.5. Construction of the Thaneermukkom barrage was meant to convert the southern segment into a fresh water body, however this lower pH indicates that although salt water inflow has stopped, water quality has deteriorated. This can affect not only the potability of the water but also the biota in the lagoon. The urban waste from Kottayam town and adjoining areas may be contributing to this lower pH. Change is also observed in the siltation patterns. Although time series hydrographic survey data are not available, existing data and spot measurements indicate that the lake is getting shallower. The central part of the lake has a depth of 2 to 2.5m with two troughs of more than 4m deep located in the northwest and southwest parts.

The ICZM Approach

The ICZM project will attempt to overcome all the area's main social and environmental problems through a series of integrated – connected and mutually supportive – interventions, emphasising the capabilities of Kerala's people as fashioned over their many years of activism and participation. The main components of the project are described in the next several sections. Many of the project activities have already been carried out successfully.

Identify the characteristics of each micro-environment and fit project goals and techniques as appropriate. The proposed zonation to be arrived at after detailed cadastral surveys will indicate areas for different developmental activities : areas for development with necessary protective measures with respect to fragility, areas of development with restrictions with respect to quality and areas for full conservation. This will provide the base for proposing all future activities related to land, water and biomass. By 2003, trained volunteers had completed participatory mapping of land use and environmental assets for all eight panchayats falling in this area. Fig. 10.3 shows a training class for the volunteers in Kanjikuzhy panchayat while Fig. 10.4 presents a sample landuse map prepared by the volunteers. Figure 10.5 shows one of the volunteers explaining a resource map to a local official.

Protect the shoreline with environmentally sustainable devices. In place of seawalls that require huge capital outlays, and which eventually break down anyway, the ICZM project will attempt to stabilise the coast with vegetative fencing, beach nourishment (dumping sand filled gunny bags or sacks) and strengthening existing sea walls. Considering the bottom topography, submerged artificial reefs with sand filled bags to break the wave energy and trap sand may also be tried.

Regenerate particular ecosystem components. The existing mangroves, the wetlands and the sacred groves will be rehabilitated. Some coastal forest sites will be identified and rehabilitated. These recreated forests will be nurtured by the Neighbourhood Groups (NHGs). Eventually limited forest products can be harvested.

Develop a strategy for sustainable ocean, pond, and lake fishing. It is proposed to build up artificial reefs along the coastlines to encourage fish breeding grounds and also as fish aggregating devices. These reefs can be made up of old tires or natural objects from the sea (Kurien 1995). A mini-fishing harbour is proposed for this region.

The main focus in the fisheries sector in the short run is going to be the promotion of pond and canal fishing. Of the nearly 30,000 ponds in the region at least 5,000 are suitable for fish cultivation. Technical assistance and financing for cleaning and stocking the ponds will be provided to households taking up fish cultivation.

Fig. 10.3 : Voluntary Training Programme in Progress
Photo by Sri Dev Das

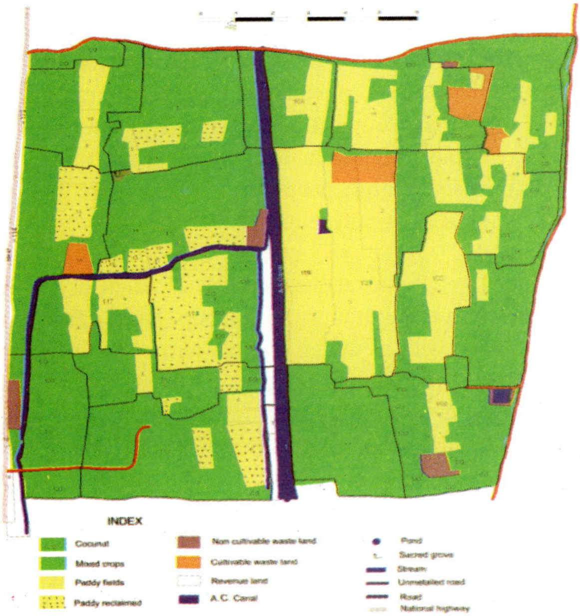

Fig. 10.4 : Portion of landuse map prepared by voluntees in Kanjikuzhy panchayat

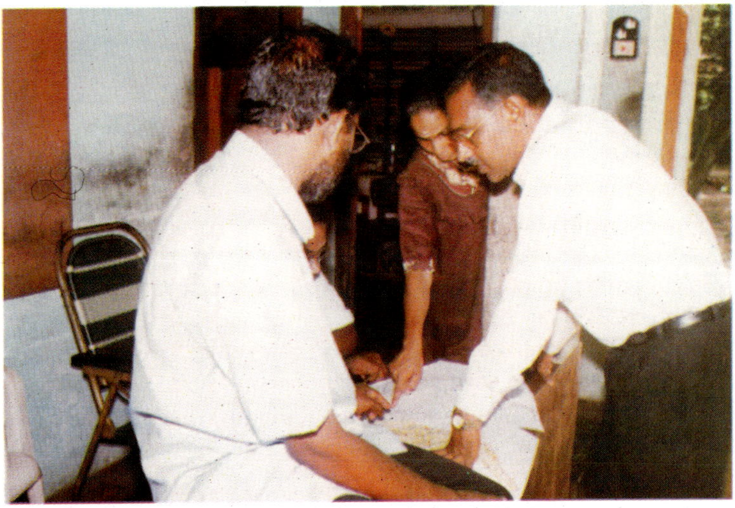

Fig. 10.5 : Woman Volunteer Explains Resource Map to Government Official
Photo by Sri Dev Das

Fishing in the Alappuzha-Cherthala canal. A major intervention will be the introduction of fishing in the Alappuzha-Cherthala Canal. The canal is to be entirely desilted and the overgrowth removed. This will also reduce the rat and mosquito populations substantially.

Creation of model villages to make more efficient use of environmental assets. People living near the shoreline are often threatened by erosion and sea surges. It is proposed to rehabilitate them in model villages, to be located in safer areas identified through land use zonation. A model village will consist of 250 to 300 households settlements in 10 to 12 clusters, each with 25 households. These 25 households will form a Neighbourhood Group (NHG). Each household will have 5 cents of land of which 2 cents will be built up area and 3 cents will be vacant. Within a cluster of 25 households there will be 75 cents (3/4 acre) of common space. Each housing unit will include dwelling space on the first floor and storage space on the ground floor. This cluster of 25 houses will have rainwater harvesting facilities, thereby its own water supply system, sanitation facilities and infrastructure for health and recreation. There will be common educational, marketing, health and recreation facilities for the model village of 250 to 300 households. The community will be entrusted with maintaining its local environment under the overall control of its grama panchayat. Because of the complex issues of ownership and finance, one cluster will be set up to begin the experiment. If it is found successful the programme will be expanded.

Sustainable, organic, community-connected agriculture. The Kanjikuzhy vegetable growing experiment described earlier in this chapter will provide the basis for an innovative agricultural expansion using sustainable practices and focused on local markets and on providing raw materials for local enterprises. At first, 1,000 acres (400 ha) of fruits, tubers, and vegetables will be cultivated, to be expanded to 3,000 acres by the fifth year of the project. NHGs will organise women to farm plots of one to five acres, with the goal of providing Rs. 50 per day in income per woman.

Some farmers will be encouraged to form Harita Sanghams (Green Groups) in conjunction with a programme sponsored by the Kerala Department of Agriculture. Farmers who join together on a minimum of 5 acres of contiguous paddy land receive a subsidy up to Rs. 50,000 for the purchase of inputs and other cultivation requirements. As a result, they are able to achieve significant economies of scale.

A new crop calendar will be prepared for the region taking explicitly into consideration seasonal price movements. Emphasis will be given to marketing by creating brand name identities for indigenous vegetables, entering into contracts with resorts and supermarkets and setting up sales outlets. The local farms will be linked to food processing micro-enterprises to be described shortly.

The vegetable-tuber-fruit cultivation programme will be based entirely on organic manure and organic insecticides. A manure collection unit will procure organic manure from outside the region and will also encourage the production of organic manure and pesticides by the local farmers. The organic waste from the neighbouring municipalities of Alappuzha and Cherthala will be transported to the compost pits owned by farmers in their fields. The coir pith waste and the waterweeds can also be recycled in this manner. A block-level composting unit will enter into agreements with the municipalities and will own sufficient vehicles for transporting the waste.

Bringing It All Together: Micro-Enterprises for Sustainable Development

Micro-enterprises allow for potentially large-scale development without the usual environmental consequences of large-scale infrastructure and large-scale businesses. In the Alappuzha-Cherthala ICZM project a complex of micro-enterprises will be built around and connected with the various environmental sustainability components just described in previous sections of this chapter. Fish, coconut, and vegetable products will be the basis for a growing economy while model villages, sustainable and organic production and recycling, and light local ecological footprints will maximise sustainability of the entire undertaking.

Products for Local Markets

The starting point for the micro-enterprises will be products such as soap and detergents that can be manufactured with local labour, requiring little formal skill, and using local materials to be sold in the local markets. Soap is ideal in this case because it is made up mostly of coconut oil, which is readily available in the area. As production expands, increasing demand for coconut oil could stimulate and support improvements in nut output which will result in more local raw materials for the coir industry as well since the same coconut that gives the oil for the soap has the fibres for the coir. Locally produced soap can compete with outside imported soap because the quality can be made practically equal and the price can be competitive. Local producers do not have to use fancy packaging or expensive advertising; they can market the product door-to-door, appealing to community solidarity and the multiplier effect of buying local products that benefit the local economy. Additional products might include school notebooks, school bags, tiffin boxes, uniforms, and umbrellas for which materials have to be bought from outside markets; and foods such as curry powder, flour, and food mixes.

Products for External Markets

Many agro-fishery-products from this area can be marketed outside the region. These include Indian style pickles *(achar)*, readymade curries, dried fish, cut-and-cleaned ready-to cook fish, mussels, vegetables, vegetables-in-bran, marinated and dried vegetables, soup powders, fried and powdered tubers, jams, fruits in syrup, cashew apple juice, and many others.

Improved Employment-Generating Coir Production

The coir industry today is organised in small-scale weaving units that sub-contract orders from exporters through middlemen. The project envisions setting up micro-production units of two to three spinning wheels each. These units will be supported by coordinating and resource supply facilities called cluster centres. The cluster centres will provide coir fibre at a reasonable price to the spinners. The cluster centres will have facilities for the storage of raw materials and finished goods, dyeing and bleaching, finishing, quality inspection, design supply and prototype development, and training facilities with master craftsmen. The cluster centres will be coordinated and funded by a Consortium of Cluster Centres. The centres will promote new product lines such as coir fibre impregnated packaging paper, coir handicrafts, products of other hard fibres, and coir pith manure – to be connected with the agricultural development described earlier. Another area would be coir geo-textiles. These have extensive application in infrastructure projects such as roads, bridges, bunds, embankment stabilisation, and watershed development activities.

Making the Production Infrastructure Environmentally Sustainable

The food products that constitute the bulk of the proposed micro-enterprises were traditionally produced within the household. To ensure uniform quality and hygiene, these production units will be organised in special work sheds that will be similar to the coir cluster units. A cluster of work sheds will have road access–thereby using only one road for the entire set of sheds. Each facility will have its own electrical hookup, drainage system, and other necessary common facilities. All cluster centres will have rainwater harvesting facilities, which would make them largely self-reliant in water. Toilets, changing and dining rooms, and a small office space are also provided for all the clusters. Each panchayat will have one or two such centres which the panchayats will own and will lease to the thrift-based micro-enterprise units that will evolve out of the NHGs. Thrift-based means that the startup capital derives from combining the small monthly savings of several households along with loans from local cooperative banks. These micro-enterprise units will be worker owned cooperatives that will federate in a larger cooperative designed to manage large-scale raw materials purchase and advertising and marketing outside the region as appropriate.

Biodiversity Programmes

In chapter 7 we noted Kerala's long struggle to protect the rich biodiversity of the Silent Valley area. But even with all the population growth and ecological change and damage that we have documented in earlier chapters, many villages still contain elements of biodiversity that can be built upon and reinforced. The project will include a major biodiversity recovery programme, emphasising the value to the inhabitants of regenerating their local biodiversity. It will focus on several areas.

Medicinal plants. Traditional medicines and herbal medicines play an important role in everyday life. Biodiversity programmes can encourage the cultivation of medicinal plants in house compound areas or within houses, by organising and developing seed kits of medicinal plants at a nominal price. This activity can generate rural employment by raising quality seeds. Some species appropriate for this area with likely medicinal value are jack tree (*Artocarpus heterophyllus*), drumstick (*Moringa oleifera),* tamarind (*Tamarindus indica)*, and papaya (*Carica papaya).* Bamboos, tuber crops, cassava and amaranthus also may have medicinal value.

Fast growing timber species in household compounds and village commons. For construction work and carpentry, every NHG can set apart areas for fast growing timber trees. These include *Artocarpus heterophyllus (Plavu or Jack), Artocarpus hirsutus (Anjili or Ayani),* and *Tetrameles nudiflora (Cheeni).*

Growing bamboos and save our artisans. Due to indiscriminate felling of bamboos for the paper industry wild bamboos have become endangered. Out of 135 species of bamboos in India, 25 species are in the endangered category. Bamboos are used in Kerala as building and scaffolding materials. They are also used by artisans for making decorative handiworks. Bamboo grows in moist conditions where water is available throughout the year but the soil is well drained. Water bodies, wastelands and fallow lands can be devoted to growing bamboo plants.

Historic farms and plant heritage sites. Since the introduction of high yielding varieties of cereals, some local rice varieties are becoming extinct. Historic farms which hold heirloom rice varieties and cultivars could be developed by the farming community. These might be sold at a high margin to plant breeders.

Wilderness gardens and sacred groves. In order to maintain biodiversity in modified habitats, the project envisions a network of wilderness gardens. Existing sacred groves associated with pilgrimage centres maintain ancient patterns of vegetation. These can be developed or rejuvenated. A survey indicated that there are more than 100 sacred groves in Kanjikuzhy grama panchayat. The majority of these kavus are owned by individual households. Family trusts and temple trusts also own sacred groves and a few of them occupy the lands administered by the Department of Revenue. Intervention in the sacred groves located on revenue lands needs government permission and interdepartmental coordination, whereas privately owned groves warrant separate institutional arrangement for intervention plans.

Some of the kavus house medicinal plants such as *Mukkutti, Narunengi, Thazhuthama, Poovamkurunil, and Kozhuppa* – plants well known to Kerala ayurvedic physicians and other traditional healers. Most of the kavus have been preserved by family or by temple trusts and are based on the old belief in snake worship. Local people also believe that ponds associated with kavus are perennial. The temple and family trusts are not always aware of the modern scientific importance of their kavus. Now panchayat councils, self help groups and nature clubs are making some conservation efforts. A detailed survey was conducted in one sacred grove, Ellanthu Kavu (Fig. 10.6). In it were found 27 floral species, seven of which are rare in the coastal tract. Some of the species are more than 100 years old (table 10.1). Two small ponds adjoin this grove.

Table 10.1 : Rare Tree Species in Ellanthu Kavu, Kanjikuzhy Panchayat

Western scientific species name	*Local name*	*No. of trees*	*Tentative age in years*
Calophyllum apetalum	Cheru punna	25	100–150
Hopea perviflora	Kambagam	9	Around 80
Vateria indica	Vella paine	1	Around 50
Lophopetalum wightianum	Vankottha	8	30–40
Holigigarna arnottiana	Charu	2	30–40
Articarpus hirsutus	Anjali	5	30–40
Garcinia gummigutta	Kodumpuli	3	20–25

Source : CESS 2003:6

These groves contain gene pools for reintroducing certain species in the locality. As part of a biodiversity conservation programme

Fig. 10.6 : Ellanthu Kavu Sacred Grove in Kanjikuzhy Panchayat. Photo by Sri Dev Das

kavus hold immense potential. Further, they can be developed as educational sites for local school and college students for training in tree identification and bird watching. Local tourism might also be developed around these sites.

A Biosphere Site at Pathiramanal Island in Vembanad Lake

A special feature of the ICZM project will be to regenerate and protect the high biodiversity site of Pathiramanal island. This island is located in the Vembanad lake and has an area of 28.5 ha. It lies about 1.5 km from the Muhamma boat jetty and about 4 km away from the Kumarakom boat jetty. The island is ideal as a pristine Backwater Island Conservation site and biosphere site. It is presently uninhabited.

The major vegetation zone of the island is a marsh comprising about 9.24 ha. Several small streams traverse the marshes containing the aquatic vegetation. The boundary of the island is a tidal marsh of 4.0 hectares. Between these two zones lies an inter-tidal flat zone of 1.99 hectares. About 2 ha of land was under paddy cultivation in the past but it has become desolate wasteland in the northwestern part of the island.

The tidal marsh zone is dominated by small mangrove trees, mangrove associates and a number of small to large climbers. These marginal littoral thickets are formed by small trees, shrubs and climbers. The composition of the island flora falls under five different groups: mangrove and mangrove associates; rare plants; medicinal plants and common plants. To create a biosphere protected area on this island, the following actions are needed:

- Proper de-weeding of climbers masking the canopy. These can easily be removed manually.
- Drawing up a strategy for managing the salvinia and water hyacinth by mechanically removing and converting it to compost for the manufacture of high fibre paper and liquid sugar for which the technology is now available. Or, the materials can become part of the compost for the agricultural development described earlier in this chapter.
- Reforesting and conserving the island margins with suitable mangrove plants. This will aid in reviving the populations of perching birds. Afforestation will enlarge the breeding and feeding grounds of prawns and fish.

The ICZM project combines environmental rehabilitation and protection with employment creation and democratic, egalitarian human development. The project represents Kerala's most advanced thinking and action so far and is the culmination of the experiences we have analysed in chapters 7 through 10. In the final chapter we shall raise the question of how Kerala's initiatives fit in with the all-India and international efforts to build an environmentally sustainable world without poverty or exploitation, and with justice and a decent life for all.

11 Consciousness Versus Collapse
Local Democracy and Our Endangered Earth

Past, Present and Future

In chapter 1, we surveyed the main endangered features of our earth. We pointed to recent studies showing that earlier societies had collapsed as a result of their failure to protect their long term resource base. The third version of the Limits to Growth study brought out frightening future scenarios of declining living standards, competition for scarce resources, and threats of violence resulting from the consequences of ecological overshoot. As we were completing this manuscript another authoritative study appeared. The *Millennium Ecosystem Assessment Synthesis Report* summarises the findings of a set of studies undertaken by 1,360 scientists from 95 countries. The *Report* warns of global warming, rising sea levels, more frequent and more devastating natural disasters, nutrient overload, collapse of fisheries, declining biodiversity and new epidemics. Like the Limits to Growth study, the Millennium Assessment projects a set of scenarios from a worst case with declining human welfare to better cases where an incipient form of sustainable growth and development becomes possible. Despite considerable uncertainty about the timing and despite substantial data gaps, both reports warn that humanity must find more sustainable means of interacting with the earth's life support system – the environment – now increasingly defined as a system of "ecological services."

Conceptual Models of Collapse

In chapter 1, we saw that the Limits to Growth approach considered and rejected a series of models of unlimited growth, sigmoid levelling off, and leaving the solution to natural processes. The Millennium Assessment Report (MA) approaches the problem with a somewhat different conceptual frame. Here collapse is interpreted to result from "ecological surprises" and "nonlinear change." Examples of recent ecological surprises include the ability of pests to evolve resistance to biocides and the eutrophication of water bodies leading to dead zones (MA 2005:130). Ecological surprises might occur from natural processes or from human activities. While no one can predict what the next surprises will be or when they will happen, "we can be certain there will be some" (MA 2005:130). Similarly, while ecosystems display significant inertia in the sense that the effects of interventions and of overshoot can be delayed for years or even decades, when gradual deterioration reaches a tipping point major changes occur rapidly. These rapid changes break the linear pattern of gradual change. Examples of nonlinear change include the sudden appearance of diseases and of epidemics, algal blooms and fish kills, fisheries collapse, coral reef species alterations, and local and regional climate changes (MA 2005:150). The Millennium Assessment Report (MA 2005:151) considers high impact nonlinear changes to be increasingly likely as the ecological footprint widens and deepens and overshoot increases. Among the most feared nonlinear events is the possibility that the atmosphere could reach a point of saturation with hydrocarbons and reactive nitrogen compounds such that it would cease its function of cleansing itself of air pollution (MA 2005:151).

Conceptual Models of Sustainability

In chapter 1 we introduced the concept of sustainability, the concept at the heart of avoiding nonlinear change and reducing the number of ecological surprises. Is there a conceptual model of sustainability from which we can derive a 21st century scenario? Following the unacceptable scenarios we outlined in chapter 1, the Limits to Growth authors identify a "Scenario 9," a computer simulation in which human welfare stabilises at a reasonable level while the ecological footprint gradually declines. Scenario 9 requires an international decision to limit family size to two children, to set modest limits for material production, to invest heavily in new technologies that increase the efficiency of resource use and that decrease pollution emissions per unit of industrial output, control land erosion, and increase yields to maintain adequate food supplies (Meadows, Randers, and Meadows 2004:245). Scenario 9 depends on population stabilisation, technological creativity, and massive amounts of political will. An occasionally stated premise is that the market based short term profit system would have to be reorganised in some way to assure that proper long term signals are given. We shall take up this premise later in this chapter.

The Millennium Assessment Report (MA) approaches the issue with a set of four scenarios with some elements quantified (MA 2005:123). All the scenarios require that humans bring about a change in the direction and/or intensity of the "drivers" that we referred to in chapter 6. The MA scenarios have names instead of numbers :

- *Global Orchestration*. Trade and liberalisation dominate, ecological problems are treated reactively, poverty and inequality are reduced, heavy investments in education and infrastructure.
- *Order from Strength*. Nations or regions focus on their own problems. Security, protection, little attention to public goods, reactive approach to ecological problems.
- *Adapting Mosaic*. Regional watershed ecosystems are the focus. Local institutions strengthened and local management strategies developed. Proactive approach to management of ecological systems.
- *Techno Garden*. Globally connected, highly managed ecosystems. Proactive approach.

It can be seen that the MA scenarios contain some of the same elements as are found in Scenario 9. The MA scenarios include both population and economic growth forecasts. The overall picture presented is complex. But the MA scenarios add two novel components to Scenario 9: drastically reducing world poverty and strengthening local institutions. These components lead us back to Kerala and what we believe is the significance of our study.

The Kerala Experience for a Sustainable World

We contend that the earlier chapters in this book show that Kerala's experience over the past several decades, and especially the last 20 years provides the world with a rich fund of experiences that can enlighten and inform activists and policy makers trying to cope with the problems of overshoot and potential collapse of ecological systems. At the same time Kerala itself faces ecological problems serious enough to warrant an acceleration of the state's current actions for sustainability. Let us consider several elements of Kerala's unusual potential profile for sustainability.

Population

Unique among India's states, and highly unusual for a non-industrial area, Kerala has experienced a steady decline in birth rates in the 20th century. Average annual population growth in Kerala between 1991 and 2001 was 0.91 per cent. This would lead to a doubling in 79 years, far longer than the 36.3 years for India overall. Total fertility in Kerala is now 1.5 births per woman compared with a worldwide average of 2.7 (GOK 2005:30; UNDP 2004:155. Doubling time computed by the authors). Indications are that Kerala's birth rate will drop still further, approaching replacement levels only. By stabilising its population, Kerala has already achieved one of the prerequisites on which all experts agree any attempt at sustainability must be based. Kerala's achievements in lowering birth rates have been the subject of numerous studies and debates that we shall not revisit here (e.g. Franke and Chasin 1994; Krishnan 1976; Ratcliffe 1978; Ramachandran 1998; Zachariah et al. 1999). For sustainability purposes we note that Kerala has met the goal.

Social Justice, Inequality, Poverty and Sustainability

Reducing inequality is seen in nearly all reports and scenarios as a formidable prerequisite for creating sustainable development. The increasingly unequal social structure of the world threatens to further endanger the ecological systems of our earth. Poverty leads to environmental harm in two major ways. Poor people have limited choices about their economic activity and are often compelled to engage in ecologically unsustainable actions such as forest destruction for firewood or planting on hillsides too steep for sustainable agriculture. At the top of the inequality spectrum wealthy consumers demand high priced and high status resources often without seeing the long term costs of their choices. Inequality can be both among and within nations. In chapter 1 we noted the excessive environmental impact of the wealth of US consumers. While the US is the world's largest single source of ecological overshoot, the entire international system of inequality facilitates ecosystem destruction even when countries such as some in West Europe make serious national efforts at environmental protection. This has led to what some observers call "the Netherlands Fallacy" (Jorgenson 2003:337; York, Rosa and Dietz 2003:287). The Netherlands has achieved remarkable success in reducing pollution and moving towards more sustainable practices domestically while acting responsible in international forums and in diplomatic circles to promote green thinking. Yet, ecological footprint analysis shows The Netherlands with a biocapacity of only 1.7 hectares per person (see chapter 1) while its wealth and position in the world inequality system allow it to purchase another 3.6 hectares of capacity (Wackernagel and Silverstein 2000:393). (The US total figure is 12.2, the world's highest – York, Rosa and Dietz 2003:282). Those 3.6 hectares are either denied in effect to poorer nations or they become part of the overshoot. Rich countries with high per capita incomes externalise their ecosystem harming consumption via the world system of international inequality (Jorgenson 2003:388), itself largely a heritage of colonial and imperial history. Sociologists Richard York, Eugene Rosa and Thomas Dietz (2003:293) found in a cross national statistical analysis that population size and affluence (measured by GDP) could account for 95 per cent of the variance in the ecological footprint among nations.

Within India as elsewhere poverty and inequality are closely related. One study (Chakraborty 1999) has specified the links between inequality and ecological decline in India's forests, grazing lands, water and soil. The author – an economist – notes the dynamic positive feedback relationship in which inequality leads to ecosystem decline which leads to further inequality. Chattopadhyay's (2004) study in selected stretches of the Kerala coast demonstrated a spatial association between the incidence of poverty and the degree of environmental degradation in certain pockets. It also indicated that environmental protection can lead to reduction of poverty. The Zero Waste Kovalam Programme, an initiative to protect the environment through waste reduction by a local NGO "Thanal" at Thiruvananthapuram, created nearly 160 jobs within Venganoor and Vizhinjam panchayats (Thanal 2005:15). Thanal trained local women to make paper bags from waste paper, cloth bags from tailoring waste and other material substitution programmes. In January 2003, 13 Below Poverty Line (BPL) women came together to start a paper products unit with the help of the panchayat, financial assistance from Kudumbasree and a bank and technical assistance from Thanal. The women involved in this programme now earn between Rs. 600 and Rs. 900 per month

(Thanal 2005:17). "Earlier we were unemployed and did not know what to do. Now we know lot of things and people also know about our products. People who used to ask 'what is the alternative for plastic bags' do not repeat their questions after seeing our products," says one woman from the group. Similarly, rejuvenation of canal segments or ponds or other water bodies is an action towards environmental protection. Involving BPL people through Women's Self Help Groups to develop fishery in these water bodies can directly help in poverty reduction.

Kerala has brought down levels of inequality and overcome substantially the scourge of poverty. Franke's study of Nadur village (1996) found that the period of intense radical reforms from 1971 to 1987 brought down the Gini Index of Inequality by 16.8 per cent. K.P. Kannan (2000:45) reports that by 1997 only 15 per cent of Kerala's people could be deemed poor using the UNDP's Human Poverty Index. By contrast, 31 per cent of all-India was poor by this measure in 2000 and among 32 countries listed as "low human development," poverty figures run from 31 per cent to 65 per cent (computed from UNDP 2004:148–49). Kannan notes further (2000:52–53) that in 1973 Kerala had a 10 per cent higher poverty rate than all-India but by 1997 the state had brought down its poverty rate to 30 per cent less than all-India. Kerala's successes in reducing both caste and rural-urban inequality and in overcoming some elements of gender inequality have been noted in several studies. In sum, Kerala has already achieved a major part of the sustainability prerequisite of overcoming poverty and inequality.

Behind Kerala's Success: Activism and Participation

Significantly, Kerala's achievements in lowering birth rates and overcoming much of its poverty result from driving forces that can be redirected towards achieving sustainable development. Kerala's achievements result from a mix of several factors—its location, environmental resource base, the history of commerce, the particular effects of colonial rule in generating an early working class, enlightened princely families, and possibly other factors. But the overarching element has been the rise of militant and creative people's movements. The modern movements began in the 19th century with caste improvement associations but were rapidly transformed into class and peasant movements; literacy, public education and health movements; land reform; workers rights; and the famous People's Science Movement (KSSP) that has been referred to in earlier chapters. These movements mobilised hundreds of thousands of people to sign petitions, write letters to newspapers, march in demonstrations, go on strike, and vote for leftist candidates. While observers debate certain details of Kerala's history, we believe it is reasonable to assert that Kerala's achievements in reducing its population growth, overcoming poverty and inequality, and raising levels of education and life expectancy can be traced primarily to these people's movements, and to the motivated and creative leaders these movements produced.

The Kerala Ecological Paradox

Kerala's achievements suggest a paradox, however. As we showed in chapters 4–6, Kerala's environment has undergone severe degradation. If Kerala's people could achieve so much in social advancement and human welfare and capacity building, why did they allow their environment to deteriorate and therefore why would they be of interest to activists and policy makers today who seek to build a sustainable world?

The answer, we believe, is contained in a famous insight by Karl Marx, as presented near the beginning of his study *The 18th Brumaire of Louis Bonaparte* (Marx 1968 [1852]:15) where he writes that people

> …make their own history, but they do not make it just as they please; they do not make it under circumstances chosen by themselves, but under circumstances directly encountered, given and transmitted from the past.

The history of Kerala's social movements illustrates Marx's idea. The caste improvement associations and the educational expansions of the 19th century in Kerala were influenced by the historical circumstances of colonialism and the interaction of Kerala's educated elite with the liberal ideas of some British thinkers such as John Stuart Mill. Much of the 20th century thinking in Kerala was influenced by the circumstances of the Russian Revolution, the depression and World War 2, the rise of the independence movement and of Gandhian thinking about a life of voluntary simplicity, all reinforced by the ideas of secular democracy and social justice that found deep roots in Kerala. Throughout these periods the dominant ideas worldwide as well as the actual experience of the people was that nature is boundless and is there for humanity to conquer, to master. Kerala was the only state in India to adopt a Science Policy Resolution in the 1970s. This policy resolution led to the creation of important scientific research centres such as the Centre for Earth Science Studies (CESS), the Centre for Water Resources Development and Management (CWRDM), and the Kerala Forest Research Institute (KFRI), whose studies provided some of the significant data and analysis used in our study.

By the 1970s Kerala's people faced new circumstances, influenced in part by their own struggle to save Silent Valley (as described in chapter 7) and in part by their now established traditions of cosmopolitan interaction with the international world of ideas. As the awareness of the fragility of the earth's life support system gained momentum, Kerala was one of the areas most deeply affected. This can be seen in the rapid development of the People's Science Movement (KSSP) and state level initiatives such as the State Institute of Languages from a focus on translating English scientific works into Malayalam and on bringing scientific thinking into the villages to a focus on organising to protect the environment. It is further seen in the rapid development of environmental consciousness among the generation of leftist activists that emerged in the 1980s to bring new ideas into the 1987–91 LDF ministry that resulted in the New Democratic Initiatives including the Panchayat Resource Mapping Programme that we described in chapter 8.

In short, Kerala's people have a history of making history by activating high levels of consciousness in response to the circumstances existing at each period. As their society has evolved through various struggles and reforms they have constantly thrown up leaders and thinkers capable of identifying the means to make history in the next set of circumstances. This process is dramatically illustrated in the People's Plan Campaign that we discussed in chapter 9 and in the various creative approaches to sustainability that have been introduced within and around the edges of the Campaign that we described in chapters 8 and 10. Furthermore, because Kerala's movements have remained vibrant and many of its people continue the traditions of activism and participation, the ideas from abroad and the ideas being generated within Kerala's own local experiments can potentially be transmitted broadly across the landscape into all the villages, and throughout the society. Few other places in the world at this time have achieved such a strategic convergence of elements necessary to attempt sustainable development.

Consciousness in Theory and in Kerala

In chapter 1 we presented several conceptualisations of sustainability. Most of them state or imply a concern for the life situation of future generations (see also Drummond and Marsden 1999:8). This concern conflicts with the short term view of modern capitalism that cultivates an orientation towards immediate material satisfaction. Indeed, the expansion and continuation of modern capitalist economies depend on selling products whether those products enhance life styles or not and whether those products help or harm the process of sustainability. Within the Western world, small numbers of individuals have begun to resist this tendency and a movement towards "voluntary simplicity" has arisen (Elgin 1993).

Kerala, by contrast, has two deeply rooted focal points of consciousness from which to establish an understanding of the connection between present and future generations. One of these is the Gandhian tradition – itself the voluntary simplicity concept par excellence. While Gandhi's ideas are sometimes more talked about than practiced, they nonetheless provide a significant backup for activists seeking to spread notions of living simply when possible. A second deeply rooted Kerala concept is that of egalitarianism, an idea fostered by the Marxist and left movements that have flourished in Kerala now for almost a century. If people can conceptualise equality among castes, classes, and genders, then it is not such a large leap to the idea of intergenerational equality – that future generations have equal rights to the bounty of the earth as does our own generation.

How widespread are Gandhian simplicity and Marxist egalitarianism in practice in Kerala? Many observers are skeptical, noting that Kerala's wealthy live heavily on the earth and arguing that poor people in Kerala use fewer resources primarily because they cannot afford to use more (e.g. Véron 2000:220–21). At a minimum we can observe that many activists make a point of engaging in life styles that set examples of simplicity. At the very least we can say that Kerala has a strong basis for expanding the consciousness of intergenerational equality along with its implications for present generations.

Another important consciousness element is cooperation. Standard neoliberal thinking today emphasises the superiority of competition over cooperation. Often it is now taken as axiomatic that cooperation leads to lower output, less achievement, and less innovation. Despite the axiomatic appeal of competition, Alfie Kohn (1986) has summarised substantial psychological research that finds cooperation superior to competition on almost every dimension from achievement to human satisfaction.

Competition is hardly absent from Kerala life and certainly not from Kerala business practices, but some estimates put the number of cooperative enterprises at 17 per cent of the state's output. Kerala's cooperatives have shown remarkable resiliency and an ability to foster an alternative business ethic in which employment generation and enhancement of local communities take precedence over simply increasing profits to share with private stockholders (Thomas Isaac, Franke and Raghavan 1998). The People's Plan Campaign we discussed in chapter 9 and several of the local initiatives described in chapters 8 and 10 make extensive use of cooperation and little use of competition. In the long term, it seems more likely that cooperation rather than competition will be axiomatic for sustainability.

Theoretical Issues: The Market *Versus* Sustainability

One of the most profound issues dividing sustainability from present day capitalist consumerism is that of markets versus common pool or public resources. Throughout the Millennium Assessment assertions are made about the need to use more market-based approaches and more international trade (MA 2005:18, 129, 131). Yet the same report notes that markets fail to provide appropriate signals that might alter investment patterns towards more sustainable resource use patterns (MA 2005:35). Ecosystem services economists Robert Costanza et al. (1997:259) put it this way: "If ecosystem services were actually paid for, in terms of their value to the global economy, the global price system would be very different from what it is today."

The Limits to Growth study is even more explicit in recognising that "market signals have not yet given the world useful information about impending physical limits" (Meadows, Randers and Meadows 2004:227). In a detailed case study (2004:223–34) they describe the failure of markets to prevent the collapse of the Newfoundland fishing industry in the early 1990s, a case now recognised worldwide as a possible harbinger for future fisheries. Furthermore, when combined with inequality, the market has compounded the collapse because well off first world consumers will pay high prices for once plentiful fish such as cod as they become more prized in the market.

There are two aspects to the problem of how markets relate to sustainability. First is the fact – implicit in the discussion just above – that markets provide pricing signals based on supply and demand but not on whether sustainable limits are being approached. In the words of ecological economists Wackernagel and Silverstine (2000:392) "there are no stabilising feedback loops working against overshoot." Thus the Newfoundland fisheries went on pushing the catch because the market provided signals telling them it was profitable. By 1970 800,000 tons were being harvested but in 1992 the catch dropped to near zero (MA 2005:50). After 13 years of attempting to regenerate the fish populations by a partial moratorium, the fisheries were closed indefinitely and it is not certain if they will ever recover (MA 2005:110). When the fish population suddenly plummeted, it was the public sector that had to absorb about $2 billion (MA 2005:92) in what some economists call the "externalities," the economic and human consequences of the market, through government welfare and retraining projects for the fisherfolk. Against this backdrop it is significant that the Integrated Coastal Zone Management project in Kerala that we described in chapter 10 aims at a scientifically managed catch that will not bring the *chagara* to the point of collapse.

Market or Capitalism? Marxist and Socialist Perspectives on Ecology

Why can't the market tell the ecological truth? The market is only an exchange mechanism and as such it reflects more fundamental economic processes. According to Marxist ecologist John Bellamy Foster, it is the capitalist accumulation process that destroys the environment because the competitive search for profits – mediated through the market mechanism but with assistance from the capitalist state – trumps all other concerns. The same processes that generate poverty and inequality in capitalism lead to destruction of the environment and the impossibility of a sustainable world. In a nutshell, "When set against the get-rich-quick imperatives of capital accumulation, the biosphere scarcely weighs in the balance" (Foster 2002:21–22). Foster (2000:163–70) argues that Marx was one of the earliest to observe capitalism's inefficiencies and misuse of natural resources. As to whether present tinkering with the market (capitalist) economy can generate the elements of sustainability, Foster (2002:67) writes that "Nothing in the history of capitalism…suggests that the system will be up to such a task."

As to the idea of using market forces to stimulate increased energy efficiency through technological advance–one of capitalism's recognised strengths, Foster draws on the work of a 19th century British economist named William Stanley Jevons. Jevons discovered a paradoxical relationship between increased efficiency and overall resource consumption: technology that results in increased efficiency in the use of a natural resource leads to increased demand for that resource and eventually increased depletion, not a slowing of depletion rates (Foster 2002:94–95). In today's world we see an example: more fuel efficient cars lead to more car purchases and more miles driven per car. The Jevons paradox is driven by market forces and therefore its resolution can only be achieved through economic planning – the opposite of the market. In the present period increasing resource efficiency is becoming less common with the Bush Administration launching a furious and sustained assault on even the modest gains of the environmental movement of the past three decades (Austin and Phoenix 2005).

Economist and libertarian socialist Robin Hahnel argues for democratic, participatory planning using workers and consumers councils in place of the profit-driven private property economy of capitalism. According to Hahnel (2005:200–207) capitalism cannot by its competitive profit-driven nature reduce pollution to "efficient" levels, nor can it function according to the necessary principle that the "polluter pays" for pollution. The system rarely compensates the main victims of its environmental destruction and it inherently hides the true costs of environmental destruction – thus obviating what economists call "incentive compatibility."

Perhaps the most theoretically ambitious alternative to date has been Joel Kovel's (2002) proposal of *ecosocialism* as the concept needed to bring classical Marxism and modern environmentalism into a needed conjuncture. Along with Foster and Hahnel, Kovel argues that capitalism has inherent properties that prevent it from ever developing a sustainable economy. He envisions ecosocialism as "a great network of productive communities, from agricultural cooperatives, to transnational scientific teams, to governing assemblies…(2002:247). How to create this great network? Hahnel (2005:382–84) envisions democratic cooperatives growing in size and number combining with anticapitalist movements to eventually create a "grand alliance." Kovel (2002:252) argues the need "to build small beachheads—liberated zones that can become the focal points of resistance and combine into larger ensembles." Kerala's experiments in democratic decentralisation and environmental activism seem consistent with both proposals.

Theoretical Issues: The Common Pool Resources Dilemma

A major problem with capitalist approaches to sustainability results from the fact that well over half of all ecological services to support human life fall outside the realm of markets. These services include gas (atmospheric) regulation, disturbance regulation, waste treatment, nutrient cycling, and the creation of aesthetically or spiritually pleasing surroundings. (Costanza et al. 1997:259; MA 2005:22). Despite their critical importance, these services have not been studied adequately (MA 2005:11). One tentative finding is that poor people are more directly dependent on ecosystem services (MA 2005:17). We have noted that the evidence from Kerala supports this view at the local level. Another finding is that developing nations are more directly dependent than are industrial nations (Sutton and Costanza 2002). Another is that the net benefits from more sustainably managed systems "are greater than those from the converted ecosystem, even though the private (market) benefits would be greater from the converted ecosystem" (MA 2005:48; cf. 148). The market and its associated short term consciousness thus compete with and undermine sustainability.

By lying outside the market, most ecosystem services automatically lie outside the system of private ownership. They are owned by everyone and no one in particular. One set of such ecosystems included "commons" lands that have been reported from nearly all parts of the globe at one or another historical time. The commons became the subject of a famous essay by Stanford biologist Garrett Hardin (1968) – "The Tragedy of the Commons." Hardin argued from a general theoretical perspective – without studying any real cases – that population pressure would inevitably lead to destruction of the commons because each individual can increase her/his wealth by intensifying individual use of the commons while eventually the community as a whole will suffer the consequences of the subsequent decline in the resource base. Therefore individual ownership is preferable. For these reasons tragedy of commons thinking lies at the basis of some privatisation arguments.

But is the tragedy of the commons approach accurate? The West

African Sahel famine of 1968 to 1974 became a classic case when it was argued that local animal herders overgrazed their pastures: each herder trying to increase output by increasing animals until the pastures succumbed to overuse and overtrampling. Franke and Chasin (1980 : esp. 120–22), however, showed that the herders had maintained collective management of the pastures for centuries. It was not the single variable population growth driver but the penetration of capitalist production systems facilitated by French colonial rule that brought on the overgrazing (Franke and Chasin 1979). The herders had in fact developed cooperative approaches to common resource use that had survived for centuries before colonial rule shattered them (Franke 1987).For India, Mike Davis (2001 : 326) found that the "El Niño" temperature oscillations of the southern Pacific Ocean generated a sequence of crop failures in the late 19th century that the British colonial regime exploited to confiscate common lands that had been community managed for centuries. In the process the colonial regime tolerated ineffective famine relief measures that appear to have left 19 million dead in a wide band from southeast to northwest India (Davis 2001 : 29, 163 and 174). Kerala seems to have avoided this disaster because of its location to the west of the Ghat mountains.

In recent years the tragedy of the commons debate has evolved into discussions of "common pool resources," (CPRs) a synonym for the commons. The Millennium Assessment (MA 2005:33) bemoans the finding that "...existing national and global institutions are not well designed to deal with the management of common pool resources, a characteristic of many ecosystem services." This should hardly be surprising in an era of neoliberal attacks on public ownership and the spread of doctrinaire and dogmatic ideas about the superiority of private ownership and competition into every sphere of life worldwide. As the prominent anti-corporate-dominated globalisation theorist and activist Naomi Klein puts it (2002:85), "The market swallows the commons."

All the more reason, then, at this historical moment, for local structures to put up as much resistance as possible by reaffirming and expanding their own age-old traditions for managing these resources. Sahelian cattle and goat herders, it turns out, are hardly unique in creating such management practices. Elinor Ostrom (1990:61–65) described successful local management of common pool resources among Swiss cattle farmers dating from AD 1224. Rural Japanese collectively managed forests and uncultivated mountain meadows for centuries. Spanish farmers managed a complex water distribution system from AD 1435 to the present and Philippine farmers created common irrigation works called *zanjeras* from at least the 17th century (Ostrom 1990:65–88). In south India, particularly in Tamil Nadu tanks are ancient, decentralised and multipurpose common property resource systems that continue to act as flood storage and can be used for irrigation purposes. On the island of Bali anthropologist J. Stephen Lansing (1991) found that Hindu temples provided not only religious activities but utilised a ritual calendar to manage the dispersal of irrigation waters in ways that corresponded to the most efficient design derived from a computer simulation of agricultural needs. Lansing (1991:48) also found that Balinese flooding and draining practices provided substantial biological pest control that got interrupted when foreign advisors from the "Green Revolution" establishment attempted to redesign the irrigation flow to fit the high yielding rice varieties, leading to pesticide pollution.

Ostrom's work on the management of common pool resources led her to identify 8 characteristics associated with successful or long enduring cases: clearly defined boundaries of the resource, appropriate rules for usage, participation by members in drafting or modifying the rules, an effective monitoring system, graduated sanctions for rule breakers, conflict resolution mechanisms, lack of interference from larger outside organisations, and effective delegation of powers and responsibilities between the local CPR group and any larger political units of which it is a part (Ostrom 1990:90). Her findings are almost an exact list of the main characteristics of the Kerala's People's Plan Campaign, assuming that the CPR unit is a local panchayat council with its associated grama sabhas and other public meetings.

But what are the common property resources in Kerala? The ocean, the sacred groves, the national parks and reserved forests, the waterways, the mountain slopes? The Karanila that we described in chapters 7 and 10 is the most clearcut example of a direct CPR institution in Kerala. Most other economic activity takes place on privately owned resources. Prior to the People's Plan Campaign left activists and policy makers had developed the group farming approach to rice production which encourages farmers to band together to manage water, fertiliser, planting schedules, and the like while retaining their privately held lands. Those lands become in effect temporary CPRs. Although the project had a mixed record (Véron 2000:224–25), in some panchayats such as Kanjikuzhy which we described in chapter 10, it led to dramatic increases in output without apparent negative environmental consequences. And the Panchayat Resource Mapping Programme we described in chapter 8 helps create a popular and public awareness of the relation between individual assets and long term community needs. In so doing, the PRM raises individual consciousness towards a virtual CPR consciousness. When combined with Kerala's already high public consciousness about social justice, the continuing influence of Gandhian and Marxist ideas, and the general effectiveness of the cooperative movement, a basis surely exists for a high level of popular understanding of sustainability and its requirements.

Political Structures for Sustainability

Our consideration of common pool resources and the institutional and consciousness requirements for making them succeed leads naturally to one of Kerala's most recent innovations: meaningful devolution of decision making power and responsibility in the context of substantial popular participation. These were hallmarks of the People's Plan Campaign as described in chapter 9 and the many related local initiatives described in chapters 9 and 10, many of them connected in theory and somewhat in practice to the Panchayat Resource Mapping Programme we discussed in chapter 8.

A key element in any plan to advance sustainability is to involve the immediate users of resources as directly as possible in the decisions about use and about long term management. René Véron (2000:224–26) has labeled this approach "community based sustainability." Its units of action are the actual stakeholders, a strategy of enlisting those persons from rural areas and poor households most impacted by ecosystem changes. We saw in chapter 10 how Perambra block could mobilise hundreds of citizens to dig contour pits, how Adat and Olavanna panchayats could set up local drinking water cooperatives, and how Thathamangalam farmers could create irrigation CPR institutions reminiscent of Ostrom's Spanish and Philippine cases. In Erimayur women emerged as key players in bringing about a community based drinking water project that has led to continuing economic benefits for a group of skilled women plumbers. The Koyilandy biological mosquito control project took place in an urban area but depended on a high level of community participation and activism. By avoiding harmful pesticides the community based programme in Koyilandy showed a creative means towards sustainable pest control.

Why Community Based Sustainability?

Many of the world's most threatening potential ecosystem failures undoubtedly require large scale international action. Why spend time then, highlighting these small scale experiments? We believe there are two major reasons. One is corporate dominated globalisation. At this moment in history private corporate expansion and rewards for short term investments for short term profits dominate the world scene. Corporate dominated globalisation is advanced by the richest and most powerful companies and individuals and is backed up by the force of the US military and the far right administration in Washington along with its minions in Britain and some other first world countries. These corporations and their wealthy stockholders are the main beneficiaries of the system of world inequality we described earlier in this chapter. While large scale international opposition to the Washington Consensus is undoubtedly a needed step towards promoting eventual sustainability, right now community based sustainability experimentation constitutes the most promising avenue for generating knowledge about what to do and for generating inspiration by showing that people can actually do it. Kerala is strategically an advanced outpost in this regard. More than that, the state constitutes a giant laboratory of creativity for which the documentation can be useful for planners and activists in other areas.

Secondly, many problems can be dealt with through a community based approach. One of the most serious sustainability issues facing humanity today is nutrient overload – the accumulation of reactive nitrogen and other chemicals in soils and water bodies. The Millennium Assessment (2005:125) identifies nutrient loading as an increasing problem for developing countries. We saw in chapter 5 how Lake Vembanad and other water bodies in Kerala are subjected to increasing levels of nitrogen and carbon that lead to eutrophication. Yet we saw in chapter 9 how Mulamkunathukavu panchayat could significantly reduce fertiliser (nitrogen) loads while simultaneously improving tillage practices and water management and increasing cropping intensity. This corresponds precisely to one of the major proposals of the Millennium Assessment as part of both the Adapting Mosaic and TechnoGarden scenarios (MA 2005:132).

Devolution and Empowered Local Democracy as Structures for Community Based Sustainability

Decentralisation can take weak or strong forms. Passing on some authority from a central office to a local line department – a form of decentralisation often called "deconcentration" does little to mobilise the strengths of a community (Thomas Isaac and Franke 2002:197). Devolution—the transfer of significant resources along with substantial decision making powers—provides a basis for genuine community participation. People participate because they have something meaningful to achieve and they can see a chance for real benefits. Thomas Isaac and Franke's study of the Kerala People's Plan Campaign brought out the theoretical basis for local democracy as a means of advancing development among a population with high literacy, political consciousness, and widespread grassroots organisations. Local democracy can facilitate transparency—reducing opportunities for corruption—improve the efficiency of implementation of development projects, draw on local skills and knowledge, make for better project monitoring and evaluation, and connect projects to the perceived needs of the people as they express them in their local assemblies (Thomas Isaac and Franke 2002:8–9 and 22–25). It can also become a mechanism for drawing out volunteer activities and contributions that in the People's Campaign of 1996–2001 appear to have added about 10 per cent on top of the government devolved resources (Thomas Isaac and Franke 2002:7). A systematic study of the health projects in the campaign revealed that health services became functionally divided among the different levels of administration, that the health projects reflected the expressed needs of the local population, that levels of equality were maintained or expanded, and that several local communities developed innovative projects (Elamon, Franke and Ekbal 2004).

A systematic study of the environmental projects in the People's Campaign remains to be undertaken, but the evidence presented in chapters 8–10 suggests that devolution and local democracy created a milieu in which protection and enhancement of resources were frequently on the agendas of local communities. Community based sustainability has an important future in Kerala.

Devolution and ITK

Most experts agree that indigenous technical knowledge (ITK) can play a major role in ecosystem maintenance. This is especially true in the era of modern science when local knowledge can be tested and enriched with the scientific method (Franke 1987). Local democracy and participation increase the likelihood of bringing ITK into the process of sustainable development as indicated by the Ernakulam Biodiversity Project we described in chapter 10. Without the energy and resources of the People's Plan Campaign it is unlikely such an outpouring of information would have occurred or that it would have been so assiduously documented and preserved. The incredible richness of the Kalliasseri biodiversity study and the Kanjikuzhy sacred grove project – also noted in chapter 10 – provide further evidence of the potential of ITK waiting to be unlocked by further experiments in empowered local democracy in Kerala and elsewhere.

The Scaling Up Problem

Development experts and ecosystem specialists have long recognised a central weakness of community based projects. They usually cannot be scaled up to higher levels (MA 2005:165). This problem is also referred to as the "small project contradiction" (Franke and Chasin 1980:235). Scaling up problems result from two different related problems. First, what works on a small scale might not be relevant or appropriate to a larger ecosystem or political unit. Second, the units themselves are not compatible: village boundaries rarely correspond to relevant ecosystem boundaries. Kerala's community based sustainability programme has developed partial solutions to both of these problems.

For scaling up to larger political units, the People's Plan Campaign devised a creative system of interaction whereby village projects were sent up to the blocks and block projects were sent up to the district councils. Each higher level of administration is to examine all the projects within its jurisdiction. Within an overall framework developed by the council at the higher level, they are to look for local projects that compete with or contradict each other and develop projects to manage that competition. In addition, they are to look for gaps in the planning at the local levels and generate block or district projects to fill in those gaps. The scaling up problem is thus handled by an increasingly integrated set of projects at higher levels – each of which bases its own planning on the planning from one level below (Thomas Isaac and Franke 2002:98–101). Ostrom refers to this theoretically as "nesting." The evidence suggests that this process worked only to a limited extent in the first five years of the People's Plan Campaign, but it remains in place for a future push to broaden its effectiveness. Even so, the People's Plan Campaign facilitated the successful scaling up of the Kudumbashree (family prosperity) antipoverty project in Kerala that is now being emulated in other Indian states (Kadiyala 2004).

A second means of overcoming the scaling up problem in Kerala was to establish the idea of watershed development plans. Watersheds rarely overlap with village boundaries. A key element is to introduce watershed consciousness from the local level up. The watershed planning process paralleled the scaling up process just described in that villages would identify their watersheds, those watersheds would be identified as parts of larger watersheds and eventually an overall watershed management programme should emerge. The success of Perambra block in creating a watershed management programme with significant people's participation that we described in chapter 10 indicates the potential of this approach.

A third means of approaching the problem of scale is to create mechanisms for knowledge and experience sharing so that local projects in one area can be adopted and adapted in other areas. Kerala made significant strides towards this goal during the People's Plan Campaign by setting up frequent district and state level seminars where representatives from one village could tell members of several others about their successes and failures (Thomas Isaac and Franke 2002:177–180). Most seminars included booklets or books printed in Malayalam – the language of Kerala – with details of the projects. This process culminated in the May 2000 International Conference on Democratic Decentralisation at which over 700 village representatives gathered in Thiruvananthapuram to share ideas and for which 6 volumes of local reports were published.

Finally, we should note that scaling up is not always an essential criterion for assessing the viability of a local initiative. Up scaling the utility of a pond beyond its capacity may turn out to be unsustainable. The concept of pond rehabilitation may be up scaled (such as through the seminars to spread information about it to other local communities) but not its physical functioning.

Do Third World People Care About the Environment?

The scaling up problem is linked to the question whether people in developing countries can be mobilised for environmental actions – indeed, whether they can understand and appreciate the problems of sustainability. At one level this question is asked because of remnants of racism and cultural chauvinism in the West. At another level, the question is based on the idea that poor people have little time or incentive to be bothered about environmental questions since they must struggle so hard just to get by and keep their families alive. Some traditional leftists have fashioned their own version of the latter view, concluding that environmentalism is a rich country fad that the victims of colonialism and imperialism can hardly afford to be sidelined by.

Available data seem to indicate that all these views are wrong. The recent Nobel Peace Prize award to Kenyan environmental activist Wangari Maathai (see chapter 1) is based on her proven success in mobilising thousands of village women to plant trees. Environmental activists from Nigeria to Brazil have attracted large followings, often leading to their execution or assassination by agents of higher classes. One of the most famous cases was that of Chico Mendes, a leader of the rubber tappers union in Brazil, who became a tireless advocate of protecting the Amazon rainforest and was gunned down in 1988 (Cappato n.d.).

A cross-national study (Brechin and Kempton 1994) of attitudes based on various opinion surveys turned up clear evidence that citizens of developing nations have environmental concerns equal to or in some cases greater than those in the wealthy nations. In the absence of strong state support for environmental preservation, people in Third World nations often take up the causes themselves, creating "decentralized, popularly supported, and often spontaneous actions to protect local natural resources" (Brechin and Kempton 1994:248). The authors (1994:247) cite studies indicating that over 100 million people in the poor countries belong to grass roots movements, many of them with an environmental focus. The authors conclude (1994:253) that "concern for the environment has become a global phenomenon." Transforming "concern for the environment" into a plan for sustainable development remains the crucial next step for which the data from Kerala that we have presented in this study could be inspiring and of practical use.

In India outside of Kerala, a few major environmental actions have become legendary, including in particular the famous Chipko tree huggers movement, the Tehri dam movement and the mass movement against the Narmada dam project. These movements involved peasants and tribal people with support from many artists, scientists and members of the middle class. In Orissa state "Save Chilika" is a major environmental movement of the local fishing community against businesses promoting prawn culture. Pani (water) panchayat activities in Maharastra is an example of sharing common property resources like water equally by all the inhabitants in the panchayat. Tarun Bharat Sangha, an NGO in Rajasthan revived a formerly perennial stream by involving local villagers in planting trees in the catchment area. There are several such movements spearheaded by the local people in various parts of the country. Environmentalism has spread to India's emerging middle class more generally as well, though middle class environmentalism often involves class prejudice as when better off people limit their perspective on environmental improvement to the removal of slum dwellers and street hawkers (Mawdsley 2004:91). India now hosts a wide range of environmental organisations including the Centre for Science and the Environment, a national research and activist centre as well as Indian branches of international groups such as Greenpeace and the World Wide Fund for Nature. The national literature on environmentalism in India has grown to the extent that a review essay could be written about it (Philip 2001). Environmental awareness in India is sufficiently high that one study found stock market prices reflected at least in the short run the Centre for Science and the Environment reports on the companies' impact on the environment (Gupta and Goldar 2005).

Even at the local level one recent study in the Western Ghats mountains outside of Kerala found that villagers stated a willingness to forego some short term income in order to protect their resources. They further expressed willingness to engage in up to 25 person days of labour annually to assist in the preservation of the elephant population that is endangered by the spread of coffee production in their region (Ninan and Sathyapalan 2005:9). All of these data tend to confirm the idea that an international consciousness is forming around the need to protect the environment. Still missing is a body of case studies showing successes and challenges for community based sustainability. Kerala has provided an initial set of these cases as we have tried to show in earlier chapters.

Scaling Up Worldwide from Community Based Sustainability

The combination of international awareness and the onset of local experiments presents the world with some starting points towards a genuine plan of action for sustainability. As more case studies accumulate more experiments are likely. Even in the first world a small scale movement has emerged to create what are now called *ecovillages* – described by one observer (Jackson and Svensson 2002:10) as

> ...communities of people who strive to lead a sustainable lifestyle in harmony with each other, other living beings and the Earth. Their purpose is to combine a supportive social-cultural environment with a low-impact lifestyle.

Recently more than 600 ecovillages and similar intentional communities have been identified in North America and more than 100 others have been identified on other continents (Fellowship for Intentional Communities 2000). Ecovillages represent a wide range of economic, social and political perspectives. Many are religiously based or have a deep spiritual orientation, but they produce nonetheless a lot of potentially valuable information about recycling, organic farming, alternative energy production and many other sustainability oriented practices. Most ecovillages encourage egalitarian relationships among their members and therefore limit displays in the differences in wealth by house size or other environmentally consequential acts (Jackson and Svensson 2002). In a parallel trend, there are now more than 1,000 registered Community Supported Agriculture (CSA) farms in the U.S. alone (Van En 2002) in which urban or suburban consumers contract with nearby farms to purchase a guaranteed amount of produce. This reduces transport costs (and greenhouse gas emissions), provides cheaper and tastier food, and insulates the farms from corporate takeovers. But it does not achieve the higher goal of permaculture in which the land produces with almost no loss of fertility through recycling of consumer wastes back on the land. The proposed recycling of urban manure and coir pith remains back to the land in Kerala's Integrated Coastal Zone Management project described in chapter 10 would constitute an advanced form of CSA approaching the permaculture model.

A small scattering of intentional communities and CSA farming setups can hardly bring about a sustainable world and to date little systematic research has been done on them. But the decentralised model can be extrapolated to a larger scale at least in theory. Australian environmentalist Ted Trainer (1995) envisions a world of "highly self-sufficient" communities in which transportation is by light rail, most foliage is fruit and nut trees, people share commonly used items such as tools in community sheds, and most people are equal or nearly equal to each other in wealth. Kerala's general demographic layout of fairly evenly dispersed population density puts it in a good position to eventually develop a version of this vision. Oxfam economist Colin Hines (2000) argues that refocusing the global economy around local markets would be the most effective form of resistance to the corporate dominated globalisation that today threatens to destroy the planet in the long run while destroying local communities in the short run. The Kerala People's Plan Campaign – especially its fourth and fifth year emphasis on local women's thrift based cooperatives producing with local products for local markets fits this idea closely. The Integrated Coastal Zone Management project we described in chapter 10 has such local cooperative enterprises at the heart of its structure. American economists Gunn and Gunn (1991) argue that local communities can own and manage productive enterprises and bend their profit orientation to fit community needs. Kerala's cooperatives furnish on-the-ground examples of this potential.

Community based sustainability can theoretically have larger political impacts. In the United States in 2005 132 city mayors led their cities to endorse and attempt to carry out the Kyoto Protocol on global warming thus bringing national attention to the recalcitrance of the Bush Administration that has refused to cooperate with the Protocol and has spread misinformation about it to the American public (Sanders 2005:A9).

Environmental Stress and Local Democracy

Our study now leads to the ultimate question: can Kerala's traditions of activism and high levels of consciousness overcome the state's steady and perhaps escalating environmental stress? Kerala faces many obstacles. As we showed in chapters 4 through 6, the environmental stress is substantial and the drivers are the same as found in other parts of the world. With a heavily rural population that is more directly dependent on ecosystem resources than are urban dwellers, Kerala's people face direct and possibly immediate consequences of the state's environmental degradation. We must add—as has been noted briefly in earlier sections of this chapter—that Kerala faces these environmental stresses in a world increasingly dominated in the present historical period by the international forces of unbridled market capitalism combined with the inherited inequality of the colonial era and the increasing inequality fostered by neoliberalism, itself an ideological "driver" of the short term profit oriented capitalist market system.

On the other hand, Kerala's people have shown remarkable resilience and spirited dedication to the principles of social justice and a willingness to act locally against apparently overwhelming power. The struggles over Silent Valley and at Plachimada for environmental justice stand along with Kerala's historic struggles for land reform, abolition of caste discrimination, workers' rights, education, and health care as landmarks in the international struggle for a better world. The recent mobilisation of hundreds of thousands of Malayalees for local democratically based development planning in the People's Campaign and the many associated experiments in identifying elements of the resource base in ways that can lead to more sustainable use of those resources put Kerala in a strategic place to advance its own programme for sustainable development even in the context of corporate dominated globalisation.

Humanity today faces the threat of a greater collapse than has any individual civilisation in the past. And yet – with the tools of modern science and the potential of local democracy we have the potential of avoiding that collapse. The power of local democracy has yet to be fully tested. Kerala's people have crafted one of the world's primary laboratories for its development. To the extent they succeed they will make their own lives better. They will also give to the rest of the world a wealth of case material to enrich the desperately needed international discussion and debate on how to use the power of human consciousness to avert collapse and build a sustainable world.

Bibliography

Achar, K. P. 1997. Documentation of peoples' knowledge and perceptions about biodiversity through PBR [People's Biodiversity Registers] at Mala Village Panchayath, Karkala. Karnataka State. Unpublished report.

Ajai Kumar, S. 2003. From Dependence to Self-Reliance. *Kerala Calling* 24(1):28–30.

Allen, Robert. 1980. *How to Save the World: Strategy for World Conservation.* Totowa, New Jersey: Barnes and Noble Books.

Andrea, M. O. and Dickinson, R. E. 1995. Sustainability and the change in the atmosphere: assessing changes in chemical and physical climate. In Munashinge, Mohan, and Shearer, Walter, eds. *Defining and Measuring Sustainability – The Biophysical Foundation.* The U. N. University and the World Bank, pp. 117–36.

Aniyan, P. V., Natarajan, C. V, and Bhageeradhan, K. R. 2002. Karappuram – Land and People. In *Seminar Papers, Decentralisation, Sustainable Development and Social Security.* Alappuzha: Joint Committee of Panchayats, Aryad and Kanjikuzhy Blocks and International Labour Organisation, pp. 31–40.

Anonymous, 1999. Prevention rather than cure. An interview with Mr. Anil Sinha, Jt. Secretary, Ministry of Agriculture. *GIS Development* 3(3):53.

Anonymous. 2005. Tsunami news. *Journal of the Geological Society of India* 65(3):273–78.

Arunachalam, N., Mariappan, T., and Somachary, N. 1992. Seasonal abundance of Culex sitiens Wiedemann 1828, in Vypeen island situations, Kochi, Kerala. Thrissur: Proceedings of the Fourth Kerala Science Congress. P. 290.

Asari, P. K. S. 1999. Silent Valley – A Candidate for World Natural Heritage Site. In Manoharan, T. M., Biju, S. D., Nayar, T. S. and Easa, P. S., eds. *Silent Valley: Whisper of Reasons.* Thiruvananthapuram: Kerala Forest Department in Association with Kerala Forest Research Institute. pp. 97–101.

Austin, Andrew, and Phoenix, Laurel. 2005. The Neoconservative Assault on the Earth: The Environmental Imperialism of the Bush Administration. *Capitalism, Nature, Socialism* 16(2): 25–43.

Baba, M., Thomas K. V., and Sreekala, Panicker. 1997. Alternate coastal engineering strategies for Kerala. Thiruvananthapuram: Proceedings of the Second Indian National Conference on Harbour and Ocean Engineering (INCHOE-97). Vol. 2:1205–1215

Balachandran, K. K. 2004. Does subterranean flow initiate mud banks off the southwest coast of India? *Estuarine,Coastal and Shelf Science* 59:589–98.

Balagopalan, M. 1995. Changes in the distribution of organic carbon and different forms of nitrogen in soils under natural forests and teak plantations. Palakkad: *Proceedings, VII Kerala Science Congress*. Pp.110–112.

Bandyopadhyay, D. 1997. People's Participation in Planning – Kerala Experiment. *Economic and Political Weekly* 32(40):2450–2454. 27 September 1997.

Barbier, Edward. 1987. The concept of sustainable economic Development. *Environmental Conservation* 14(2):101–110.

Barlow, Maude, and Clarke, Tony. 2002. *Blue Gold: The Fight to Stop the Corporate Theft of the World's Water.* New York: The New Press.

Basu, S.R. and Main, H.A.C. 2001. Calcutta's water supply: demand, governance and environmental change. *Applied Geography* 21(1):23–24.

Bhatt, D. M. 1990. Litter production and seasonality in tropical moist forest ecosystem of Uttar Kanada District, Karnataka. Bangalore: *Proceedings of the Indian Academy of Science* (plant science) Vol.100:139–152.

Biswas, A. B. 1995. Water supply in Calcutta. In S. Chaudhri, ed. *Calcutta: The Living City*, Vol. 2: The Present and Future. Oxford: Oxford University Press. pp. 160–66.

Biswas, A. K. 1993. Sustainable water development: a global perspective. *Water Resources Journal*. Bangkok: Economic and Social Commission for Asia and the Pacific – (ESCAP). March 1993: 1–12.

Blakie, P. M., and Brookfield, H. C. 1987. *Land Degradation and Society*. London: Methuen and Co. Ltd.

Blanford, H.F. 1888. Influence of forests on rainfall. In *Rainfall of India, Indian Meteorological Memoirs* 3. Calcutta: India Meteorology Department.

Bose, A. 1978. *India's Urbanisation, 1901–2001*. New Delhi: Tata McGraw Hill Publishing Co. Ltd.

Boserup, E. 1965. *The Condition of Agricultural Progress*. London: Allen and Unwin.

Brechnin, Steven R., and Kempton, Willett. 1994. Global Environmentalism : A Challenge to the Postmaterialism Thesis? *Social Science Quarterly* 75(2):245–69.

Bright, Chris. 2003. A History of Our Future. In Gardner, Gary, et al. eds. *State of the World 2003: A Worldwatch Report on Progress Toward a Sustainable Society*. New York: W. W. Norton and Company. pp. 3–13.

Brown, Becky. J., Hanson, Mark. E., Liverman, Diana. M., and Merideth, Robert W. Jr. 1987. Global sustainability: Toward definition. *Environmental Management* 11(6):713–19.

Brown, Lester R. 1981. *Building a Sustainable Society*. Washington, D.C.: Worldwatch Institute.

Brown, Lester R. 2001. *Eco-Economy—Building an Economy for the Earth*. Earth Policy Institute. New York: W. W. Norton and Company.

Brown, Lester R. 2003. *Plan B: Rescuing a Planet under Stress and a Civilization in Trouble*. New York: W. W. Norton and Company.

Brown, Lester R. 2004. *Outgrowing the Earth: The Food Security Challenge in an Age of Falling Water Tables and Rising Temperatures*. New York: W. W. Norton and Company.

Brown, Lester R. 2006. *Plan B 2.0 : Rescuing a Planet under Stress and a Civilization in Trouble*. New York : W. W. Norton and Company.

Buttel, F. and Taylor, P. 1994. Environmental sociology and global environmental change: a critical assessment, In M. Redelht and T. Benton, eds. *Social Theory and the Global Environment*. New York: Routledge. pp. 228–55.

Cappato, Jorge. N.d. Who Was Chico Mendes? Global Forum: United Nations Environment Programme. *http://www.global500.org/feature-6.html*.

Carpenter, R.A. and Harper, D. E. 1989. Towards a science of sustainable upland management in developing countries. *Environmental Management* 13(1):43–54.

Biswas, B.C. 1980. Effect of forest cover on rainfall distribution at Andaman and Car – Nicobar Islands. *Mausam* 31:178–180.

Centre for Earth Science Studies (CESS). 1997. Analysis of the Environmental Impact of the Lowland to Highland Migration in the Western Ghats Region, Kerala. Thiruvananthapuram: Unpublished project report.

Centre for Earth Science Studies (CESS). 1998. Catchment Conservation/

Development Plan for Sasthamcotta lake, Kollam district. Thiruvananthapuram: Unpublished project report.

Centre for Earth Science Studies (CESS).2000. Conservation of Vellayani lake in Trivandrum district as drinking water source— Detailed study. Thiruvananthapuram: Unpublished project report.

Centre for Earth Science Studies (CESS). 2000. GIS Application for Urban Space Management, Thiruvananthapuram city region, Kerala. Thiruvananthapuram: Unpublished Project report.

Centre for Earth Science Studies (CESS). 2000. *Panchayat Level Resource Mapping: Final Technical Report*. Thiruvanan-thapuram.

Centre for Earth Science Studies (CESS). 2000. Terrain characteristics and soil nutrient studies of the Rattendon valley annex reserved forests of Periyar Tiger Reserve. Thiruvananthapuram: Unpublished report.

Centre for Earth Science Studies (CESS). 2003. Regional Technical Assistance for Coastal and Marine Resources Management and Poverty Reduction in South Asia. India (Kerala Component). Final Report. Thiruvananthapuram: unpublished.

Centre for Science and Environment (CSE). 1990. *State of India's Environment – A Citizens Report: Floods, Flood Plains and Environmental Myths*. New Delhi.

Chakraborty, D., Biswas, B. K., Roy Chowdhaury, T., Basu G. K., Mandal, B. K., Chowdhury, U. K., Mukherjee, S. C., Gupta, J. P., Chowdhury, S. R. and Rathore K. C. 1999. Arsenic ground water contamination and sufferings of people in Rajnandgaon district, Madhya Pradesh, India. *Current Science* 77(4):502–504.

Chakraborty, Rabindra Nath. 1999. Linkages between Income Distribution and Environmental Degradation in Rural India. In Madsen, Stig Toft, ed. *State, Society and the Environment in South Asia*. Oslo: Nordic Institute of Asian Studies. pp. 165–99.

Chakraborty, S. C. 1988. Extended metropolitan areas: their relevance for understanding urban process in India. Paper submitted for the conference on the "Extended Metropolis in Asia." Honululu: Environment and Policy Institute, East-West Centre.

Chambers, Robert. 1994. Participatory rural appraisal: challenges, potentials and paradigm. *World Development* 22:1437–59.

Chapman, D., and Kimstach, V., 1997. Selection of water quality variables. In Chapman, D., ed. *Water Quality Assessments*. London: UNESCO, WHO and UNEP. Pp 59–126.

Chasin, Barbara H. 2004. *Inequality and Violence in the United States:*

Casualties of Capitalism. Amherst, NY: Humanity Books. Second edition.

Chathukulam, J., and Gireesan, K. 2004. Survey and Documentation of Individual Household Based Micro Hydel Power Ventures in Idukki District. Thiruvananthapuram: Project Report Submitted to the Kerala State Committee on Science, Technology and Environment; and Kottayam: Centre for Rural Management.

Chathukulam, J., and Gireesan, K. 2005. People's Power – Idukki District in Kerala Shows that Small Hydel Power Plants Can Be Profitable. *Down To Earth* 13(21):53.

Chattopadhyay, M., Sajikumar, S. and Saritha. D. 1996. Integrated Resource Evaluation of the Vamanapuram River Basin. Project report submitted to the State Committee on Science, Technology and Environment, Govt. of Kerala. Thiruvananthapuram: Centre for Earth Science Studies.

Chattopadhyay, M., Sakunthala, C., Ouseph, P. P., and Soman, K. 1998. Basin approach in pollution mitigation measures – A case study on Killi Ar, Thiruvananthapuram district. Kochi: Proceedings of the VIII Kerala Science Congress. P. 84.

Chattopadhyay, S. 1985. Deforestation in parts of Western Ghats region Kerala, India. *International Journal of Environmental Management* 20:219–230.

Chattopadhyay, S. 1986. Land classification – an ecological approach. Examples from Kerala. In Richter, Hans and Schonfelder, Gunther, eds. *Landscape Synthesis – Foundation, Classification and Management Part-II. Landscape Classification and Management.* Martin Luther University Halle-Wittenberg. Halle(Saale), pp.87–99.

Chattopadhyay, S. 1988. Urbanisation in Kerala. *Geographical Review of India* 50(2):8–25.

Chattopadhyay, S. 1995a. Environmental impact on urbanisation: a case study of Kerala, India. *Transactions*. Institute of Indian Geographers 17(1):15–27.

Chattopadhyay, S. 1995b. Infrastructural Facilities' Role in Emergence of Kerala Model. *Productivity* (National Productivity Council) 36(2):236–245.

Chattopadhyay, S. 1996a: Coastal zone management. A methodo-logical approach with example from Kerala. Maharashtra Association of Engineers (MAEER's), Maharashtra Institute of Technology *(MIT) Pune Journal* 4(15 and 16):14–22.

Chattopadhyay, S., 1996b: Rubber cultivation in Kerala: Some issues in environmental perspective. In *Proceedings of the Interna-tional Seminar on Natural Rubber - An Eco Friendly Material.* Trivandrum, pp. 1–14.

Chattopadhyay, S.1998. Application of sustainable development concept: Some issues. In Srikumar Chattopadhyay and H. S. Sharma, eds. *Sustainable Development – Issues and Case Studies.* New Delhi: Concept Publishing Co.

Chattopadhyay, S. 1999. Traditional rainwater harvesting system in Kerala: Present status and future research requirements. Paper presented at the international symposium on "Making Water Everybody's Business." Organised by the Centre for Science and Environment, New Delhi, 1999.

Chattopadhyay, S. 2000. Floods in Kerala, India. An overview. *Proceedings of the Euro-Asian Workshop on Flood.* Department of Flood Control and Dyke Management of Vietnam. Hanoi, Vietnam.

Chattopadhyay, S. 2003: Towards Gram Swaraj: West Bengal and Kerala experiments. *ANASAKTI* (Journal of the International Centre for Gandhian Studies) 1(1):93–115.

Chattopadhyay, S. 2004. Poverty and environmental degradation in selected areas of Kerala coast. In Surya Kant, Singh, N., Singh, Ja., and Mukherji, A. B., eds. *Reinventing Regional Develo-pment.* New Delhi: Rawat Publications. Pp. 377–395.

Chattopadhyay, S., and Aniyan, P. V. 2000. Spatial Planning: Scope of Kerala's Experiment in People's Campaign for Decentralised Planning. Paper presented at the International Conference on Democratic Decentralisation, 23–27 May 2000. Thiruvananthapuram: Kerala State Planning Board.

Chattopadhyay, S., and Carpenter, R. A., 1990. A Theoretical Framework for Sustainable Development Planning in the Context of a River Basin. Unpublished report, East-West Environment and Policy Institute, Hawaii, USA.

Chattopadhyay, S. and Chattopadhyay, M., 1995. Terrain Analysis of Kerala. Kerala State Committee on Science Technology and Environment, Technical monograph no. 1/95. Thiruvananthapuram : Government of Kerala.

Chattopadhyay, S., and Chattopadhyay, M. In press. Impact of terrain characteristics on catchment yield – a case study of Kabbini river basin, Kerala. In Thakur, B., ed. *Natural Resource Management.* New Delhi: Concept Publishing Co.

Chattopadhyay, S., and Jayaprasad, B. K. 1991. Evaluation of paddy field in Thiruvananthapuram city. Report submitted to the Trivandrum Development Authority. Thiruvananthapuram: Centre for Earth Science Studies.

Chattopadhyay, S., Sangeetha P. Viswan and Asa Rani. L. 2001. Landuse/landcover change and its environmental and human dimension: A case of Chalakudy river basin, Interim project report. Thiruvananthapuram: Centre for Earth Science Studies.

Chattopadhyay, S., and Sasidharan, C. K. 1994. Changing pattern of agricultural land use in Kerala: A study in environmental perspectives. In Singh, R. B., ed. *Disasters, Environment and Development*. New Delhi: Oxford and IBH Publishing Co. Pp. 479–84.

Chattopadhyay, S., Velayudham, S., and Salim, M. B. 1986. Trends of Deforestation in Kerala. In Radhakrishna, B. P., and Ramachjandran, K. K., eds. *India's Environment: Problems and Perspectives.* Bangalore: Geological Society of India. Memoir 5, pp. 289–298.

Chin, A. and Gregory, K. J. 2001. Urbanisation and adjustment of ephemeral stream channels. *Annals of the Association of American Geographers* 91(4)595–608.

Chinnamani, S. 1965. Runoff and soil loss studies under different vegetative cover. *The Indian Forester* 91:672–682.

Chisholm, M. 1962. Rural Settlement and Landuse: An Essay in Location. London: Hutchinson University Library.

Christian Aid. Nd. *Behind the Mask (pp. 45–49): Living Its Values – Coca-Cola in India http://www.christian-aid.org.uk/indepth/0401csr/csr-casestudy3india.pdf*

Coca-Cola. 2003. Frequently Asked Questions About Plachimada: *http://www.coca-colaindia.com/press/faqs-kerala.html*

Cohn, Bernard S. 1971. *India: The Social Anthropology of a Civilization.* Englewood Cliffs, NJ: Prentice-Hall.

Coomer, James. C. 1979. The Nature of the Quest for a Sustainable Society. In Coomer, James C., ed. *Quest for a Sustainable Society.* New York: Pergamon Press. Published in Cooperation with The Woodlands Conference.

Costanza, Robert, d'Arge, Ralph, de Groot, Rudolf, Farber, Stephen, Grasso, Monic, Hannon, Brue, Limbur, Karin, Naeem, Shahid, O'Neill, Robert V., Paruelo, Jose, Raskin, Robert G., Sutton, Paul, and van den Belt, Marjan. 1997. The value of the world's ecosystem services and natural capital. *Nature* 387(15 May 1997):253–260.

CWRDM (Centre for Water Resources Development and Management) 1989. *Tanks and Ponds of Kerala*, Kozhikode: Centre for Water Resources Development and Management.

Dabral, B. J., Yadav, J. S. P. and Sharma, D. R. 1965. Soil Moisture Studies in Chirpine, Teak and Sal Plantations at New Forest, Dehra Dun. *The Indian Forester 91(10):701–13.*

Daly Herman E. 1986. Thermodynamic and Economic Concept as Related to Resource-Use Policies: Comment. *Land Economics* 62(3) : 319–22.

Daly, Herman E. 1987. The Steady State Economy: Alternative to Growthmania. Washington, D. C.: Monograph published by Population-Environment Balance. Pp. 1–8.

Daly, Herman E., 1990a. Sustainable development: From concept and theory towards operational principles. *Population and Development Review* 16:25–43. Supplement: Resources, Environment, and Population: Present Knowledge, Future Options.

Daly, Herman E. 1990b. Towards some operational principles of sustainable development. *Ecological Economics* 2(1):1–6.

Daly, Herman E. 1991 [1977]. *Steady-State Economics*. Washington D. C.: Island Press. Second edition with new essays.

Damodaran, A. 2003. What happened in Wayanad. *Down To Earth* 12(21):53.

Dasmann, Raymond F. 1984. An introduction to world conservation. In Thibodeau, Francis R. and Field, Hermann H., eds. *Sustaining Tomorrow: A Strategy for World Conservation and Development.* Hanover, New Hampshire: University Press of New England. pp. 16–24. Bangkok.

Davis, Mike. 2001. *Late Victorian Holocausts : El Niño Famines and the Making of the Third World.* London : Verso.

Davidson, D.A .1982. Assessment of landuse capability. In *Proceedings of the 1st International Symposium on Soil, Geology and Landforms: Impact on Landuse Planning in Developing Countries*, pp. G201–G257.

Davis, Mike, 2001. *Late Victorian Holocausts : El Niño Famines and the Making of the Third World.* London : Verso.

Diamond, Jared. 2005. *Collapse: How Societies Choose to Fail or Succeed.* New York: Viking.

Down To Earth. 2003. India's Sacred Groves. *http://www.cseindia.org/dte-supplement/forest20031231/what_sacred.htm*

Drummond, Ian, and Marsden, Terry. 1999. *The Condition of Sustainability*. London and New York: Routledge.

Dunn, S. and Flavin, C. 2002. Moving the climate change agenda forward. In Flavin, Christopher, French, Hilary, and Gardner, Gary, eds. *State of the World, 2002*. The Worldwatch Institute. New York: W. W. Norton and Company, pp. 24–50.

Elamon, Joy, Franke, Richard W., and Ekbal, B. 2004. Decentralization of Health Services: The Kerala People's Campaign. *International Journal of Health Services* 34(4):681–708.

Elgin, Duane. 1993 [1981]. *Voluntary Simplicity: Toward a Way of Life That is Outwardly Simple, Inwardly Rich*. New York: William Morrow.

Ernakulam District Panchayat. 1999. People's Biodiversity Register– A District Panchayat Project Ernakulam District Biodiversity Committee. Kochi : Civil Station.

FAO. 1995. Planning for sustainable use of land resources : Towards a new approach. FAO Land and Water Bulletin-2, p. 50.

Fellowship for Intentional Community. 2000. *Communities Directory: A Guide to Intentional Communities and Cooperative Living*. Rutledge, MO: Fellowship for Intentional Community.

Fisher, H. B. 1979. Growth Centre Planning in India. In Misra, R. P., and Sundaram, K. V., eds. *Rural Area Development: Perspective and Approaches*, New Delhi: Startling Publishers Pvt. Ltd. pp. 92–121.

Foster, John Bellamy. 2000. *Marx's Ecology: Materialism and Nature*. New York: Monthly Review Press.

Foster, John Bellamy. 2002. *Ecology Against Capitalism*. New York: Monthly Review Press.

Franke, Richard W. 1987. Power, Class and Traditional Knowledge in Sahel Food Production. In Markovitz, Irving L., ed. *Studies in Power and Class in Africa*. New York: Oxford University Press. pp. 257–85.

Franke, Richard W. 1995. Promoting Participation–Seeking Sustainability: The ITK Dilemma. *Reviews in Anthropology* 24(3):159–68.

Franke, Richard W. 1996 [1993]. *Life Is a Little Better: Redistribution as a Development Strategy in Nadur Village, Kerala*. Delhi: Promilla.

Franke, Richard W., and Chasin, Barbara H. 1979. Peanuts, Peasants, Profits, and Pastoralists: The Social and Economic Background to Ecological Deterioration in Niger and Its Implications for Current Development Programs. *The Journal of Peasant Studies* 8(3):1-30.

Franke, Richard W., and Chasin, Barbara H. 1980. *Seeds of Famine: Ecological Destruction and the Development Dilemma in the West African Sahel*. Montclair, N. J.: Allanheld Osmun and Company.

Franke, Richard W., and Chasin, Barbara H. 1997. Power to the Malayalee people. *Economic and Political Weekly* 32(48):3061–68.

Franke, Richard W., and Chasin, Barbara H. 1994. *Kerala: Radical Reform as Development in an Indian State*. Oakland, CA: Food First and Delhi: Promilla. Second edition.

Freeman A. M. 1991. Valuing Environmental Resources under Alternative Management Regimes. *Ecological Economics* 3:247–256.

Friedrich, G., Chapman, D., and Beim, A. 1992. The use of biological material. In Chapman, E., ed. *Water Quality Assessment*. UNESCO, WHO and UNEP. London: E. and F. N. Spon. pp. 175–242.

Gardner, Gary, Assadourian, Erik, and Sarin, Radhika. 2004. The State of Consumption Today. In Halweil, Brian, and Mastny, Lisa, eds. 2004. *State of the World 2004: A Worldwatch Institute Report on Progress Toward a Sustainable Society*. New York: W. W. Norton and Company. pp. 3–21.

Gaussen, H, Legris, P. and Mehir Homji V. M. 1986. International Map of Vegetation and Environmental Conditions - India Sheet-Cape Comorin. Pondicherry: French Institute.

Global Environmental Change Programme (GECP). 2001. Earth System Science: an integrated approach. *Environment* 43(8):21–27.

George, P. S. and Chattopadhyay, S. 2001. Population and Landuse in Kerala. In *Growing Populations – Changing Landscapes: Studies from India,China and the United States.* Indian National Science Academy, Chinese Academy of Sciences, and U. S. National Academy of Sciences. Washington, D. C.: National Academy Press. pp. 79–106.

George, P. S. 1982. Agricultural Price Movements in Kerala. In Pillai, P. P., ed. *Agricultural Development in Kerala.* New Delhi: Agricole Publishing Academy. pp. 142–57.

George, T. K, Harida, V. and Sreekumar, B. 1988. Role of Government and Structural Changes in Rubber Plantation Industry. *Economic and Political Weekly* 23:M108–M116. November 1988.

Ginsberg, N. 1990. *The Urban Transition: Reflections on the American and Asian Experience*. Hong Kong: The Chinese University Press.

Goldsmith, Edward, Allen, Robert, Allaby, Michael, Davoll, John, and Lawrence, Sam (Editors of *The Ecologist*). 1972. *Blueprint for Survival.* Boston: Houghton Mifflin.

Goodland, Robert, and Ledec, George. 1987. Neoclassical economics and principles of sustainable development. *Ecological Modelling* 38:19–46.

Gopalan, U. K., Meenakshikanjamma, P. P., and Vengayil, D. T. 1987. Macrobenthos of Vembanad estuary in relation to the deposition of degraded water fern Salvinia and other macrophytes. *Proceedings, National Seminar on Estuarine Management*, Trivandrum, pp.410–414.

Gopi, K. C. 2000. Freshwater Fishes of Kerala State. In Ponniah, A. G. and Gopalakrishnan, A., eds. *Endemic Fish Diversity of Western Ghats*. NBFGR-NATP Publication. Lucknow: National Bureau of Fish Genetic Resources. pp. 56–76.

Gopinathan, C., and Sundaresan, C. S. 1990. *Cropping Pattern Changes and Employment Effects in Selected Districts of Kerala*. Thiruvananthapuram: Centre for Management Development.

Gould, John and Nissen-Petersen, Erik. 1999. *Rainwater Catchment Systems for Domestic Supply: Design, Construction and Implementation*. London: ITDG (Intermediate Technology Development Group) Publishing.

Government of Kerala. 1974. *Report of the Committee on Agro-climatic Zones and Cropping Patterns*. Thiruvananthapuram: Department of Agriculture.

Government of Kerala. 1984. *Western Ghat Development Programme: Status paper for area subgroup.*

Government of Kerala. 1992. *Kerala Fisheries: An Overview*. Thiruvananthapuram: Department of Fisheries.

Government of Kerala. 1994. *Agricultural Statistics of Kerala, 1992–93*.Thiruvananthapuram : Department of Agriculture.

Government of Kerala. 1995. *Farm Guide,* 1994. Thiruvananthapuram : Department of Agriculture.

Government of Kerala. 1995. *Land Resources of Kerala State*. Thiruvananthapuram: Kerala State Land Use Board.

Government of Kerala. 1997. *Kerala Fisheries: An Overview,* 1997. Thiruvananthapuram: Department of Fisheries.

Government of Kerala.1997. *Report of the Task Force on Drainage and Flood Management*. Thiruvananthapuram: State Planning Board.

Government of Kerala. 1997. *Report of the Taskforce on Fisheries and Agriculture*. Thiruvananthapuram: State Planning Board.

Government of Kerala. 1998. *Forest Statistics,* 1998. Thiruvananthapuram: Department of Forests.

Government of Kerala. 1999. *Natural Resource Management: Community Based Action Research Programme for Kerala. Project Profile.* Thiruvananthapuram: Kerala State Planning Board.

Government of Kerala. 2005. *Economic Review 2004*. Thiruvananthapuram: Kerala State Planning Board.

Government of West Bengal. 2004. West Bengal Human Development Report, 2004. Kolkata: Development and Planning Department, Government of West Bengal.

Gray, Andrew. 1991. The Impact of Biodiversity Conservation on Indigenous Peoples. In Shiva, Vandana, Anderson, Patrick, Schücking, Heffa, Gray, Andrew, Lohmann, Larry and Cooper, David, eds. *Biodiversity: Social and Ecological Perspectives*. London: Zed Books; and Penang: World Rainforest Movement. pp. 59–76.

Gunn, Christopher, and Gunn, Hazel Dayton. 1991. *Reclaiming Capital: Democratic Initiatives and Community Development*. Ithaca: Cornell University Press.

Gupta, G. R. 1990. A meso scale study of unusually heavy rainfall over south Kerala. *Mausam* 41(1):109–114.

Gupta, Shreekant, and Goldar, Bishwanath. 2005. Do Stock Markets Penalize Environment—Unfriendly Behaviour? Evidence from India. *Ecological Economics* 52(2005):81–95.

Hahnel, Robin. 2005. *Economic Justice and Democracy: From Competition to Cooperation*. New York: Routledge.

Halweil, Brian. 2002a. Farming in the Public Interest. In Flavin, Christopher, French, Hilary, and Gardner, Gary, eds. *State of the World,* 2002. The Worldwatch Institute in Cooperation with the United Nations Environment Programme. New York: W. W. Norton and Company. pp.51–74.

Halweil, Brian. 2002b. Farmland Quality Deteriorating. In *Vital Signs: The Trends That Are Shaping Our Future*. The Worldwatch Institute in Cooperation with the United Nations Environment Programme. New York: W. W. Norton and Company. pp. 102–103.

Halweil, Brian. 2003. Grain Production Drops. In Renner, Michael, Sheehan, Molly, et al., eds. *Vital Signs: The Trends That Are Shaping Our Future*. The Worldwatch Institute in Cooperation with the United Nations Environment Programme. New York: W. W. Norton and Company. pp. 28–29.

Hardin, Garrett. 1968. The Tragedy of the Commons. *Science* 162:1243–48.

Harris, Marvin. 1974. *Cows, Pigs, Wars and Witches: The Riddles of Culture*. New York: Random House.

Harris, Marvin. 1977. *Cannibals and Kings: The Origins of Culture*. New York: Random House.

Higgins, B. 1981. Planning Regional Development: Art, Science or Philosophy. In Mather, O. P., ed. *Training for Regional Development Planning*. Nagoya: Maruzen Asia. United Nations Centre for Regional Development Planning Regional Development Series Vol. 9. pp. 137–75.

The Hindu. 2002. Call to evolve methods to harvest rainwater. *The Hindu - Southern States*, India, Online edition, Feb. 25.

Hines, Colin. 2000. *Localization: A Global Manifesto*. London: Earthscan Publications Ltd.

Holmgren, David. 2002. *Permaculture: Principles and Pathways Beyond Sustainability*. Hepburn, Australia: Holmgren Design Services. *http://www.holmgren.com.au.*

Indian Council of Agricultural Research (ICAR). 1991. Agroclimatic Zone Specific Research. New Delhi: ICAR.

Ipye, George. 2003. "Sludge at Coke Factory is Deadly." (*http://www.rediff.com/money/2003/aug/05spec2.htm*)

Jackson, Hildur, and Svensson, Karen. 2002. *Ecovillage Living: Restoring the Earth and Her People*. Devon, UK: Green Books Ltd. *http://www.greenbooks.co.uk* in association with Gaia Trust, *http://www.gaia.org*

James, A. and Evison, Lillian, eds. 1979. *Biological Indicators of Water Quality*. Chichester and New York: John Wiley and Sons.

James, E. J., Nambudiripad, K. D., and Sreedharan, K. E. undated. Sedimentation processes in the catchments of Western Ghats in the Indian Peninsula, unpublished seminar paper. Kozhikode: Centre for Water Resources Development and Management.

Jananeethi. 2002. *Jananeethi Report:On the Amplitude of Environmental and Human Rights Ramification*. Thrissur: Jananeethi. *http://jjwoodworld.port5.com/jananeethi/archives/cocacola.pdf*

Jansson, A. 1977. Micronectinae as indicator of water quality in two lakes in southern Finland. *Ann. Zool. Fennici* 14:118–24.

Jarrett, H. R. 1977. *Tropical Geography*. Plymouth: Macdonald and Evans Ltd.

Jayan, T. V. 2004 Long Lives the Drink. *Down To Earth* 12(21):21.

Jayarajan, M. 2004. Sacred Groves of North Malabar. Thiruvananthapuram:Kerala Research Programme on Local Level Development. Discussion Paper No. 92.

Jeena, T. S. 1997. Internal Rural Migration and the Impact of Deforestation on Cropping Patterns in Kerala. New Delhi: Jawaharlal Nehru University. Upublished M. Phil. Thesis.

Jeffrey, Robin. 1993. *Politics, Women and Well-Being: How Kerala Became 'A Model.'* Delhi: Oxford University Press.

Jorgenson, Andrew K. 2003. Consumption and environmental degradation: A cross-national analysis of the ecological footprint. *Social Problems* 50(3):374–394.

Joseph, K. V. 1998. Migration and Land Use: A Case Study of Migration from Travancore to Malabar. Thiruvananthapuram: Centre for Development Studies. Population and Land Use Studies Background Paper No. 4.

Joy Sankar, B. V. 2003. *Laws on Land in Kerala*. Old Faridabad, Haryana: Navin Publications.

Kadiyala, Suneetha. 2004. Scaling up Kudumbashree—Collective Action for Poverty Alleviation and Women's Empowerment. Washington D. C.: International Food Policy Research Institute. Food Consumption and Nutrition Division Discussion Paper No. 180.

Kannan, K. P. 1988. *Of Rural Proletarians: Mobilisation and Organisation of Rural Workers in South-west India*. Delhi: Oxford University Press.

Kannan, K. P. 2000. Poverty Alleviation as Advancing Basic Human Capabilities: Kerala's Achievements Compared. In Parayil, Govindan, ed. *Kerala: The Development Experience – Reflections on Sustainability and Replicability*. London: Zed Press, pp. 40–65.

Kannan, K. P., and Pushpangadan, P. 1988. Agricultural stagnation in Kerala: an exploratory analysis. *Economic and Political Weekly* 23(39):A120–A128.

Karthikakutty Amma, M., Suresh, P. R., Aiyer R. S., and Valsa, G. 1996. Effects of continuous cultivation of rubber *Hevea Brasiliensis* on total and available nitrogen status of soil. Kochi: Proceedings of the VIII Kerala Science Congress, pp.80–82.

Karthikakutty Amma, M., Suresh P.R, Valsa, G., Aiyer R. S, and Mathew, M. 1995. Comparative evaluation of soil organic matter status of rubber growing soils in relation to forest ecosystem. Palakkad: Proceedings of the VII Kerala Science Congress, pp.100–101.

Kerala Engineering Research Institute (KERI). 1977. Sedimentation studies in Malampuzha reservoir. Peechi: Unpublished report.

Kerala Engineering Research Institute (KERI). 1982. Sedimentation Studies in Peechi reservoir. Peechi: Unpublished report.

Kerala State Land Use Board (KSLUB). 1995. *Land Resources of Kerala State*. Thiruvananthapuram: Government of Kerala.

Kerala Statistical Institute. 1994. *Conversion of Paddy Land, Kerala*. Thiruvananthapuram: Kerala Statistical Institute.

Klein, Naomi. 2002. *Fences and Windows: Dispatches from the Front Lines of the Globalization Debate*. New York: Picador.

Kohn, Alfie. 1992 [1986]. *No Contest: The Case against Competition—Why We Lose in our Race to Win*. Boston: Houghton-Mifflin.

Kolavalli S. L., and Kerr J. 2002. Mainstreaming participatory watershed development. *Economic and Political Weekly* 37(3):225–42.

Köppen, Wladimir Peter. 1936. *Das geographische System der Klimate*. In *Handbuch der Klimatologie*. Vol. I. Berlin: Gebrüder Bomtraeger.

Kovel, Joel. 2002. *Enemy of Nature: The End of Capitalism or the End of the World*. London: Zed Press.

Krishnakumar, R. 2003. Testing Times for Cola Firms. *Frontline* 20(17):33–37. 29 August 2003. *http://www.flonnet.com/fl2017/stories/20030829003803300.htm*

Krishnakumar, R. 2004a. A Judicial Intervention. *Frontline* 21(2):34–36. 30 January 2004. *http://www.flonnet.com/fl2102/stories/20040130003203400.htm*

Krishnakumar. 2004b. Water Resources: Resistance in Kerala. *The Hindu* 24 January 2004. *http://www.thehindu.com/2004/01/24/stories/2004012402321300.htm*

Krishnan, T. N. 1976. Demographic Transition in Kerala: Facts and Factors. *Economic and Political Weekly* 26(26):A1203–1224.

Kumaran, P. 1994. Urban pollution in Kerala with specific reference to Kochi. Trivandrum : International Congress on Kerala Studies, Vol. 5. Abstracts pp.175–176.

Kurien, John. 1995. Collective Action for Common Property Resource Rejuvenation: The Case of People's Artificial Reefs in Kerala State, India. *Human Organization* 54(2):160–68.

Kurien, John 2005. Securing the future against tsunamis. *Economic and Political Weekly* 40(1):89–99.

Kurien, John. and Sebastian, M. 1980. Technological Change in Fishing and its Impact on Fishermen. Thiruvananthapuram: Centre for Development Studies Review of Research Agenda for ICSSR (Indian Council for Social Science Research).

Kurien, John, and Vijayan, A. J. 1995. Income spreading mechanisms in common property resource : Karanila system in Kerala's fishing. *Economic and Political Weekly* 30(26):1780–1785.

Lafferty, William M., and Langhelle, Oluf. 1999. Sustainable development as concept and norm. In Lafferty, William M., and Langhelle, Oluf, eds. *Towards Sustainable Development: On the Goals of Development – and the Conditions of Sustainability*. New York: St. Martin's Press, Inc. pp. 1–29.

Lansing, J. Stephen. 1991. *Priests and Programmers: Technologies of Power in the Engineered Landscape of Bali*. Princeton: Princeton University Press.

Lee, R. 1980. *Forest Hydrology*. New York: Columbia University Press.

Lefebvre, Henri. 1976. Reflections on the Politics of Space. *Antipode* 8(2):30–37.

Lele, S. 1991. Sustainable Development – a Critical Review. *World Development* 19(6):606–21.

Lin, G. C. S. 2001. Evolving Spatial Form of Urban-Rural Interaction in the Pearl River Delta, China. *The Professional Geographer* 53(1):56–70.

Logan, William. 1995 [1887]. *Malabar*. 2 volumes. New Delhi: Educational Services.

Maathai, Wangari. 2004. Nobel Lecture on Receiving the Nobel Peace Prize. Oslo, Norway: 10 December 2004. *http://nobelprize.org/peace/laureates/2004/maathai-lecture-text.html*

Maathai, Wangari. 2005. Nobel Peace Laureate Wangari on the Environment, the War in Iraq, Debt and Women's Equality. New York: "Democracy Now" Radio Interview, 8 March 2005. *http://www.democracynow.org/article.pl?sid=05/03/08/1444212*

McDonald, Mia, and Nierenberg, Danielle. 2003. Linking Population, Women, and Biodiversity. In Gardner, Gary, et al. *State of the World 2003: A Worldwatch Report on Progress Toward a Sustainable Society*. pp. 38–61.

McGee, T. G. 1987. Urbanisasi or Kotadesasi, The emergence of new regions of economic interaction in Asia. Hawaii: East-West Centre Environment and Policy Institute (EAPI). Working Paper No. 87–7.

McGee, T. G. 1989. Urbanisasi or Kotadesasi? Evolving Pattern of Urbanisation in Asia. In Costa, F. J., Dutt, A. K., Ma, L. J. C., and Noble, A. G., eds. *Urbanisation in Asia*. Honolulu: University of Hawaii Press. pp. 93–108.

McGee, T. G. 1991. The Emergence of Desakota Regions in Asia: Expanding a Hypothesis. In Ginsberg, N., Koppel, B., and McGee, T. G., eds. *The Extended Metropolis: Settlement Transition in Asia*. Honolulu: University of Hawaii Press. pp. 3–25.

McGinn, Anne Platt. 2003 Combating Malaria. In Gardner, Gary (Project Director). *State of the World 2003: A Worldwatch Institute Report on Progress Toward a Sustainable Society*. The Worldwatch Institute, Washington, DC and Norton Books, pp. 62–84.

McKibben, Bill. 1995. *Hope, Human and Wild: True Stories of Living Lightly on the Earth.* St. Paul, Minnesota: Hungry Mind Press.

Madhav, Gadgil. 1996. *New Meanings for Old Knowledge: The People's Biodiversity Register Programme.* Bangalore: Centre for Ecological Sciences, Indian Institute of Science.

Mahadevan, T. M. 2005. General Path of the Tsunami of 26th December. *Journal of the Geological Society of India* 65 (3):373–74.

Mahendrarajah, S., Warr, P. G., and Jakeman, A. J. 1993. Optimal Extraction of Small Scale Surface Water Storage in Asia. *Water Resources Journal.* Bangkok: United Nations Economic and Social Commission for Asia and the Pacific. pp. 50–64.

Mahesh, R. 1999. Causes and Consequences of Change in Cropping Pattern: A Location Specific Study. Thiruvananthapuram: Kerala Research Programme on Local Level Development (KRPLLD). Discussion Paper Number 11.

Mandal, B. K., Roy Chowdhury, T., Samanta, G., Basu, G. K., Chowdhury, P. P, Chanda, C. R., Lodh, D., Karan, N. K, Dhar, R. K, Tamili, D. K, Das, D., Saha, K. C. and Chakraborti, D. 1996. Arsenic in ground water in seven districts of West Bengal, India: the biggest arsenic calamity in the world. *Current Science* 70(11):976–85.

Marx, Karl. 1968 [1852]. *The 18th Brumaire of Louis Bonaparte.* New York: International Publishers.

Mary Kutty, A., Joy, C. M., and Joseph, T. 1996. Potential effect of fertiliser residues on algae of Kuttanad. Thiruvananthapuram: Proceedings of the VIII Kerala Science Congress. pp. 59–60.

Mathew, G. 1994. *Panchayati Raj: From Legislation to Movement.* New Delhi: Concept Publishing Co.

Mathew, K. J., Gopinathan, C. P., Rao, D. S., and Murty, A. V. N. 1984. Ecology of mudbanks-zooplanktons. Central Marine Fisheries Research Institute (CMFRI) Bulletin 31:35–39.

Mathur, H. N., Babu, R., Joshie, P., and Singh, B. 1976. Effect of clear felling and reforestation on runoff and peak rates in small watersheds. *The Indian Forester* 102(4):219–226.

Mawdsley, Emma. 2004. India's middle classes and the environment. *Development and Change* 35(1):79–103.

Maya, S., Prameela Kumari, S., and Sarojini, M. V. 2001. Quality Profile of Selected Temple Tanks of Southern Kerala: A Case Study. Thrissur: Proceedings of the XIII Kerala Science Congress. pp. 511–17.

Meadows, Donella, Randers, Jorgen, and Meadows, Dennis. 2004. *Limits to Growth: The 30-Year Update*. White River Junction, Vermont: Chelsea Green Publishing Company.

Meher-Homji, V. M. 1980. The link between rainfall and forest clearance: case studies from Western Karnataka, *Transactions of the Institute of Indian Geographers* 2(1)59–65.

Mencher, Joan. 1966. Kerala and Madras: a comparative study of ecology and social structure. *Ethnology* 5(2):135–71.

Menon Sarojinji,V. 1997. Sacred groves. In Balachandran Thampi, K., Nayar, N. M., and Nair, C. S., eds. *The Natural Resources of Kerala*. Thiruvananthapuram: World Wild Life Fund for Nature-India. Kerala State Office. pp. 159–167.

Menon, Sreedhara A. 1965. *Kerala District Gazetteers, Ernakulam*. Thiruvananthapuram: Government of Kerala.

Meybeck, M., and Helmer, R. 1989. The quality of rivers: From pristine stage to global pollution. Paleo Geography. Paleo-climatology. Paleo-ecology (Global and planetary change section) Vol. 75, pp. 283-307.

Meybeck, M. 1998. The International Geosphere Biosphere Programme (IGBP) water group: a response to a growing global concern *Global Change Newsletter* 36:8–12.

Millennium Assessment. 2005. *Millennium Ecosystem Assessment Synthesis Report: Strengthening Capacity to Manage Ecosystem Sustainability for Human Well-Being.* Washington D. C.: Island Press. *http://www.millenniumassessment.org*

Miller, Arthur Austin. 1971 [1949]. *Climatology*. London: Methuen.

Mitra, D. 1995. Subterranean blues. Calcutta: *The Telegraph*, May 22.

Mjøs, Ole Danbolt. 2004. The Nobel Peace Prize 2004. Oslo: Norway. Nobel Peace Prize Presentation Speech. 10 December 2004. *http://nobelprize.org/peace/laureates/2004/presentation-speech.html*

Mohanan, C. N. 1997. Mangroves. In Thampi, K. Balachandran, Nair, N. M., and Nair, C. S., *The Natural Resources of Kerala*. Thiruvananthapuram: World Wide Fund for Nature-India, Kerala State Office. pp 149–158.

Mukherjee, N., and Bandopadhyay, D. 1993. New Horizons for West Bengal Panchayats. Kolkata : Government of West Bengal. Unpublished report.

Munashinge, M., and McNeely, J. 1995. Key concepts and terminology of sustainable development. In Mohan Munashinge and Walter Shearer, eds. *Defining and Measuring Sustainability: The Biogeographical Foundation*. Washington, D. C.: The United Nations University and the World Bank.

Muralikrishna, M. 2000. Environmental Impact Assessment of Human Activities in Ashtamudi Estuary, Kollam, Kerala. Thiruvananthapuram: Univeristy of Kerala. Unpublished Ph.D. thesis.

Nair, K. K. N., Khanduri, S. K., and Balasubramanyan, K. 2001. *Shola Forests of Kerala: Environment and Diversity*. Thiruvananthapuram: Kerala Forest Department and Kerala Forest Research Institute.

Nair, K. N., and Chattopadhyay, S. 2005 : *Water Resources of Kerala-Issues and Case Studies.* Thiruvananthapuram : Kerala Research Programme on Local Level Development, Centre for Development Studies.

Nair, K. S., Thampi, K. B., and Babu, N. V. T. 1999. Silent Valley National Park – A Historical Perspective. In Manoharan, T. M., Biju, S. D., Nayar, T. S. and Easa, P. S., eds. *Silent Valley: Whisper of Reasons.* Thiruvananthapuram: Kerala Forest Department in Association with Kerala Forest Research Institute. pp. 79–87.

Nair, M. M. 1987. Coastal geomorphology of Kerala. *Journal of the Geological Society of India* 29(4): 450–458.

Nair, N. J. K., Reghunathan Pillai, K., and Nalinakumar S., 1987. Environmental Geomorphic Atlas, Kerala coastal zone. Final report submitted to the Department of Environment, Govt. of India. Thiruvananthapuram: Centre for Earth Science Studies.

Nair, Ramachandran P. V., Gopinathan, C. P., Balachandran, V. K., Mathew, K. J., Regunathan, A., Rao, D. S., and Murty, A. V. S. 1984. Ecology of mudbanks: Phytoplankton productivity in Alleppey mudbank. Central Marine Fisheries Research Institute (CMFRI) Bulletin 31:28–34.

Narayanan, Vasudha. 2001. Water, wood, and wisdom: ecological perspectives from the Hindu traditions. *Daedalus* 130(4):179–206.

Nayak, N. 1993. *Continuity and Change in Artisanal Fishing Community – A Study of Socio-economic Conditions of Artisanal Fishing Communities on the South-west Coast of India Following Motorisation of Fishing Crafts.* Thiruvananthapuram: Programme for Community Organisation and South India Federation for Fishermen Societies.

Nayar, M. P., and Sastry, A. R. K. 1987–90. *Red Data Book of Indian Plants.* 3 Volumes. Howrah: Botanical Survey of India.

Ninan, K. N. and Sathyapalan, Jyothis. 2005. The economics of biodiversity conservation: a study of a coffee growing region in the Western Ghats of India. *Ecological Economics.* In press.

Nirmala, E., Jalaja, T. K., Nirmala, Stephen, and Nair, S. R. 1994. Assessment of water pollution in Connolly canal. Thiruvananthapuram: Proceedings of the VI Kerala Science Congress. P. 98.

NOW. 2003. Interview with Vandana Shiva; Letter from Perumatty Panchayat, and Response from Hindustan Coca-Cola India. *http://www.pbs.org/now/transcript/transcript_shiva.html*

O'Neill, R. V., Hunsaker, C., and Levine, D. 1992. Monitoring challenges and innovative ideas. In Mckenzie, D. H., Hyatta, D. E., and McDonald, V. J., eds., *Ecological Indicators*. New York: Elsevier. pp.1443–1460.

O'Neill, R. V., Hunsaker, C. T., Klopatek, D. J. M., Dale, N. H., Turner, M. G., Gardener, R. H., and Graham, R. 1995. Sustainability at landscape and regional scale. In Munasinghe, Mohan, and Shearer, Walter, eds. *Defining and Measuring Sustainability: The Biogeophysical Foundations*. Washington, D. C. and Tokyo: The United Nations University and the World Bank. pp.137–43.

O'Neill, R. V., Johnson, A. R., and King, A. W. 1989. A hierarchical framework for the analysis of scale. *Landscape Ecology* 3:193–205.

Oommen, M. A. 1994. Land Reforms and Economic Change: Experience and Lessons from Kerala. In Prakash, B. A., ed. *Kerala's Economy: Performance, Problems, Prospects*. New Delhi: Sage Publications. pp. 117–40.

Ostrom, Elinor. 1990. *Governing the Commons: The Evolution of Institutions for Collective Action*. Cambridge: Cambridge University Press.

Padmalal, D., and Maya, K. 2000. Sand from Kerala rivers. Mruthika: A World Wildlife Fund of India Kerala State Office Publication. Aug-Sept.

Pandurangan, A. G. 1995. A study on anthropogenic activities causing rarity of flowering plants in Idukki reservoir project area, Kerala. Proceedings of the VII Kerala Science Congress, pp.66–69.

Panikar, P. G. K. 1980. Recent Trends in Area under Production of Rice in Kerala. Thiruvananthapuram: Centre for Development Studies. Working Paper No.116.

Pathiyoor, G., Gopalkrishnan, R., Parameswaran, M. P., Narayanan, N. and Babu Ampat, 1990. Group Approach for Locally Adapted and Sustainable Agriculture (GALASA): A Blue Print for Decentralised Agricultural Development. Palakkad: IRTC and KSSP.

Pearce, D., Markandya, A., and Barbier, E. B. 1989. *Blue Print for a Green Economy.* London : UK Department of Environment.

Pennyback, Mindy. 2004. Behind the Scenes: Antibacterial Soap. In Halweil, Brian, and Lisa Mastny, eds. *State of the World 2004: Special Focus: The Consumer Society*. Washington, D.C.: The Worldwatch Institute. New York: W. W. Norton and Company. pp. 66–67.

Pezzy, John. 1989. Economic analysis of sustainable growth and sustainable development. Washington, D.C. : World Bank Environmental Department Working Paper No. 15.

Philip, Kavita. 2001. Seeds of Neo-colonialism? Reflections on Ecological Politics in the New World Order. *Capitalism Nature Socialism: A Journal of Socialist Ecology* 12(2):3–47.

Prabhakaran, P. V., Balakrishna Pillai, P. and Karmachandran, K. M. 1991. Climatic classification of Kerala. *Mausam* 43(4):434–36.

Prakasam, V. R. and Joseph, M. L. 1991. Lake "Sasthamcotta" : Water sediment interaction and productivity. Kozhikode: Proceedings of the III Kerala Science Congress, pp.5–6.

Prasad, M. K. 1999. The Silent Valley Crusade. In Manoharan, T. M., Biju, S. D., Nayar, T. S., and Easa, P. S., eds. *Silent Valley: Whisper of Reasons*. Thiruvananthapuram: Kerala Forest Department in Association with Kerala Forest Research Institute. pp. 43–50.

Prasad, S. N., Vijayam, L., Balachandran, S., Ramachandran, V. S. and Varghese, C. P. A. 1998. Conversion planning for the Western Ghats of Kerala - A GIS approach for location of biodiversity hotspots. *Current Science* 75(3):211–219.

Pretty, J. N. 1995. Participatory learning for sustainable agriculture. *World Development* 23(8):1247–1263.

Prugh, Thomas, Flavin, Christopher, and Sawin, Janet L. 2005. Changing the Oil Economy. In Renner, Michael, French, Hilary, and Assadourian, Erik, eds. *State of the World 2005: A Worldwatch Institute Report on Progress Toward a Sustainable Society*. New York: W. W. Norton and Company, pp. 100–119.

Puri, G. S., Meher-Homji, V. M., Gupta, R. K., and Puri, S. 1983. *Forest Ecology - Phytogeography and Forest Conservation*. New Delhi: Oxford and J. B. H. Publishing Company.

Raj, K. N., and Michael Tharakan, P. K. 1983. Agrarian reforms in Kerala and its agrarian impact on the rural economy—a preliminary assessment. In Ghose, Ajit Kumar, ed. *Agrarian Reform in Contemporary Developing Countries*. New York: St. Martin's Press. pp. 31–90.

Ramachandran, K. K., Chattopadhyay, S., and Salim, M. B. 1987. A Comparative Study of Land Use Pattern in Two Selected River Basins (Mangalam-Gayatripuzha and Aralampuzha-Bavalipuzha) of Western Ghats – Kerala. Thiruvananthapuram: Centre for Earth Science Studies.

Ramachandran, V. K. 1998. On Kerala's Development Achievements. In Drèze, Jean, and Sen, Amartya, eds. *Indian Development: Selected Regional Perspectives*. Delhi: Oxford University Press. pp. 205–356.

Ramalingom Pillai, A., Rema Devi, V., Saravanan, V. and Ouseph, P. P. 1994. Distribution of heavy metals and nutrients in the nearshore sediments and waters along the southwest coast of India. In Proceedings of the VI Kerala Science Congress, Thiruvananthapuram, pp. 91–93.

Rao, C. H. Hamunantha. 2000. Watershed Development in India: Recent Experience and Emerging Issues. *Economic and Political Weekly* 34(45):3943–47.

Ratcliffe, John. 1978. Social justice and the demographic transition: Lessons from India's Kerala State. *International Journal of Health Services* 8(1):123–44.

Rayner, S., Bretherton, F., Buol, S., Fosberg, M., Grossman, W., Houghton, R., Lal, R., Lee, J., Lonergan, S., Olson, J., Rockwell, R., Sage, C., and Von Imhoff, E. 1994. A Wiring Diagram for the Study of Land-Use/Land-Cover Change: Report of the Working Group A. In Meyer, W. B., and Turnet, B. L. II, eds. *Changes in Land Use and Land Cover: A Global Perspective*. Cambridge: Cambridge University Press. pp. 13–53.

Redclift, Michael R. 1987. *Sustainable Development: Exploring the Contradictions*. London: Methuen and Co. Ltd.

Renner, Michael, Sheehan, Molly, et. al. 2003. *Vital Signs: The Trends That Are Shaping Our Future*. The Worldwatch Institute in Cooperation with the United Nations Environment Programme. New York: W. W. Norton and Company.

Resamma Stephen 1991. Variations in zooplankton density in the Cochin backwaters consequent to environmental change. Kozhikode: Proceedings of the III Kerala Science Congress. pp.57–58.

Reshmi, P. B. 1996. Environment and sustainable development: The case of brick industry in Kerala. New Delhi: Jawaharlal Nehru University, M. Phil. Thesis.

Roy, P. 1973. Micro regional planning for the management of agriculture. *Social Change* 1(1):6–19.

Roy Burman, B. K. 2003. Interview given to Sopan Joshi. *Down To Earth* 12(21):36–37.

Rukmani, R. 2004. Participatory Ecorestoration Activities at Micro Watershed Level in Mulamkunnathukavu panchayat : A Retrospective Analysis. Thiruvananthapuram: Kerala Research Programme for Local Level Development. Unpublished Project Report. *http://www.krpcds.org/reportleft.htm*

Sabesan, S., Kumar, N. P., Krishnamoorthy, K. and Panicker, K. N. 1991. Seasonal abundance and the biting behaviour of Mansonia annulifera, M. uniformis and M. indiana and the relative role of in the transmission of Malayan filariasis in Shertalla (Kerala state). *Indian Journal of Medical Research* (A)2:1–253.

Sanders, Eli. 2005. Rebuffing Bush, 132 Mayors Embrace Kyoto Rules. *The New York Times*. 14 May 2005:A9.

Sanderson, S. 1994. Political-Economic Institutions. In Meyer, W. B., and Turnet, B. L. II., eds. *Changes in Land Use and Land Cover: A Global Perspective*. Cambridge: Cambridge University Press. pp. 329–55.

Sarin, Radhika. 2003. Behind the Scenes: Computers. In Halweil, Brian, and Mastny, Lisa, eds. *State of the World 2004: A Worldwatch Institute Report on Progress Toward a Sustainable Society*. New York: W. W. Norton and Company. Pp. 44–45.

Saseendran, S. A., Rakhecha, P. R., and Dhar, O. N. 1983. Highest Observed 1 to 10-day Point and Areal Rainfall over Kerala State. *Indian Journal of Power and River Valley Development* 33(12):438–43.

Saseendran, S. A., Singh, K. K., Rathore, L. S., Singh, S. V., and Sinha, S. K. 2000. Effects of Climate Change on Rice Production in the Tropical Humid Climate of Kerala, India. *Climatic Change* 44:495–514.

Satheesan, K. V. 1991. Water requirements of rubber plantation in humid tropical areas. Kozhikode: Proceedings of the III Kerala Science Congress, pp.5–6.

Satheesan, K. V., Gururaja Rao, G, Sethuraj, M. R. and Reghavendra, A. S. 1984. Canopy photosynthesis in rubber *Hevea Brasiliensis*: Characteristics of leaves in relation to light interception. In Sybesma, C., ed. *Advances in Photosysthesis Research*. Belgium: Proceedings of the Sixth International Conference on Photosynthesis. Vol. IV. The Hague, The Netherlands: Martinus Nijhoff Dr. W. Junk, N. V. Publishers. pp.125–128.

Satheesan K. V., Jayakumar, M., Varadan, K. M., and Unni, P. N. 1993. Effect of conversion of natural forests into plantation crops on water use in the Western Ghats. Kottayam: Proceedings of the V Kerala Science Congress. pp. 71–73.

Sawin, Janet. 2004. Making Better Energy Choices. In Halweil, Brian, and Mastny, Lisa, eds. *State of the World 2004: A Worldwatch Institute Report on Progress Toward a Sustainable Society*. New York: W. W. Norton and Company. pp. 24–43.

Sawin, Janet. 2005. Climate Change Indicators on the Rise. In Worldwatch Institute. *Vital Signs: The Trends That Are Shaping our Future*. New York: W. W. Norton and Company. pp. 40–41.

Schücking, Heffa, and Anderson, Patrick. 1991. Voices Unheard and Unheeded. In Shiva, Vandana, Anderson, Patrick, Schücking, Heffa, Gray, Andrew, Lohmann, Larry and Cooper, David, eds. *Biodiversity: Social and Ecological Perspectives*. London: Zed Books and Penang: World Rainforest Movement. pp. 13–41.

Sen, L. K., Wanmali, S., Bose, S., Misra, G. K., and Ramesh, K. S. 1971. Planning Rural Growth Centres for Integrated Area Development: A study in Miryalguda Taluk. Hyderabad: National Institute of Community Development. National Institute of Rural Development.

Shaji, V. M. M., and Abraham, T. K. 1994. Litter production in forest ecosystems of Kallar and Palode, Kerala. Thiruvananthapuram: Proceedings of the VI Kerala Science Congress. P. 94.

Sheehan, Molly O. 2003. Carbon Emissions and Temperature Climb. In Renner, Michael, and Sheehan, Molly O., eds. *Vital Signs: The Trends That Are Shaping Our Future*. New York: W. W. Norton and Company. pp. 40–41.

Shiva, Vandana. 1991. *The Violence of the Green Revolution: Third World Agriculture, Ecology, and Politics*. London: Zed Books and Penang: Third World Network.

Shiva, Vandana. 2002. *Water Wars: Privatization, Pollution, and Profit*. Cambridge, Mass.: South End Press.

Shiva, Vandana. 2004. Building Water Democracy: People's Victory against Coca-Cola in Plachimada. ZNet Commentary May 13, 2004. *http://www.zmag.org*; also available at *http:/www.awakenedwoman.com/wbw_shiva_plachimada.htm*

Shravankumar V., Chattopadhyay, S. and Nair, N. J. K. 1988. Bio-climatic characteristics of Kerala. *The Indian Geographical Journal* 63(2):88–99.

Silas, E. G. 1984. Mudbanks of Kerala-Karnataka : need for an integrated study. *Central Marine Fisheries Research Institute* (CMFRI) *Bulletin* 31:2–7.

Singh, H. 1997. Unsustainable water use in agriculture and its socio-economic ramifications : A case study of Punjab, India. *Asian Thought and Society* 22(65):134–145.

Singh, J. S., Singh, S. P., Saxsena, A. K., and Rewat, Y. S. 1984. India's Silent Valley and its Threatened Rain Forest Eco-System. *Environmental Conservation* 11(3):223–33.

Singh, N., and Soman, M. K. 1990. Some aspects of hydroclimatic fluctuations in Kerala, India. *Indian Journal of Power and River Valley Development*, 40(5–6):47–52.

Sinha, V. K., Uniyal, V. K., and Rennenson, G. 1999. Status paper on Silent Valley national park : The management perspective. In Manoharan, T. M., Biju, S.D., Nayar, T.S., and Easa, P.S., eds. *Silent Valley : Whisper of Reasons*. Thiruvananthapuram: Kerala Forest Department in Association with Kerala Forest Research Institute. pp. 89-95.

Sivanandan, P. K. 1985. Kerala's Agricultural Performance, Differential Trends and Determinants of Growth. New Delhi: unpublished M. Phil thesis Jawaharlal Nehru University.

Skole, D. 1994. Data on Global Land-Cover Change: Acquisition, Assessment, and Analysis. In Meyer, W. B., and Turner, B. L. II, eds. *Changes in Land Use and Land-Cover: A Global Perspective*. Cambridge: Cambridge University Press. pp. 437–71.

Soman, K., Chattopadhyay, Mahamaya, and Ouseph, P. P. 1992. Status of riverine pollution in South Kerala and its relation to physiography and landuse. *Proceedings of the IX Kerala Science Congress*, pp.93–95.

Soman, K. 1997. *Geology of Kerala*. Bangalore: Geological Society of India.

Soman, K. 2002. *Geology of Kerala*. Bangalore: Geological Society of India.

Soman, M. K. Krishnakumar K. and Singh, N. 1988. Decreasing trend in the rainfall of Kerala. *Current Science* 57(1):7–12.

Sooraj, S., Preeja, N. B, Baji, S. and Viswan, Sangeetha P. 2000. Spatial variation of nutrient content and water quality parameters in Vembanad lake, south of Thaneermukkom bund. *Proceedings of the XII Kerala Science Congress*. pp.640–43.

Sreedharan, T. P. 2004. Biological Diversity of Kerala: A Survey of Kalliasseri panchayat, Kannur district. Thiruvananthapuram: Kerala

Research Programme on Local Level Development. Discussion Paper No. 62.

Subramanyan, A. R., and Murthy, N. S. 1982. Climatic zones of Kerala state and their water balances. *Decan Geographer* 20(21):285–293.

Sugatha Kumari. 1999. Silent Valley – A Case Study. In Manoharan, T. M., Biju, S. D., Nayar, T. S. and Easa, P. S., eds. *Silent Valley: Whisper of Reasons*. Thiruvananthapuram: Kerala Forest Department in Association with Kerala Forest Research Institute. pp. 11–20.

Sundar, Aparna. 1999. Sea change: Organizing around the fishery in a South Indian community. In Barker, Jonathan, ed. *Street Level Democracy: Political Settings at the Margins of Global Power*. West Hartford, Connecticut: Kumarian Press. pp. 79–114.

Sundaram, K. V. 1997. Area Development: Trends and Perspectives. First Professor V. L. S. Prakash Rao Memorial Lecture. Delivered at the 19th Indian Geography Congress, Thiruvananthapuram, 29–31 December, 1997.

Surendranath, C. 2003a. Water Contamination by Coke in Kerala. *CorpWatch*. July 10, 2003. *http://corpwatch.radicaldesigns.org/article.php?id=7548*

Surendranath, C. 2003b. Coca-Cola: Continuing the Battle in Kerala. *CorpWatch*. July 10, 2003. *http://corpwatch.radicaldesigns.org/article.php?id=7528*

Surendranath, C. 2004. Silent Valley: Threatened Again. *http://www.indiatogether.org/2004/jul/env-silentval.htm*

Sushanth, C. M., and Jayakumar, M. undated. Erosion hazards under different landuse in Western Ghats. Kozhikode: unpublished seminar paper, Centre for Water Resources Development and Management.

Sutton, Paul C., and Costanza, Robert. 2002. Global estimates of market and non-market values derived from nighttime satellite imagery, land cover, and ecosystem service valuation. *Ecological Economics* Tainter, Joseph. 1988. *The Collapse of complex societies.* Cambridge : Cambridge University. 41:509–27.

Tainter, Joseph, 1988. *The Collapse of Complex Societies.* Cambridge : Cambridge University Press.

Tamta, S. R., Thambi, D. S., Subburaj, A., and Shaji, E., 2004. Rain water as a source for recharging ground water. In *Water Resource Development through Rain Water Harvesting – A Compendium of Selected Papers on Rain Water Harvesting*. Thiruvanan-thapuram: Western Ghat Cell, Planning and Economic Affairs Department, Government of Kerala. pp 190–199.

Thanal. 2005. International POPs (Persistent Organic Pollutants) Elimination Project: Case Studies of Zero Waste Kovalam –A progressive waste management programme with focus on the best available technology options and material substitution. Thiruvananthapuram: Thanal. *http://www.thanal.org*

Tharakan, P. K. M. 1976. Migration of Farmers from Travancore to Malabar, 1930 to 1960: An Analysis of Economic Causes. New Delhi: Jawaharlal Nehru University. Unpublished M. Phil. Thesis.

Thomas Isaac, T. M., Franke, Richard W., and Parameswaran, M. P. 1997. From Anti-feudalism to Sustainable Development: The Kerala Peoples Science Movement. *Critical Asian Studies [Bulletin of Concerned Asian Scholars]* 29(3):34–44.

Thomas Isaac, T. M., Franke, Richard W., and Raghavan, Pyaralal. 1998. *Democracy at Work in an Indian Industrial Cooperative: The Story of Kerala Dinesh Beedi*. Ithaca, N.Y.: Cornell University Press.

Thomas Isaac, T. M., and Franke, Richard W. 2000. *Local Democracy and Development: People's Campaign for Decentralised Planning in Kerala*, New Delhi: Leftword. Also 2002. Boulder, Colorado: Rowman and Littlefield.

Thomson, K. T. 2002. Coastal and Marine Resource Management and Poverty Reduction in South Asia: India (Kerala): Legal and Institutional Issues. Unpublished report prepared for the Asian Development Bank, Manila and submitted to the Centre for Earth Science Studies, Thiruvananthapuram.

Thornthwaite, C. W. 1948. An approach towards a rational classification of climate. *Geographical Review* 38(1):56–94.

Trainer, Ted. 1995. *The Conserver Society: Alternatives for Sustain-ability*. London: Zed Press.

Turner, B. L. 1990. Two types of global environmental change – definitional and spatial scale issues in their human dimensions. *Global Environmental Change* 1(1):14–22.

Turner, B. L. 1997. Spirals, Bridges and Tunnels: Engaging Human-Environment Perspectives in Geography. *Ecumene* 4:196–217.

Turner, M. D. 1999. Merging local and regional analyses of landuse change: The case of livestock in the Sahel. *Annals of the Association of American Geographers* 89(2):191–219.

United Nations Development Programme, 2004. *Human Development Report 2004: Cultural Liberty in Today's Diverse World*. New York: UNDP.

Unni, J. 1983. Changes in cropping pattern in Kerala: Some evidence on substitution of coconut for rice. *Economic and Political Weekly* 18(39):A100–A107.

USDA. 1981. *World Agricultural Trends and Indicators, 1970–80.* Washington D. C.: United States Department of Agriculture. pp. 19-56.

Van En, Robyn. 2002. Robyn Van En Center for Community Supported Agriculture. *http://www.csacenter.org*

Varadan, K. M. and Raghunath, B. 1985. Hydrological properties of latosols for water management in humid tropics. *Journal of Indian Society of Soil Science* 33:497–506.

Varadan, K. M., Madhava Chandran, K. and Lakshmanan, K. 1990. User Guide on Farm Irrigation Scheduling for Upland Crops of Kerala. Kozhikode: Centre for Water Resources Development and Management.

Varma, Ajay Kumar, Resy George, A. S. K. Nair, Krishnakumar, and Aniyan, P. V. 2000. Watershed Master Plan: An Approach to Reinforce Local Level Development. Paper Presented at the International Conference on Democratic Decentralisation, 23–27 May, 2000. Thiruvananthapuram: Kerala State Planning Board.

Venugopal, P. 2003. Toxicity in Plachimada Sludge. *The Hindu* 27 July 2003. *http://www.thehindu.com/2003/07/27/stories/2003072703401000.htm*

Véron, René. 2000. Sustainability and the 'New' Kerala Model. In Parayil, Govindan, ed. *Kerala: The Development Experience – Reflections on Sustainability and Replicability*. London: Zed Press. pp. 212–29.

Viederman, Stephen. 1995. Knowledge for sustainable development: what do we need to know? In Trzyna, Thaddeus C., ed. *A Sustainable World: Defining and Measuring Sustainable Development.* Sacramento and Claremont CA: Published for The World Conservation Union by the International Center for the Environment and Public Policy of the California Institute of Public Affairs. pp. 37–43.

Vijayalakshmi, E. 2003. Calling the shots. *Down To Earth* 12(12): 15–19.

Wackernagel, Mathis, and Silverstein, Judith. 2000. Big things first: focusing on the scale imperative with the ecological footprint. *Ecological Economics* 32:391–94.

Wackernagel, Mathis, Schulz, Niels B., Deumling, Diana, Linares,

Alejandro Callejas, Jenkens, Martin, Kapos, Valerie, Monfreda, Chad, Loh, J. Monathan, Myers, Norman, Norgaard, Richard, and Randers, Jørgen. 2002. Tracking the ecological overshoot of the human economy. *Proceedings of the National Academy of Sciences* 99(14):9266–71. *http://www.pnas.org/cgi/reprint/99/14/9266*

Walker, R.T., and Solecki, W. D. 1999. Managing landuse and landcover change: The New Jersey pinelands biosphere reserve. *Annals of the Association of American Geographers* 89(2):220–237.

Watkins, K. 2002.Making globalization work for the poor. *Finance and Development* 39(1): 24–26.

Weaver, James. H., Rock, Michael. T., and Kusterer, Kenneth. 1997. *Achieving Broad-Based Sustainable Development: Governance, Environment, and Growth with Equity*. West Hartford, Connecticut: Kumarian Press.

Wiegers, E. S., Robert, J. H., Dominique, H., and Fresco, L. O. 1999. Landuse intensification and disintensification in the Upper Canete Valley, Peru. *Human Ecology* 27(2):319–339.

Wilbanks, T. J, 1994. Sustainable development in geographic perspective. *Annals of the Association of American Geographers* 84(4):541–556

Wolman, M. G.1967. A cycle of sedimentation and erosion in urban river channels. *Geografiska Annaler* 49A. pp. 385–395.

Wood, W. B. 2004. American geography and international research: A sustainable-development agenda. *The Professional Geographer* 56(1):53–61.

World Commission on Environment and Development. 1987: *Our Common Future*. Oxford: Oxford University Press.

World Conservation Union (IUCN). 1991. *Caring for the Earth: A Strategy for Sustainable Living*. Gland, Switzerland: The World Conservation Union, The United Nations Environment Programme, and the World Wide Fund for Nature.

World Scientists. 1992. World Scientists' Warning to Humanity. Cambridge, MA: Union of Concerned Scientists. *http://www.ucsusa.org/ucs/about/page.cfm?pageID=1009*

Wramner, Eva. 2004. *Fighting Cocacolanisation in Plachimada: Water, soft drinks and a tragedy of the commons in an Indian village*. Lund University Human Ecology Division Paper. *http://web.forumsyd.se/Arkiv/Faktabanken/uppsats/006DA285-000F6CFB-006DA2A7.0/Fighting_Cocacolanisation_i.pdf*

Wright, Ronald, 2004. *A Short History of Progress*. New York : Carroll and Graf Publishers.

York, Richard, Rosa, Eugene A., and Dietz, Thomas. 2003. Footprints on the Earth: The Environmental Consequences of Modernity. *American Sociological Review* 68(2):279–300.

Youth, Howard. 2003. Watching Birds Disappear. In Gardner, Gary et al. *The State of the World 2003: A Worldwatch Institute Report on Progress Toward a Sustainable Society*. New York: W. W. Norton and Company. pp. 14–37.

Zachariah, K. C., Rajan, S. Irudaya, Sarma, P. S., Navaneethan, K., Nair, P. S. Gopinathan, and Mishra, U. S. 1999. Demographic Transition in Kerala in the 1980s. Thiruvananthapuram: Centre for Development Studies. Second edition.

Zachariah, Mathew, and Sooryamoorthy, R. 1994. *Science in Participatory Development: The Achievements and Dilemmas of a Development Movement: The Case of Kerala.* London: Zed Books.

Index